*Techn*Ozarks:

Essays in Technology, Regional Economy, and Culture

*Techn*Ozarks:

Essays in Technology, Regional Economy, and Culture

Edited
by

Thomas A. Peters

Paul L. Durham

Foreword
by

Greg Burris

Former Springfield City Manager

President and CEO of the

United Way of the Ozarks

The Ozarks Studies Institute of
Missouri State University
Springfield, Missouri
2019

Published by the Ozarks Studies Institute, an initiative of the Missouri State University Libraries. Copyrights of materials remain in possession of the authors, artists, owners, and first publishers. For inquiries, contact:

Duane G. Meyer Library
901 South National Library
Springfield MO 65897
417.836.4525

Book cover and layout design by Jesse Nickles.
Cover photo: The Chadwick Flyer stopping in Ozark, Missouri (*ca.* 1909).
   Courtesy of the Richard Crabtree Collection.
Half-title page photo: Neosho-built engines in a Saturn V rocket (*ca.* 1967).
   Courtesy of Larry James.
Interchapter I image (p.lii): Streetcar in the Town Square (1919).
   Courtesy of the History Museum on the Square, Springfield, Missouri.
Interchapter II image (p.92): Lily Tulip Cup Corporation (1951).
   Courtesy of the History Museum on the Square, Springfield, Missouri.
Interchapter III image (p.184): X-ray photoelectron spectrometer, JVIC (2016).
   Starboard & Port Photography, courtesy of Allen D. Kunkel.
Image facing Notes on Contributors (p. 282): Panorama at West Shops in Springfield, Missouri (1927). *Thomas Railroad Collection, MSU Special Collections and Archives.*

ISBN-13: 978-1-7321222-2-2
Library of Congress Cataloging-in-Publication Data:
   Names: Peters, Thomas A. 1957- editor. | Durham, Paul L., editor.
   Burris, Greg, writer of foreword.
   Title: TechnOzarks : essays in technology, regional economy, and culture / edited by Thomas A. Peters, Paul L. Durham ; foreword by Greg Burris.
   Description: Springfield, Missouri : The Ozarks Studies Institute of Missouri State University, 2019.
   Notes: Includes bibliographical references.
   Identifiers: ISBN 978-1-7321222-2-2
   Subjects: LCSH: Technology—Ozark Mountains Region. | Technology—Social aspects—Ozark Mountains Region. | Technology—Economic aspects—Ozark Mountains Region. | Ozark Mountains Region—Social life and customs. | Ozark Mountains Region—History.
   Classification: LCC T14.5.T43 2019 | DDC 303.483—dc23

Introducing The OSI Publications Series in Ozarks History and Culture
The Ozarks Studies Institute (OSI) of Missouri State University seeks to preserve the heritage of the Ozarks, its culture, environment, and history by fostering a comprehensive knowledge of Ozarks' peoples, places, characteristics, and dynamics. The Institute promotes a sense of place for residents and visitors alike and serves as an educational resource by collecting existing—and discovering new—knowledge about the Ozarks and by providing access to that knowledge.

Following *Living Ozarks: The Ecology and Culture of a Natural Place* (2018), *TechnOzarks* is the second volume in the OSI series. Along with its companion journal, *OzarksWatch*, the series aims "to introduce the Ozarks to the world," and vice versa.

What readers have said of the first volume, *Living Ozarks*:

> Authors in this anthology are aware of tourism's fantasies that overlay geology's reality and that the Ozarks' fragile natural landscape requires stewardship. We know that the environment shapes all creatures that live within it, including us. We must be prepared to address our presence as part of the natural—what is the cost to absorb our footprint?
>
> —Lynn Morrow, editor, *Ozarks in Missouri History: Discoveries in an American Region*

> Any discussion of sustainability in the Ozarks must involve not only the natural environment, but also elements not commonly thought of as natural resources: the history, the heritage, and the people. These are key elements that make this region unique and attractive to outsiders and tourists and give the Ozarks its unique identity. *Living Ozarks: The Ecology and Culture of a Natural Place* brings this point home in a decisive and definitive work.
>
> —Paul W. Johns, author, *Unto These Hills: True Tales of the Ozarks*

OZARKS STUDIES INSTITUTE

Broadcast tower (undated).
Ozark Public Television Collection,
MSU Special Collections and Archives.

Foreword

It's all just zeros and ones ... how complicated can it all be? Very complicated, as it turns out.

I'm deeply honored to have been asked to write this foreword, but it is quite intimidating. While the half-life of technological knowledge is probably measured in months now due to the increasing speed of change, I have been blessed to have an opportunity to view technologies through a variety of lenses during my various careers. That's one of the advantages of not being able to keep a job and having to continually reinvent yourself: You get exposed to a lot of smart people along the journey. And you get to see the world from different corners of the room.

My background is in Information Technology (IT). I didn't really discover IT until my junior year at Southwest Missouri State University (SMSU—now Missouri State University). SMSU's Computer Information Systems program had a great reputation, and employers came from throughout the Midwest and beyond, recruiting our graduates to become application computer programmers and systems analysts.

Those were the days of mainframes. Personal computers were things you built from a kit at home. Thus, my classmates and I lived in computer labs. We carried decks of punch cards and cringed when someone dropped their assignment, as cards skittered everywhere. We later programmed while staring into green-screen cathode ray tube (CRT) monitors.

I was hired by the university right before I graduated and planned to stay a few years to get some work experience while earning my MBA. This was a time when we wrote almost all of the administrative software used at the university; we didn't purchase much. Armed with some in-the-trenches IT work experience and an MBA, I would head out to conquer the world.

Thirty-five years later, I'm still in the Ozarks. I love a lot of things about living here. I worked twenty-five years at the university in a variety of roles—from computer programmer to vice president.

When I left Missouri State University to become the city manager of Springfield, Missouri, the "systems thinking" I had learned working at the university continued to serve me well. A city is a series of systems, many of them interconnected. As an applications systems analyst, I learned to start at the end, then work backwards. I applied this process many times when attacking community issues: Start by getting everyone to coalesce around the vision of what you want to achieve, then work backwards. And within that process, things will collide.

I've always been fascinated by where things intersect and collide. I believe life gets most interesting where things intersect—ideas, opinions, trends, philosophies, sabers, lives, railroads, demographics, musical influences, technologies, and cities (e.g., the merger of two towns that created Springfield in 1887).

If you throw a frog in boiling water, it'll jump out, but if you put it in tepid water and slowly raise the heat, the frog won't notice until it's too late—or so the story goes. Just like the boiling frog, we too become accustomed to the changes, slowly adapt, and no longer notice their impact on our lives. How many technologies influence your life just during your morning commute? Your vehicle is now a shiny metal box full of computer systems. Are you listening to traditional radio, satellite radio, or a CD in the car? Intelligent traffic management systems optimize traffic light functions to minimize delays and maximize throughput. See any digital billboards? I hope you're not texting and driving, so we won't count the mini-super-computer you have in your pocket. And on and on.

But work ethic is work ethic. And the Ozarks are full of smart, hard-working people. The result is cutting-edge research, tech startups, and a palpable "buzz" in a community with 45,000 college students. Work ethic is energy, and the work ethic of the Ozarks continues to pay dividends in a global economy. Innovation is energy, and there is a serious entrepreneurial spirit in the Ozarks. Most places would love to have that mix of ingredients in their community stew.

In an increasingly fierce competition for talent—that sound you hear in the background is the vacuum being created by Baby Boomers exiting the workforce at the pace of 10,000 per day—the Ozarks' high quality of life gives us an advantage. The Ozarks, once geographically isolated to some degree, are on a level playing field with any other community. Talent attracts talent. If it's true that people today select a place to live and then select a job—which is just the opposite of my generation: we would chase a job to just about any location—we hold a winning hand in the Ozarks.

Technologies have the ability to thrill us, frustrate us beyond belief, captivate us, and terrorize us. Technologies are tools. And, like any tool, they can be used for good and evil. Yin and yang.

Drones, for example, have the ability to enhance our lives in a variety of ways. Drones can also be used to invade your privacy and cause harm in ways we've not yet even considered. Technologies like this will force elected officials and others to wrestle with incredibly complex policy issues … just like past technologies have done.

Cryptocurrency will devastate our global financial systems. Or it will reinvent and revolutionize our financial systems. Or neither. Or both.

Wireless technologies will continue to bring conveniences to our lives. Or they will interrupt our sleep patterns and cause other health issues. Or both.

The mining of "big data" (instead of lead, zinc, and iron) that reveal our digital footprints as we tramp through the ether (as others scoop them up and archive them) will bring convenience to our lives or wreak havoc in our lives. Or both.

Digital security breaches will continue to drain bank accounts, devastate financial credit scores, and cause victims to feel violated. Or biometric security advances and other tools will greatly reduce our reliance on computer IDs and passwords. Or both.

Technologies will be used to rape our environment. Or technologies will allow us to reverse the damage we've done to our planet. Or neither. Or both.

Artificial intelligence will eliminate a lot of existing jobs. Or AI will help us cure diseases. Or both.

Technology can be used to level the playing field. Or it can be used to amplify injustices. Or both.

Moore's Law originally referenced the ever-increasing rate at which transistors could be placed on integrated circuits, doubling every year or two. It is sometimes applied more generally to describe the speed at which broader advancements in digital electronics occur.

I hope you will slow down long enough to enjoy this collection of inspiring essays.

And even though it seems that things continue to move faster and faster, if the speed of future innovation continues at a rate anything close to Moore's Law, then enjoy today … because things will never move this slowly again. Can we keep up?

Greg Burris
Springfield, Missouri

Table of Contents
TechnOzarks: Essays on Technology, Regional Economy, and Culture

Editors' Introduction and Acknowledgments

Thomas A. Peters and Paul L. Durham

We take Missouri State University's Public Affairs Conference theme for 2019, "The 21st Century Digital World," as an occasion for this anthology, which explores the role of technology in building the Ozarks. In its collection of essays and images, *Techn*Ozarks follows a rough chronology, charting where Springfield and the Ozarks came from, where they've arrived, and where they're going with respect to technology, economy, and culture. These terms, by the way, are inseparable: To speak of the region's entrepreneurship and "culture of innovation," for example, is to raise implications for its economy—and for its ecology, since the Ozarks has relied on its rich natural resources and faces the need to conserve these for future use. Much like the interconnectedness of these terms, the book's division into a technological-economic-cultural "past," "present," and "future" remains arbitrary. Many of the essays that follow find themselves equally at home in two or more timeframes: Historical discussions lead to a consideration of present conditions, which enables predictions of our "digital future."

*Techn*Ozarks aims at an educated general readership, readers whose expertise might lie in one or more of the fields covered but who will find ours a useful synthesis, a "big picture" view of the components—technological, economic, cultural—that create a region like our own. Our work is not "authoritative" in any scholarly sense: We aim, rather, to start and extend conversations over the present state and future possibilities facing Springfield and the Ozarks. If anything unites the authors and audiences of this book, it's an appreciation of the region and its gifts. Though in varying degrees, most of the book's contributors "belong to" the Ozarks, as will many if not most of its readers. As Ozarkers, we share a common history; we

shall also, most likely, share a common fate. In the 21st century, our future remains tied to regional resources in technology, economy, and culture; at the same time, our fate belongs increasingly to the nation and, indeed, to the planet. Once isolated geographically, we are no longer; technology has contracted time and space, rendering us all Ozarkers and global citizens simultaneously.

Collaboration in writing and editing is like singing a duet; both voices join in harmony, though the notes sung differ. Through the next two sections, we give our solo versions of the book and its intentions.

Tom Peters goes first:

Why *Techn*Ozarks?

We offer the portmanteau word, "*Techn*Ozarks," in homage to the RadiOzark, a techno-savvy radio transcription service from the late 1940s into the 1950s affiliated with KWTO-AM 560 radio station in Springfield, Missouri. The name was jingly, inventive, and very much in the tradition of Ozarks place names and businesses.[1]

---

1. Ironically, "Ozarks" is itself a portmanteau combined from the French *aux Arks*. A poem (admittedly mediocre) stands on the first page of the 1909 *Ozarko*, the first year of the yearbook of Fourth District Normal School:

> Midst the forests of the Ozarks where the scenes with beauty glow, / Where the brooklets clear as crystal seem to carry as they flow / Happy greetings from the mountains to the valley green below: /... Here the people, peaceful neighbors, seem to gather as they go / Common sense with education, 'til they soon begin to know / That the grandest land in all the world is that of "OZARKO."

While the yearbook is playfully inventive in its naming, it turns out that Ozarkers have long used regional wordplay as a selling

The early broadcast media, first radio and then television, were revolutionary, because they used the "ether," not wires like the telegraph and telephone did, to carry content—signals—from one source to many listeners and viewers. Wireless broadcasting was a "disruptive" technological breakthrough, though it had temporal limitations. This type of broadcasting worked in real time only.

As a child growing up in Iowa, if I wanted to watch at least the beginning of *Bonanza*, I had to cajole my parents into letting me stay up until 9 p.m. Central Time on Sunday evening—a school night—because that's when *Bonanza* aired. This was decades before TiVo. A radio transcription service like RadiOzark used recording technologies akin to making phonograph recordings to enable a radio program to be distributed on acetate media—records—to be used by over 650 radio stations in far-flung geographic locations whenever they wanted to broadcast the pre-re-

corded program themselves. It was an exciting high-tech industry after the Second World War, attracting young people with vision. In 1953, when Don Richardson wrote an article about RadiOzark, the principals were local young entrepreneurs: Si Siman was 32 and John Mahaffey only 27.[2]

Television added real-time moving images to radio sound broadcasts, and radio transcriptions solved the time-shifting challenges. It also proved to be a lucrative business, helping KWTO and its sister companies attract and retain talented performers. Transcriptions helped lure the famous Red Foley to Springfield. On April 10, 1954, Foley signed a five-year contract with RadiOzark and Top Talent (a talent management and booking agency), sibling companies of KWTO, for a series of fifteen-minute radio transcriptions, personal appearances, a thirty-minute radio network show and perhaps, as reported in an industry update in *Billboard*, a thirty-minute filmed TV program (*Billboard*, April 17, 1954, p. 16).

In this story above, note the impact of an innovative, indeed "disruptive" technology upon regional economy: RadiOzark made its owners rich while attracting nationally famous radio and TV personalities to Springfield. And note the TV technology's impact upon regional culture: By means of nationally syndicated radio and television broadcast, Springfield became, for a time, the epicenter of "country music" entertainment in America. This pattern—of local entrepreneurs taking advantage of technological innovations that would impact regional economy and culture—repeats itself in the essays following.

I've called RadiOzark's transcription service a "disruptive" technology, and with good reason: Technological innovations easily disrupt existing personal, social, economic, political, and cultural norms. As if to maintain some semblance of continuity, we often make

point. "ADMAN LAUDS OZARK REGION," reads a headline of the June 12, 1927 *Springfield Leader and Press*:

> Leo P. Bott, Jr. of the Bott Advertising Agency of Little Rock, specialists on Ozarks and other community and commercial advertising,... was originator of the words, "Ozarkansaw" and "Mozarks," which he has released to posterity....
>
> When God made "Ozarkansaw," the land of the Ozarks in Arkansas, and "Mozark," the land of the Ozarks in southern Missouri, He gave us mountains that He had molded by fire, chiseled and carved by streams and the elements. For a country without mountains is a country without grandeur....

Bott's wordplay caught on quickly: There was the Mozark Baseball League, which began play in 1927 and continued through 1930, being done in by the Depression. In the May 1, 1929 issue of the *St. Louis Post-Dispatch*, an editorial proposed Mozark Lake "in the naming of the projected lake on the Osage River," adding, "it can be readily seen that the word combines Missouri and the Ozarks. It has geographical value, is easily understood and is also euphonious." (Compare Lake Taneycomo, another regional coining.) If not a lake, a movie house received the moniker: In 1934, the Isis—a smallish theater tucked into the southwest corner of the Springfield town square—was renamed the Mozark. These are a few of the moniker set-ups to RadiOzark.

2. Don Richardson, "Ozarks Studios Produce Nationwide Radio Programs," *Ozarks Mountaineer* (November 1953), p. 9.

sense of the new in terms of older, established technologies, ideas, and processes. When presented with an innovation, our typical first impulse is to wonder how it can be adapted to do the work of older technologies. In the 1970s, for example, the screen-mouse-keyboard combination of early desktop computers resembled a typewriter. When word-processing first emerged, it was understood as typing on screen. Automatic spell-checking, hyperlinking to other information objects, group-authoring processes, and writing in the cloud came later: What fueled the revolution in personal computing was the capacity to type—in electrons rather than in ink—to correct misspellings without erasers or "Wite-Out," and to store one's words in retrievable "memory."

Even the names we give to new technologies reveal our impulse to understand the new based on the known older technologies. Thus the railroad locomotive was called the iron horse. The automobile was called the horseless carriage. And the radio was called the wireless—as it was understood in the context of older wired technologies, such as the telegraph.

Each technology has its own set of inherent "affordances" or possible uses/applications. Entrepreneurs and startups are quick to understand these inherent affordances, which disrupt existing markets and established corporations. Netflix, for example, essentially marginalized Blockbuster and the entire VHS rental sector. Online travel websites marginalized the established travel agency model.

Our collective task as adopters and users of such technologies is to understand and exploit their affordances. For example, one affordance of automobile technology is that, compared to the older locomotive, it is much easier to change course, start, and stop. This basic affordance of the automobile has had a profound impact on the American landscape, both as it is configured and as it is imagined and understood.

It's Paul's turn now:

Watching *The Wizard of Oz* on Black-and-White TV

Do you remember the first time that you realized the impact of a new technology in your life? Like many of you, I have a fond memory of how a particular technology changed the way I viewed the world. It was in the late '60s when I first saw "the yellow brick road" as a *yellow* road in the movie, *The Wizard of Oz*. Our family happened to be visiting my uncle, and my cousins were all excited about their new color television set. At that time, our family owned a single large black-and-white Magnavox television. (This was before solid-state electronics, cable feeds, and remote controls: Back then, your TV had a heavy cathode ray picture tube and smaller vacuum tubes that regularly blew out, requiring house calls by the TV repairman, "rabbit ear" antennas to fiddle with the reception, and manual channel dials—one for the VHF "network" stations, one for the UHF stations.) On my uncle's new TV, *The Wizard of Oz* started as it had always begun, with black-and-white scenes of Dorothy and Toto in their Kansas farmhouse before being lifted up by a tornado and falling from the sky. To their credit, my cousins gave us no clue as to what was about to happen: When Dorothy opens the farmhouse door and gazes out, the world of Oz bursts into color. It must have been a magnificent spectacle in the "big screen" movie theaters back in 1939, being the first major film released in Technicolor; thirty years later on the color TV "small screen," its effect was no less wondrous.

Seeing that yellow brick road and, later in 1969, images of the Earth and moon as seen from the Apollo 11 mission evoked a sense of awe and an awareness that we had entered a golden age of technology. Little did I know that this was just the beginning of technological advances that would continue to impact my personal and professional life, allowing me to see and experience the world in more colorful and profound ways. On a personal level, my grateful wife appreciates not having to be the "navigator and map reader" on our trips anymore, because she has been replaced by "Hazel"—the name that we affectionately gave to our voice-activated GPS—so when something goes wrong I can blame "Hazelnut" for the misinformation. On a professional level, I am a cell and molecular neuroscientist whose career began in the early '80s, when technological advances in DNA analysis and digital microscopy were changing how we could see inside cells to better understand the causes of human disease and discover new treatment strategies. This path led me to Missouri State University, where I now have the privilege of serving as Director for the Center for Biomedical and Life Sciences at the Jordan Valley Innovation Center.

I am honored to be the Provost Fellow for the 2019 Public Affairs Conference, whose theme focuses on the impact of the digital world in the 21st century. Technological advances have been influencing human evolution for thousands of years; however, the pace of technological change has greatly accelerated in the past century, impacting all aspects of human society. Every niche of our existence, from economics to politics, science, medicine, education, transportation, agriculture, entertainment, and media, has been morphed by technological breakthroughs. We now live in a digital world, in which information lies at our fingertips and communication is possible from even the most remote places on Earth.

We are more connected than ever before—or are we? With all our labor-saving devices, we have more time to enjoy life—or do we? As illustrated in the Ozarks region, technological advances have greatly improved many aspects of our lives; but they have done harm in some respects, as well. As a biology professor, I tend to view the world from a naturalist's point of view. Based on my understanding, it is by means of technology that we have taken control of our own evolution. In the essays and images that follow, you will learn of resilient and creative Ozarkers who have pioneered and/or adapted to the changing economic, ecological, cultural, and technological landscape; and, in so doing, have helped shape the past, present, and future of the Ozarks region. I, personally, have hope for the future and look forward to the sort of *Wizard of Oz* wonderments that our own children, and our children's children, will experience in their lifetime.

Joining Voices

We have many to thank, for many have contributed to our enterprise. We thank Dr. Frank A. Einhellig, Provost of Missouri State University, for his support of the Ozarks Studies Institute and its Publications Series in Ozarks History and Culture. His unwavering commitment makes projects of this sort possible. We thank Mary Ann Wood and Candace Fisk, Conference Program and Content Coordinators for the MSU Public Affairs Conference, for including our project in the Fall 2019 PAC activities. As we've noted, *Techn*Ozarks was compiled in response to the 2019 PAC theme, "The 21st Century Digital World."

We thank Dr. John F. Chuchiak IV, Director of the MSU Honors College, and Honors College Assistant Director Scott Handley for approving our "Forum on the Future" as a student/faculty collaboration. Enrolled in Dr. Baumlin's Spring 2019 Honors section of English 310: Writing for Graduate/Professional School, the following students took the "Forum" as a team project, conducting and editing interviews: Allee Armitage, Sarah Crain, Jasmine Crawford, Hannah Fox, Cara Hawks, Arylle Kathcart, Jacob Miles, Hanna Moellenhoff, Austin Sams, Grace Sullentrup, Nicholas Stoll, Emma Sullivan, and Jesse Walker-McGraw. (We're pleased to include Hanna Moellenhoff's own original contribution, as well.)

We thank Dr. Baumlin for lending us his students and for serving as the book's textual editor. (We also forgive him for nagging over deadlines.) We thank Dr. William B. Edgar, MSU Clinical Associate Professor of Library Science, for serving as project manager, coordinating duties and communications and keeping us on schedule. And we thank Jesse Nickles, an alumnus of the MSU Art and Design Department, for his artistry in layout and book design.

We give special thanks to Joan Hampton-Porter, curator at Springfield's History Museum on the Square, for her support of our project. The History Museum's photographs make a vital contribution to our enterprise. We also thank the museum's photographer, Palmer Johnson, for preparing images for our use. As with other OSI publication projects, we thank Shannon Mawhiney and Tracie Gieselman-Holthaus, our colleagues in Meyer Library Special Collections and Archives, for researching, gathering, preparing, and captioning images. We also thank Nathan Neuschwander of the Library Dean's office for photography and preparation of images.

We have others to thank for providing images and access to equipment and archival holdings, including Dr. Toby J. Dogwiler, Head of the MSU Department of Geography, Geology, and Planning; Dr. Robert S. Patterson, MSU Professor of Astronomy; and Allen D. Kunkel, MSU Associate Vice President for Economic Development and Director of the Jordan Valley Innovation Center.

Our understanding of regional history remains a work-in-progress, and we are pleased to acknowledge our debts to previous scholarship. Included in this present collection, Lynn Morrow's essay, "Gasconade Mills: The Ozarks' First Timber Boom," draws materials from his previous work, "Old Pulaski: A Lumbering and Rafting Legacy," Part I in *Old Settlers Gazette* 34 (2016), pp. 30-44, and Part II in *Gazette* 35 (2017), pp. 38-58. Richard Schur's essay, "Walter Majors, African American Inventor and Entrepreneur," draws from his "Memories of Walter Majors: Searching for African American History in Springfield," in *Springfield's Urban Histories: Essays on the Queen City of the Missouri Ozarks*, ed. Stephen L. McIntyre (Springfield, MO: Moon City, 2012), pp. 119-125. And Tom Dicke's essay, "'Red Gold' of the Ozarks: Tomato Canneries in the Ozarks, 1890-1960" draws from "Red Gold of the Ozarks: The Rise and Decline of Tomato Canning, 1885-1955," in *Agricultural History* 79 (2005), pp. 1-26.

We thank Karen Craigo for her skillful proofing, which has saved us from embarrassments in grammar, citation, and spelling. We thank our readers, Dr. Tita French Baumlin, Rachel M. Besara, Lynn Morrow, Lynn Cline, Elaine Stuart, Rira Zamani, and Dr. Craig A. Meyer; their suggestions have improved the text, helping minimize the inevitable errors. And we thank our contributors, some seventy-two in total, who represent the diversity of interests, disciplines, and communities in Springfield and the Ozarks. Given this diversity, our contributors have used a variety of citational styles and formats. We have let these stand, aiming instead for fullness and accuracy in citation.

Following an overview by our colleague, Dr. Baumlin, we proceed to the essays and photo albums, allowing these to speak for themselves.

Inventing the Ozarks (I):

On the Confluence of Technology, Regional Economy, and Culture
James S. Baumlin

In his *Journal of a Tour into the Interior of Missouri and Arkansaw* (1821), Henry R. Schoolcraft—the region's first English-speaking "tourist"—records his impressions of Finley Creek and the environs south of modern-day Springfield. His diary entry for January 4, 1819 follows:

> The prairies, which commence at the distance of a mile west of this river, are the most extensive, rich, and beautiful, of any which I have ever seen west of the Mississippi river…. The lands consist of a rich black alluvial soil, apparently deep, and calculated for corn, flax, and hemp. The river-banks are skirted with cane,… and the lands rise gently from the river for a mile, terminating in highlands, without bluffs, with a handsome growth of hickory and oak…. Taking these circumstances into view, with the fertility and extent of soil, its advantages for water-carriage, and other objects, among which its mines deserve to be noticed, *it offers great attractions to enterprising emigrants,* and particularly to such as may consider great prospective advantages an equivalent for the dangers and privations of a frontier settlement. The junction of Findley's Fork [*sic*] with James' River, a high, rich point of land, *is an eligible spot for a town,* and the erection of a new county … would soon give the settlers the advantages elsewhere enjoyed in civil communities…. A water communication exists with the Mississippi. Steamboats may ascend White River to the mouth of its Great North Fork. Keelboats of twenty tons burthen may, during the greater part of the year, ascend to the mouth of James' River; and boats of eight tons burthen may ascend that to the junction of Findley's Fork,… to which the navigation may be continued in smaller boats, thus establishing a communication by which the peltries, the lead, and the agricultural products of the country, could be easily, cheaply, and at all seasons, taken to market, and merchandize brought up in return. (pp. 58-59; emphasis added)

Like most tourists or "prospectors" of his age, Schoolcraft sees the present *but looks to the future.* Resources—mineral, wildlife, and agricultural especially—when matched with adequate transportation, conspire to make this "an eligible spot for a town, and the erection of a new county." *There was money to be made* by those "enterprising emigrants" who'd be willing to endure "the dangers and privations of a frontier settlement." Equally important, their collective work would transform wilderness into a "civil communit[y]," a place where the amenities of culture and the good life could be pursued. I'm willing to declare Schoolcraft's *Journal* the region's first piece of published Ozarks "boosterism." I'd also declare Schoolcraft's predictions, made 200 years ago, to have proved true.

The following is an essay in the history of a place *and an idea*: specifically, of technological innovation and its role in creating a progressive, future-oriented "booster" image for early Springfield and the Ozarks. In recent years, historians have focused on the region's economy and politics; on its shifting demographics; on ecology and shifting practices in land-use; on industry—agriculture, mining, and manufacture—and on technologies under discussion here: the railroad (arriving in 1870), the automobile (in 1901), and radio broadcast (in 1921).[3] From the start, let me state that this is *not* a local history of the railroad *per se*, or of the motorcar, or of radio telecommunication; it is, rather, an exploration of attitudes and aspirations of Springfield *as a town* and of the Ozarks *as a region* whose growth rested in the possession and exploitation of such technologies as these.

"The booster spirit was strong in Springfield," writes Charles K. Piehl of the decades following the Civil War (p. 89). The first big "boost" came in 1870, when the region's "bright and happy future, the subject of our wishes for many long years,… arrived" with the railroad.[4] A second boost or "boom" came in 1887, when Springfield merged with "New Town" or North Springfield. As the August 19, 1887 issue of the *Springfield Daily Leader* reads, the newly consolidated town "booms and booms and keeps on booming. It is a perpetual motion boom." This booster-boomer spirit lasted well into the 20th century; arguably, vestiges of it remain to this day.

Though the Chamber of Commerce organized in 1919, prominent businessmen had long supported "the Springfield Club" and the northside "Commercial Club." By these and other "business fraternities,"

boosters sought to attract outside investment and settlement through the promise of cheap land, civic order, scenic beauty, and urbane culture.[5] Natural resources abounded; needful were the tools to extract, refine, transport, and sell the same. Enter the technologies—trains, autos, radios—that promised a healthful, prosperous, comfortable Ozarks lifestyle. In his article, "The Small City in American History," Timothy R. Mahoney distinguishes towns from small cities from major metropolises, particularly as these evolved in the Middle West.[6] Typical of the small town, Mahoney notes, "is that at one time, many, if not most, of its citizens imagined themselves to be living in a 'future metropolis' or at least a significant regional center" (p. 316), the economy and culture of such towns having been "constructed within the framework of a 'booster ethos'" (Mahoney, p. 316):

> In towns across the country a predominantly American-born middle-class elite articulated

3. See Lynn Morrow and Linda Myers-Phinney, *Shepherd of the Hills Country: Tourism Transforms the Ozarks, 1880-1930s*, and Brooks Blevins, *A History of the Ozarks, Volume 1: The Old Ozarks*. See also essays gathered in several collections: Lynn Morrow, *The Ozarks in Missouri History: Discoveries in an American Region*; Stephen L. McIntyre, *Springfield's Urban Histories: Essays on the Queen City of the Missouri Ozarks*; and William B. Edgar, Rachel M. Besara, and James S. Baumlin, *Living Ozarks: The Ecology and Culture of a Natural Place*.

4. "Speech of Hon. John S. Phelps," from *Opening of the Atlantic and Pacific Railroad* (p. 8). For discussion of the arrival of the railroad and the rivalry between "Old Town" and North Springfield, see Piehl's "Race of Improvement: Springfield Society, 1865–1881," in Morrow (pp. 71-100).

I take this opportunity to thank colleagues Craig A. Meyer, Elaine Stuart, Lynn Morrow, and Cathie English for help in improving early versions of this essay.

5. "I believe that a man should be proud of the city in which he lives and that he should so live that the city will be proud he lives in it." This anonymous quote stands on the back page of the Chamber's first official publication, *Springfield Greets You* (1919). We can take it as the Chamber's boosterish civic motto.

6. "In the urban history of the United States," writes Mahoney, "two predominant narratives have emerged: that of the metropolis and that of the small town" (p. 314). He continues:

> The former is the story of regional and national centers of economic development that enjoyed steady, even rapid, growth and became focal points of the emergence of the modern nation. The latter is the story of local or regional centers that played peripheral, secondary, or reactive roles in the national economy.… What is missing, of course, is a story line for those urban places in between the small town and the metropolis: small cities of America. (pp. 314-15)

From the Civil War through the Second World War, Springfield fit fairly neatly into the small-town economy and ethos, as Mahoney describes it. After World War II, Springfield evolved into one of those "small cities of America," making for new opportunities and challenges (some of which will be described below).

a local boosterism. According to this view one achieved success in work and enjoyed the satisfactions of family life through self-control, hard work, and religious faith. Middle-class citizens built a successful town by developing the town economy, establishing a system of law and order, founding institutions, creating a civic life, and formulating booster policy. (p. 316)

In Springfield from the 1870s and 1880s through the 1920s, Mahoney's description is spot-on: for, "at the core" of its booster ethos "was a strong entrepreneurial impetus that distinguished most towns in the Midwest from those in the upland South and New England" (p. 317). This ethos is on full display in *A Booming City* (1887), a pamphlet published by the local real estate firm, Lapham and Bro. Though its "commercial importance" came "by slow stages, covering a generation," Springfield "awoke a few mornings since, to find itself confronted by that young Samson of the West, the 'Boom,' which, with his magic wand, *makes towns of villages and cities of towns*" (p. 8; emphasis added).

Within this booster-boomer ethos resided an optimism over the future that Springfieldians pursued, not just as businessmen but as entrepreneurs—investors speculating in land, industry, and technology. The local newspapers spurred inventors on, reporting on their innovations and calling for more; booster advertisers spread the word, "letting the world know that Springfield is up and coming—not going or standing still" (January 19, 1927 *Leader and Press*). In its practice of entrepreneurship, the Ozarks yoked economy, ecology, technology, and culture together. Exploitable resources proved the region's great attraction: Enter the entrepreneur. Geographic isolation (caused by the daunting terrain) proved the region's great challenge: Enter technology—specifically, the innovations in manufacture and transport that would bring region-

ally produced goods to markets state- and nationwide.[7] In local history, the inventor needed the investor, and *vice versa*. With every innovation in transport, communication, or manufacture, *someone* had to buy it, bring it to Springfield, and adapt it to local conditions. Springfield's booster ethos enabled this synergy by uniting capital and labor, exploiting the regional ecology while growing the regional economy.

Before automated assembly lines and prefabricated components, most items of local manufacture were assembled manually from machine-tooled parts. If an engine or some item broke, it would be repaired in a local shop by skilled labor. Parts might be tooled on site and the item rebuilt—and even, perhaps, "improved" by some adjustment, however minor. In this manner, as Pagan Kennedy notes, the local factory "turned workmen into inventors" (*Inventology*, p. 5). Tools and technologies that served in one

---

7. We need to remember the 210 miles separating Springfield from St. Louis; the 155 miles separating Springfield from Kansas City; the 195 miles separating Springfield from Tulsa; the 260 miles separating Springfield from Wichita; and the 295 miles separating Springfield from Memphis. Even as railways arrived from each compass point, the region remained in relative isolation, given these distances. If a train broke down or needed service, it would be fixed here, with parts at hand. One cannot overstate the innovative energies of "the Frisco" and other local machine shops: As their engineers and machinists proved time and again, necessity is the "mother of invention."

After the railroad came the automobile; but it, too, was hampered by terrain. Begun for bicycles and taken up by the automobile, the national Good Road Movement made crawling progress through southwest Missouri: From the 1870s through the 1920s, Ozarks roadways were notoriously bad. Writing in 1915, Jonathan Fairbanks and Clyde Edwin Tuck see little improvement from pioneer days:

If the reader will take a map of Missouri, and trace the route of that little caravan of pioneers, he will find that they covered probably 250 miles of the roughest hill country in the Ozarks, a route which even today, with all the improvements in roads and bridges that have been made in eighty-four years, would put any automobile on wheels out of business. (p. 685)

terrain underperformed or broke down in another. Local conditions—the difficulty of transport particularly—made practical problem-solving part of one's job. A survey of patents registered to Springfieldians from the 1870s through the mid-1920s shows that most were transportation- or work-related, aimed at improving safety, speed, and efficiency, reducing costs, or increasing profits.

So, while future-oriented boosterism remains an abiding theme, local innovation provides its twin thesis: The region's achievements follow predictable patterns in what might be called entrepreneurial technoculture. Local industries faced local problems tied to local conditions of terrain that demanded local solutions. While Springfieldians innovated for health, hearth, and home—medicines, clothing, cookware, stoves, and other domestic items were registered with the U.S. Patent Office—most innovations served industry (mining and agriculture especially), power supply, and transportation (the railroad and, by the 1910s, the automobile). At the time, these were "emerging technologies" of national import. From the turn of the 20th century to the Great Depression, Springfield enjoyed decades of industrial/mechanical/technological innovation. The twin themes of boosterism and entrepreneurship played their part in creating a distinctively modern Springfield: that is, an urbanized, industrialized Springfield whose Ozarks hinterland provided resources and markets as well as recreation. Though this present survey takes 1929 as an endpoint, the booster-boomer promise of future prosperity was fulfilled, in large part, by Springfield's growth through the latter half of the 20th century.

It's worth asking whether Springfield of the 2020s will prove as innovative, in its own way, as Springfield of the 1920s. Futuristic in their time, the "big machines" of previous decades yield to today's nanotechnologies, which are carrying us headlong into the "digital futures" of 21st century technoculture.

Does the region's need to attract new entrepreneurs return us to boosterism? The question is worth asking, though it's not yet time to answer. We need first to finish outlining the thesis and underlying technocultural assumptions, not just of this present essay, but of the volume it serves to introduce.

We are not here to dwell on the past—we are to consider the present and the future.
—"Speech of Hon. John S. Phelps," from *Opening of the Atlantic and Pacific Railroad, and Completion of the Southwest Pacific Rairoad to Springfield, Mo.* (1870)

Technology in the modern episteme is meant to bring the future under human control.
—J. Macgregor Wise, *Exploring Technology and Social Space* (1997)

In *Living Ozarks*—first of the OSI Publications Series in Ozarks History and Culture—the focus lay in intersections of culture and ecology: in the role that nature (in its rich resources of land, water, and wildlife) has played in creating, and sustaining, the Ozarks as we experience it today. In this second volume, *TechnOzarks*, the focus lies in intersections of culture and technology: in the role that innovation (in agriculture, transportation, communication, and commerce) has played *in building* the Ozarks, adjusting its tools and industries to the region's unique features (and, in the process, reshaping its landscape). We have learned to enjoy the land and conserve its resources; such was the message of *Living Ozarks*. We continue to learn how to use the land and its resources wisely; such is a message of *TechnOzarks*.

What is "the Ozarks," as explored in this present anthology? Increasingly urbanized; no longer isolated geographically; having brought some natural resources (mining, logging) to near-exhaustion while expanding to newer, "renewable" sources (hydroelectricity, solar- and wind-power); having survived the transition from a primarily production-based to a service-based economy; seeking its share in an expanding global market whose prized commodity is *not* mined or grown or manufactured goods, but is *information*. Such describes the current state of the Ozarks generally, and of Springfield in particular. *Techn*Ozarks offers essays in the history of innovations that have built the Ozarks into a vibrant culture and economy. *Techn*Ozarks also—in the spirit of the region's first English-speaking explorer, Henry R. Schoolcraft— aims to see the future already contained, in germ, in the present.

The future poses its challenges; we can name several already. Will a fully modernized, globalized Ozarks lose its character as a unique "natural" environment, a place of healthful recreation and refuge? (When climate change takes its seemingly inevitable toll, will we have kept a sufficient supply of water and arable land?) Will Ozarkians celebrate, or lament, the region's assimilation into the global economy and, by extension, into the "global village"? (As time and space continue to shrink, will events occurring "around the world" and "around the block" affect us equally?) These and other challenges are posed as questions whose probing belongs not to science or technology alone, nor to business or government, but to an informed citizenship whose future health and prosperity lie in the balance. And it is, indeed, the Ozarks' future that we seek, in that "technology," as J. Macgregor Wise tells us, "is meant to bring the future under human control."

In saying that the Ozarks today is transiting from a "modern" to a "postmodern" culture and economy, we're compelled to police our terms, starting with modernism. What does that mean or entail? In hazarding an answer, we look to four markers of modernity,

each implicated in technology. A "modern" Ozarks is *urbanized*, with housing, businesses, entertainments, workplaces, schools, government offices, and other services concentrated into major city centers encircled by suburbs and exurbs; it is *industrialized*, connected to (and participating in) a regional, national, global production-economy; it is integrated into *networks of transportation* carrying goods (and people) quickly and efficiently across expanses of land and sea and air; and it is integrated into *networks of communication* carrying information accurately and instantaneously across the globe. Though the region remains primarily rural, its natural resources in land, water, agriculture, and minerals have long been exploited: grown and harvested, extracted, refined, machined, packaged, shipped, and traded. And there's a further resource upon which a modernized Ozarks depends: Beyond production and transportation of goods, the region's development rests in an abundant, accessible *supply of energy*. Out of these markers—urbanization, industrialization, transportation, and communication, with energy as an underlying resource—our modernized version of the Ozarks has been "built."

Geographically, the Ozarks describes a place on a map with defining features of topography; geologically, the Ozarks has its distinctive features above and below ground; ecologically, the Ozarks has its flora and fauna, though these have changed over time through human intervention; ethnographically, the Ozarks has seen its share of displacements and migrations. As an inhabited space, the region's greatest changes have come through technology: that is, through industries and machines and tools and techniques (and the socialized/institutionalized knowledge of their uses) that have transformed the landscape. Driven by technology, the Ozarks has turned from "wilderness" to "pioneer settlement" to "timberland and mining land" to "vacation land." This last development—a.k.a. the tourism industry—remains central to the region's self-promotion today. To many outsiders, the Ozarks is sold as a nostalgic image, an old-time Hill Country marked by lakeside condos, rentable bass boats, and commodified pop culture, Branson style. For most Ozarkians, the situation "on the ground" is more complex. While capital wealth and trained labor concentrate in urban centers, many of the region's historic smaller towns (often premised on single industries: mining, logging, livestock, textile and clothing manufacture, etc.) have fallen on hard times. Agribusiness has largely replaced the rural subsistence farming that, for more than a century, sustained families on smallish plots of land. But industry, economy, ecology, and culture intertwine: Where any one of these goes into decline, the rest suffer.

A vast literature has grown around the history and sociology of technoscience.[8] Compared to this literature, ours is a pencil sketch of the technologies that have shaped, and will continue to shape, the Ozarks in its economy, ecology, and culture. We lack space to explore adequately the ways that technology commands each aspect of contemporary life, down to the very definition of our humanness. (Are we the masters or servants of technology? Has technology and its "built spaces" replaced nature as our *habitus* or

dwelling place?) For now, we're content to focus on the markers of modernity described above—that is, on energy supply, communication, transportation, industrialization, and urbanization—and on the roles these have played in "boosting" Springfield and the Ozarks, building the region into its current recognizable form.

1. The Iron Horse Arrives

On May 3, 1870, the railroad had at last arrived in a depot north of Springfield, carrying dignitaries from St. Louis, Jefferson City, and other points along the way. Disembarking to cheering crowds, cannons firing, and "flags fluttering in the breeze," they spent the day celebrating and speechifying. First to speak was Springfieldian and future Missouri governor, John S. Phelps (1814-1886):[9]

> Many of you perhaps have had business relations for years with some of the people of this city ... yet, as this is your first visit to our beautiful country, you can hardly appreciate the difficulties under which we have labored without an easy and expeditious connection with the other portions of the State. We were almost in an isolated condition; access to our country could only be obtained by days of tiresome and weary travel over rough and rugged roads, and through a hilly and mountainous country, whilst for years you have been in the enjoyment of railroad communication. (pp. 7-8)

Such were the region's past circumstances, dictated by terrain. As for the future, it had just arrived—by train:

---

8. See, for example, the rich gathering of materials in David M. Kaplan's *Readings in the Philosophy of Technology* (2009). Though this present essay ends by questioning the social, political, and ethical implications of contemporary technoscience, my approach remains social-constructionist: a fancy phrase, but not too difficult to apply. The social-constructionist model assumes a two-way street between material technology and human society. Within the social-constructionist model, writes Kaplan, "society simultaneously shapes technology as technology shapes society" (p. xviii). He explains:

> Far from being applied science, technology on this model is more like *embodied humanity*. Technologies are part human, part material, and always social.... The advantage of viewing technology in this way is that it calls attention to the way that humanity, technology, and the environment are bound up together in a relationship of mutual constitution. (Kaplan, p. xviii; emphasis in original)

9. I quote from the commemorative pamphlet, *Opening of the Atlantic and Pacific Railroad, and Completion of the Southwest Pacific Railroad to Springfield, Mo., May 3, 1870.* The title page gives "Springfield / 1870" as the place and date but lists no author or press. The Southwest Printing Office of North Springfield—publisher of *The Springfield Republican*—is a likely local candidate, with the railroad serving as underwriter.

Everything which can be produced in the United States can here be produced in superabundance, except the ice of Alaska, the cotton and rice of Carolina, and the tropic fruits of Florida.... The bright and happy future, the subject of our wishes for many long years, has just arrived upon us. No longer shall we be compelled to travel by stage on bad and dangerous roads, over a broken, hilly and mountainous country, to reach the commercial emporium of our State.... [A]nd though I have spoken of hills and mountains between this city and St. Louis as objects we dreaded in our journey, yet those hills and mountains are rich in minerals, and will soon greatly contribute to swell the volume of wealth of our State. (pp. 8-9)

Phelps overstates the region's "superabundance," but he was right about the role of rail transport.[10]

The epoch of modernism begins with mechanized production—a.k.a. "the Industrial Revolution," which (as we've all been taught) began in coal- and iron-rich England in the 18th century and, crossing the Atlantic, exploded through Yankee ingenuity and entrepreneurship, moving steadily westward across the North American continent as rivers and railroads allowed. Its

10. While repeating the typical futurist tropes, the rest of Phelps's speech is remarkable in calling for global commerce and immigration. The laying of tracks, he knew, would continue south to the Gulf and west to the Pacific:

> But let us remember that we are seeking to extend and enlarge our commerce with China, Japan, and the East India trade which is rapidly increasing. As our business relations with the people of China and Japan shall become more extended, these nations, with their abundant population, will furnish many emigrants to this country. And why shall they not come among us, if they shall desire to do so?... Why shall they not, by their industry, add to the wealth of this nation; and why shall they not become citizens, if such shall be their wish? Shall we repel laborers from coming amongst us? We say let them come.... (p. 13; emphasis added)

Now *that's* globalism, expressed in 1870s Springfield.

slowed arrival into the Ozarks—lamented by Phelps, above—is explained by the lack of efficient transport, whether by river or by rail. During its pioneer days, amenities of modern culture dribbled into rather than flooded the Ozarks; how quickly things changed can be gauged by a report in *The Springfield Daily Leader* for May 26, 1870, some three weeks after the formal opening of the depot of the Southwest Pacific Railroad on Commercial Street in that "new town" to the north of old Springfield:

> Boonville Street, from early morning to late in the evening, is crowded its entire length with wagons and teams hauling goods from the depot. In their new relation to the markets of the country, our merchants are no longer kept "waiting for the wagon." Goods that were formerly from ten days to three weeks in transit from St. Louis now arrive in twenty-four hours.... Our merchants ... can now largely increase their stocks on the capital invested, and assort their stocks to please customers. There is no longer danger of goods becoming old and unsalable on their hands. Country merchants coming in find the stocks in our market all that they could wish,... and when we take into consideration that goods of all descriptions are offered at St. Louis prices,... we cannot see why any should go beyond us for their supplies.

No longer a group of pioneers "waiting for the wagon" to bring in goods, the townspeople became suppliers to the larger Ozarks region. By this report, the railroad transformed both Springfields, the "old town" and the "new" together.[11]

11. Indeed, much of the story of post-Civil War Springfield revolves around iron rails, beginning with the building of two separate, incorporated towns: "Old town" Springfield, whose town square (intersected by Boonville-South Street and College Street-St. Louis) served as its business center, and "new town" North Springfield, centered around Commercial Street. In 1878,

✳ ✳ ✳

Many appellations have been applied to the present epoch ... such as the electric or steam age; none of the terms, it seems, being broad enough. But if we should christen it the age of invention, we would evidently not go far amiss.... If we look at the far-reaching effects of the inventions of only a few such wizards as Edison, Tesla, Bell, and Maxim, we would see the appropriateness of the last-named phrase to this the greatest age since the dawn of the world's history.
—Jonathan Fairbanks and Clyde Edwin Tuck, *Past and Present of Greene County, Missouri* (1915)

In the epigraph above, Fairbanks and Tuck offer to christen their own "present epoch" as "the age of invention." And rightly so. But even as they list the age's great inventors—Edison, Tesla, Bell—they give credit to those lesser-known names who, "by mere commonplace hard work," have improved the lives and labor of their fellows. These include inhabitants of the Ozarks:

Here and there, in every civilized nation may be found someone ... who has by his genius or talent or, perchance, by merely commonplace hard work, produced some device that has lightened or facilitated man's work, and therefore added his little quota to the great aggregate force that is lifting from humanity's shoulders "the burden of the world." (p. 1923)

---

a second track located on Main Street (just north of the town square) brought the St. Louis-San Francisco Railroad into downtown Springfield. Though fierce cultural-economic rivals at first, the two towns merged in 1887.

Here's the point: Transportation technology—here, the placement of railroad tracks—shaped the city's map-grid as we know it today. Road construction has continued this gridwork, cutting and dividing (and defining) neighborhoods, shopping and entertainment districts, industrial zones, and so on.

This description holds for the region's early inventors: In the main, they were farmers, shop men, and machinists—laborers who, by "some device" of their own making, sought to lighten or facilitate their own daily labor. In celebrating the lives of Greene Countians, Fairbanks and Tuck singled out these sorts of men.

Thus far, we've considered what the train did for Springfield; reversing the terms, we can consider *what Springfieldians did for the train*—and for other technologies. Just as the railroad demanded its accommodations of terrain, so the local technologies of farming, mining, logging, and rail-less transport demanded their own adaptations. Local newspapers took pride in reporting on inventors and their innovations. A survey of inventions from the 1870s through the 1920s attests to the interrelatedness of local needs and available technologies. The rocky, root-riven Ozarks soil put farmers and their implements to the test: Tired of repairing broken coulters and ploughshares, back-sore from piling up rocks and pulling out stumps manually, the region's farmers began improvising.[12] All forms of transport were studied, but rail received the most attention, leading to local innovations in ground-leveling and the laying of track; in strengthening car couplers for train safety; and in improving tools and techniques to make engine maintenance/repair quicker, safer, and more efficient.[13] These were developed by

---

12. Writing in 1915, Fairbanks and Tuck note the evolution of regional agriculture and technology:

In pioneer days when farming implements and machinery were of the crudest kind, requiring a goodly supply of both muscle and grit,... brawn, more than brains, was needed ... in order to rescue the fertile soils from the wilderness of forest and prairie growth. In these modern days of worn and worn-out soils and the abandoned farm, with the most improved labor-saving farm machinery, the business of farming needs brains more than brawn, that our soils may be rescued from the wilderness and desert or wasted fertility that has stifled and depleted them. (p. 1002; emphasis added)

13. In surveying patent records for the years 1870 through 1929, I'd

the railroad employees themselves. It was their own daily labor that they sought to make quicker, safer, less repetitive, and more efficient: They saw a need, took the materials at hand, and adapted them to local conditions.

Today's innovator-entrepreneur tends to look beyond "merely local" needs, applications, and markets. Still, the problem-solving model remains more or less unchanged. In today's parlance, might we call Springfield Wagon Company the region's first successful "startup"? From 1872 to its closing in 1951, the company's innovations in manufacturing wheel hubs and related components allowed its wagons to conquer the rugged Ozarks fields and roads. After dominating markets in Missouri, Arkansas, Texas, and Oklahoma, Springfield Wagon came to monopolize this aspect of rural rail-less transport, becoming sole provider to the U.S. Army. By 1925, virtually all wagons produced commercially in the United States were produced by the Springfield Wagon Company. But, while impres-

---

estimate that some 300 were registered to addresses in Springfield, Missouri (Annual Reports for the years 1872-1876 and 1926-1929 are unavailable online.)

The forty-eight patents in agriculture included a harvester-rake (1870); a cultivator (1878); a hand corn-planter (1879); a hedge trimmer (1880); a horse hay-rake (1882); a combined plow and harrow (1884); a wire and-picket-fence-making machine (1888); a post-driver (1891); a hand planter (1900); and a machine for cultivating orchards (1901).

Of the ninety-four patents in transportation, the following brief selection served the railroad: a train chimney (1870); a clamp to hold ratchet drills for drilling railway rails (1877); a car-coupler (1883); a railway-joint (1892); a railway bridle-rod (1904); a ditching machine (1904) for road excavation; a frogless railway switch (1907), being a rail section where trains cross over and change tracks; a brake-shoe brace (1908); concrete tie and rail-fastening (1909); a railway crossing (1911); a stop for railway switches (1915); a rail anticreeper (1915) to slow lateral displacement of track; a guard for railway frogs (1920); and a whistle-operating mechanism (1925).

By the 1910s, patents servicing autos included a device for raising and supporting automobiles (1916); a fuel supply to internal combustion engines (1917); and an acetylene-gas mixer (1917) for headlights.

sive in themselves, the Wagon Company's successes were overshadowed by the individual achievements of its co-founders and early shareholders—F. J. Underwood and H. F. Fellows above all.

Singlehandedly, Flavius J. Underwood (1831-1914) "secured about twenty patents," note Fairbanks and Tuck:

> [H]e built the first successful two-horse cultivator, which has revolutionized agricultural work, especially in the corn producing states. He enjoys the distinction of being the first person to advocate and demonstrate the circulation of steam for the purpose of heating buildings, which method is now so universally employed. Among his many inventions is a coal chute which he patented in 1904 and which is widely used. He believes his best invention is a machine for boring out hubs in which to insert boxes. (p. 1083)

While the last item above pertains to wagons, his other inventions served other purposes: tool manufacture, home heating, and farming. In fact, Underwood's body of work demonstrates the regional interconnectedness of industrial agriculture, tool manufacture, and efficient rural transportation. And though his inventions were used nationwide, each supplied a local need or solved a local problem—for which reason "his name," declare Fairbanks and Tuck, "is deserving of a high place among the successful inventors of his day" (p. 1083).

If Underwood represents the entrepreneur-inventor, Homer F. Fellows (1831-1894) represents the business-entrepreneur who anticipated—and promoted, purchased, invested in, or managed—virtually every "emerging technology" of his lifetime, bringing them to Springfield. In 1859, "he was one of the stockholders of the first telegraph line through Springfield" (Fairbanks and Tuck, p. 1366). He also "built the first telephone line that came into Springfield,... which connected his office and residence" (p. 1366). In 1870, he was among the first to open business in North Springfield. In 1871, he built Springfield's first grain elevator. During the Panic of 1873, he rescued Springfield Wagon from bankruptcy "and remained manager of the wagon factory the rest of his life" (p. 1367). In 1881, he was "the chief promoter of the Springfield street railway system" (p. 1366), serving for years as company president. He was a chief shareholder in "the Kansas City, Ft. Scott & Memphis railroad, which was made a part of the Frisco System in 1900" (p. 1367). And "he was one of the organizers of the Springfield Water Works" (p. 1367), serving for years as its president. Twice elected mayor, he was a longtime member of the school board. Upon his passing, he was properly eulogized. As Homer Barlow Stevens writes,

> Homer F. Fellows was an esteemed and valued citizen—public-spirited, strong in courage, clear in judgement, unimpeachable in character, and faithful to every trust reposed in him.... His character as a man of enterprise and genius is quite apparent. He was broad in his conceptions as he was upright in his methods. He was a public benefactor, the results of whose life have been a prominent factor in the development of this city and community .... (p. 312)

The words "public" and "citizen" resonate in the passage above. Together, they embody the American Midwest "booster ethos," as Mahoney describes it: "Citizens" like Fellows "built a successful town by developing the town economy,... founding institutions, creating a civic life, and formulating booster policy" (p. 316). Fellows was lionized in his lifetime, not for creating private wealth, but for creating "a city and community." As a business-entrepreneur, the "innovation" to which he made real contributions was Springfield itself.

2. Cars and Roads

Fellows' wagons were custom-made for the Ozarks terrain; nonetheless, the challenge of rail-less transport *lay beneath*, not upon or above, the wagon's wheels. By the turn of the 20th century, iron rails had tamed the region in part; where trains could not reach, one still relied "on bad and dangerous roads," as Phelps described them, roads that traversed "a broken, hilly and mountainous country." Invention had its incentives: "Invent it and you are wealthy for life," declares an article in the October 9, 1908 *News-Leader*. And that "it," for which "the wealth of a Rothschild is waiting," was "the invention of a satisfactory paving material." One might note that tarmacadam—a paving mixture of petroleum-based asphalt and sand— had been put into mass production just the year prior. But the surfacing material was difficult to transport and had yet to reach southwest Missouri. Besides, the article's author lacks vision of the future use of road surfacing, as his follow-up sentence suggests:

> At present what is good for the wheels is bad for the hoofs, and vice versa. That is to say, where the road is smooth and the wheels run easily there is no grip for the hoofs; and where it is rough the vehicle is hard to drag.... What is wanted is a smooth, hard, absorbent surface, with at the same time the perfect grip.

Today's inventologist might point to the author's "design fixation" (Kennedy, p. 246), which cannot see beyond the horse-drawn carriage. It's for a *horseless* carriage—the automobile—that the Ozarks' roads would eventually be improved.[14]

---

14. Another example of design fixation comes from the January 11, 1900 *Springfield Leader and Press*: "If an automobile can be invented to navigate some of our bad streets, our people would no doubt invest heavily." Put baldly, the problem lay not with the automobile, but with the streets.

At the turn of the 20ᵗʰ century, few Ozarkians had as yet seen a working automobile, though most had heard of it and many recognized its potential.[15] A technology in itself, the gasoline-powered internal combustion engine provided the energy source that would drive the industries of rail-less transportation. To this day, road construction remains a work in progress; still, the decades ensuing (from the 1920s through the '60s) turned the Ozarks into a spiderweb of asphalt-concrete roadways, with motor vehicles—cars, buses, trucks—carrying goods and passengers far beyond the reach of rail. By the 1970s, passenger rail transportation had left the region, unable to compete.

✳ ✳ ✳

"We are approaching the age of the automobile," declares an article in the August 4, 1899 issue of the *Springfield Leader and Press*:

> In this age of applied science, our old equine friend is passing away. That he may still be seen ambling unapprehensively up and down the streets of our different American cities is quite true, but now that the automobile has passed out of the experimental stage of its existence and is firmly established in popular favor, it is simply a matter of time till the merchant and the millionaire, the drayman and the doctor, will all "mote" about the face of

this earth for business or pleasure, as the case may be.

In 1899, apparently, a "drayman" or beer brewer was sufficiently well-heeled to be mentioned alongside a doctor or merchant. For the common man, however, price remained an issue: "During the last year or two great improvements have been made in the building of automobiles, and the only problem now ... is the question of reducing the cost of construction."

A group of local entrepreneurs thought they had found a solution. "THE AUTOMOBILE: A Company Forming to Operate This Latest Fad in Springfield," reads a headline in the August 21, 1899 issue of the *Springfield Leader and Press*:

> Springfield will soon be decidedly in fashion. Springfield is to have the latest fad on record—the automobile. A company is now being formed having for its object the purchasing and operating of this new vehicle....
>
> It is intended to purchase the patterned portions of the machine and have them put together and the balance manufactured in the city. By this procedure it is claimed a saving of at least one-third may be made.... The company will start operations with two carriages, two hacks or buses, and one baggage wagon. As business demands it other vehicles will be added to their stock.

Here's a typical Ozarks-style improvisation: Buy what you can't make for yourself and manufacture the rest at home. The plan fell through, however, leaving the town car-less. A year later, in its April 21, 1900 issue, the *Springfield Leader and Press* gave the headline, "Automobile for Springfield." "The Pickwick Livery Company has ordered an automobile," the article notes, "which will run from the depots to the hotels. It will cost about $4,000, and it will be here just as soon as the factory can turn it out. It will be the first

---

15. It's easy to forget the impact that one-time revolutionary technologies have on our worlds, since these tend "to become invisible" during daily use, retreating into the white-noise background of our lives and environs. (As a rule, technology calls attention to itself only when it breaks down or needs human management.) John Sellars, director of Springfield's History Museum on the Square, tells of an oldster living at the turn of the previous century, when the technologies of modernism were making their way into the Ozarks. Asked which "first sight" of which invention most impressed him, it was not the high-flying airplane but the dust-throwing automobile that caused the greatest wonder. For, "if that were possible," said the old Ozarker, "anything could follow."

automobile in Springfield." Did it arrive, in fact? In its newspaper ads through 1901, the Pickwick Livery & Transfer Co. makes no mention of an automobile—which, surely, would have been a "draw" for business.

It was on April 7, 1901, that local history was made. In its April 8 issue, the *Springfield Leader-Democrat* reported "a strange vehicle on the streets yesterday":

> People gathered about to make a close inspection and see how it was made and horses shied at it. It was no more or less than an automobile, propelled by gasoline and made by a young colored man of Springfield. The trip of this first horseless carriage made in Springfield was not entirely successful but the vehicle moved and could be steered and stopped at will. It did get a rapid move on it and there are some glaring faults in its construction[, but] the young colored man has the right principle and he can perfect the machine so it will carry him on smooth streets at a rapid rate.

The "young colored man" abovementioned was Springfield's own Walter Majors, an African American who built the town's first workable car. Surely cars had driven through town, given their use in promotion and product advertising. But this one, made at home in Majors' garage, drove up Commercial Street, came to an idle, and drew a crowd. The auto had arrived and was here to stay.[16]

By 1902, Springfieldians could buy cars from Martin Howard & Co. Other dealers entered the market. By 1903, the automobile was a regular downtown sight, sharing the road with streetcars and horse-drawn carriages. By 1904, cars were racing at the Fairgrounds. And causing accidents: "J. M. DOLING WAS HURT," reads a headline in May 25, 1904 *Springfield News-Leader*, whose subject-line adds, "Horse Became Frightened At Automobile And Buggy Overturned."

> Hon. J. M. Doling, ex-member of the legislature and owner of Doling Park, was the victim of a serious runaway accident yesterday afternoon.
>
> About five o'clock Mr. Doling, who was returning from the public square, met an automobile, which caused his horse to make a sudden turn. The buggy was overturned, but fortunately did not fall upon the occupant. Mr. Doling however was thrown violently to the ground…. The patient has

---

16. The automobile stayed, but Majors didn't: In 1907 or '08, he left for St. Louis. Still, he often returned to a hero's welcome, having "spent the greater part of his life in Springfield" and remaining "widely known among the older railroad men," with whom he had worked. I quote from the September 24, 1916 issue of the *News-Leader*, whose headline reads, "'Duck' Majors Here on 4,000-Mile Auto Tour." The article continues:

> Walter L. Majors, colored, better known as "Duck" Majors, who built and operated the first gasoline car in Springfield in 1896, and [is] now head of the Oxford College of Hair and Beauty Culture at St. Louis, arrived in Springfield last night in a specially equipped "Speedwell" six-cylinder 70-horse-

power automobile. Majors will leave Friday for St. Louis, completing a 4,000-mile tour of the Middle West….

In this particular visit, "he came … to demonstrate the advantages of his college to colored residents of Springfield." We're told, too, that he "will speak at the negro churches here before leaving." Surely his lay-sermon covered more than the college. For Springfield's African American community, Majors exemplified success in innovation and business. He was a walking (rather "motoring") advertisement for the St. Louis college, the Springfield community—and for his fancy Speedwell "Six," a five-seat touring car whose 1913 version cost a whopping $2,850 fully equipped. And, on this trip, the local roads posed more problems than racial prejudice:

> Roads in Michigan, Illinois, Iowa, Kansas and Oklahoma are in excellent condition compared with Missouri roads, Majors said. With good roads leading into Springfield, he said, scores of tourists would pass through here daily. At every point he stopped, Majors encountered unusual hospitality and at many points was assisted in changing tires by white men who disregarded color prejudices.

For further discussion of Walter Majors' achievement, see the essay by Richard Schur, included in this present volume.

sustained a severe gash on the left cheek and ... six stitches were taken....

He does not blame any person for the accident, saying that the horse jumped suddenly and it could not be avoided. He expects to be all right in a few days and is thankful that it was no worse.

It was a clash between technologies—between a horse-drawn and a horseless carriage. And, in Doling's case, the newer technology "won."[17] By 1906, one could rent a chauffeured car in town by the hour. And Ozarkians were out testing the technology's limits. The September 1, 1906 issue of the *Springfield News-Leader* reports on a "LONG AUTOMOBILE TRIP." The paper's readers would smile at the news that "Mr. and Mrs. F. T. Snapp left from their home in Joplin in their automobile" (a Buick, we're told), and "made the trip to Springfield in nine hours, covering ninety-eight miles." From Joplin to Springfield in nine hours: *Not bad!*

In 1915, Fairbanks and Tuck give the following assessment: Though "it has not been so very many years ago since the first automobile made its appearance in Springfield," the auto business "has grown with perhaps greater strides than any other line in the twentieth century" (p. 972). They continue:

These autos are not only to be found in the larger cities, but in almost every city and town in the

Union, and even on the wide plains of the West and in mountainous districts. One finds them in many of the rough, poor sections of the Ozarks. People not only enjoy riding in them, but they realize that they are time savers and thus in many instances money makers. Those engaged in this line of business, whether in manufacture, selling or repairing, are making a success. (p. 972)

Thus the automobile and its technologies (in manufacture, sale, and repair) contributed mightily to Midwest entrepreneurship: By the 1920s, "the age of the automobile" had arrived, scratching its way through "rough, poor sections of the Ozarks." It was an age, not just of the automobile, but of roadbuilding; and it was an age dominated by the likes of John T. Woodruff (1868-1949). His story has been told elsewhere, so what follows is a summary.[18]

Though he began as a "railroad man," working as an attorney for the Frisco, Woodruff saw the Ozarks' future in its rail-less roadways. By the 1920s, the Ford Motor Company (among other industry brands) was reinventing American transportation—and American lifestyle, which became increasingly dependent upon the automobile for work *and leisure* (including that newfangled urban-American middle-class practice of "Sunday drives" and vacationing). Springfieldians needed easy ways to get in and out of town, and

---

17. I'm uncertain as to the first Springfield fatality, though Fairbanks and Tuck give a gruesome account of Albert N. Hanson:

The death of Mr. Hanson occurred on April 16, 1915, as the result of an accident. He was driving across the street in his automobile when a street car crashed into his machine, hurling him from his seat a distance some twenty feet, his head striking the curbing. Burning oil from the gasoline tank of the automobile was scattered over him and the oil took fire, igniting his clothing. Help reached him immediately, but he remained unconscious to the end which came a few hours later, as a result of injuries to the head. (p. 1078)

18. For an authoritative source, see Thomas A. Peters, *John T. Woodruff*. Perhaps ironically, the automobile's energy source—gasoline—depended on the railroad for its supply chain. In delightfully concatenated prose, Lynn Morrow writes, "autos could never have been in the Ozarks without the railroad tank cars that docked at rail towns and pumped gasoline into large petroleum tanks by the railroad for over-the-road tank trucks that loaded up and then drove to gas stations to fill smaller tanks whose contents were then transferred again to consumers driving autos with smaller tanks yet. The tank truck drivers and their overland routes were fed by national franchises, like Standard Oil, Sinclair, Conoco, etc. And the local agents, often in county seat towns,... drove the routes and provided petroleum products to gas stations" ("The Auto").

Woodruff lobbied—successfully—to put the town on the map of the nation's first great east-west motorway, eventually to be named Route 66. As Woodruff writes in his memoir,

> The fever to build common things is intermittent. It comes by fits and spurts. Not so as to roads. Once you come down with it, the fever and fervor continue. Travel on horseback, by wagon, buckboard, buggy, or stagecoach required roads of course, and there were then roads of a kind. But the advent of the motor car propelled by the internal combustion engine called for more roads, good all-weather roads.
>
> The feverish anxiety to gain these took root in the Ozarks as early as anywhere else. The result was the enactment by the Missouri Legislature of laws authorizing the creation of the Special Road District in the country and the "Eight Mile" District in and around the city of Springfield, Missouri. We were not remiss in employing the plans thus provided, in any respect.... ("John T. Woodruff," p. 54)

Whereas the Ozarks lagged behind in other transportation technologies, it strove "to keep pace" in this definitively modern development.

The history of Route 66—the nation's "Mother Road," which passed through Springfield's town square—has been told numerous times. What we'd call attention to here is the "feverish" entrepreneurial spirit that Woodruff describes, one that saw a specific technology, "the internal combustion engine," fed by an accessible power supply (gasoline), turned into a means of transport that would reshape the terrain, *forever changing the Ozarks*. In 1926, when Woodruff opened the Kentwood Arms Hotel on St. Louis Street, he prophesied the role that the automobile would play in building modern Springfield. Kentwood Arms would be the town's first hotel built specifically for motor-tourism. (Previously, hotels were built near the train depots in service of rail passengers. Other motor-hotels—a.k.a. motels—would follow, though typically less impressive, along Route 66.)

Local entrepreneurs—boosters all—have left their mark. Before there was John Q. Hammons (1919-2013) and C. Arch Bay (1909-1993), there was John T. Woodruff (1868-1949); before him, there were H. F. Fellows (1831-1894), Sempronius H. "Pony" Boyd (1828-1894), John S. Phelps (1814-1886), and others leading back to Joseph Rountree (1782-1874) and John Polk Campbell (1804-1853). They built the region by attracting settlers, investment capital, and technologies of transport. (Even Campbell contributed in this regard, being a horse-trader by profession.)

Beyond the story of Midwest town-building, the history that we've told declares the slow, steady triumph of modernism—specifically, of the "conquest of time and space" via technology. The story of local transport is one not just of access, but of speed: Distances and terrains that took pioneer settlers weeks, even months to traverse could now be measured in hours. So long as it's moveable, there's nothing nowadays that can't be brought into, or taken out of, the Ozarks.

We have concentrated thus far on transportation as one of four markers of modernity. We turn now to a second marker, communication: for it, too, is implicated in the modernist "conquest of time and space."[19]

---

19. Note that any discussion of radio telecommunication must take account of its energy source: electricity. If wood and coal powered the 19th century, and gasoline (among other petroleum products) powered engines in the 20th century, then this most modern of all energy supplies—bolstered in 1913, with the completion of Powersite Dam in Forsyth, Missouri—has carried the Ozarks reliably into the 21st century. It can be said, without much exaggeration, that our lives (and certainly our economy and lifestyle) depend on an interconnected regional/national power grid. *Sans* electricity, the "information age" is inconceivable.

For a discussion of Powersite Dam, see the essay by Thomas A. Peters, included in this present volume.

3. Boosting by Radio

"In the last twenty years, neither matter nor space nor time has been what it was from time immemorial. We must expect innovations to transform the entire technique of the arts, thereby affecting artistic invention itself." Paul Valéry, French artist-philosopher, wrote these words in 1928. While his subject is aesthetics, his claims hold for all modes of electronic telecommunication/transmission. Soon, Valéry suggests, "it will be possible to send anywhere or to re-create anywhere a system of sensations, or more precisely a system of stimuli, provoked by some object or event in any given place." He continues:

> Just as water, gas, and electricity are brought into our houses from far off to satisfy our needs in response to a minimal effort, so we shall be supplied with visual or auditory images, which will appear and disappear at a simple movement of the hand, hardly more than a sign. Just as we are accustomed, if not enslaved, to the various forms of energy that pour into our homes, we shall find it perfectly natural to receive the ultrarapid variations or oscillations that our sense organs gather in and integrate into all we know. I do not know whether a philosopher has ever dreamed of a company engaged in the home delivery of Sensory Reality.

Composed some ninety years ago, Valéry's triumphalist vision seems to describe today's technologies of cable television and movies on-demand, of the internet and iPhone, of video game consoles and virtual reality goggles. From this distance, it's hard to imagine that radio, "mere" radio, was the source of Valéry's rapture.

Though radio came first, its development makes these future technologies seem unsurprisingly inevitable. For the radio had already conquered time and space as *problems of mass communication*. As Valéry

writes, it made "a piece of music audible at any point on the earth, regardless of where it is performed." Further, it preserved "live" events for future performance, allowing its engineers "to reproduce a piece of music at will, anywhere on the globe and at any time."[20] Again, his focus rests in aesthetics, but the technologies of transmission range through all informational content, from broadcast news to encrypted military messaging.

In fact, an ad in the December 13, 1925 issue of the *News and Leader* had already touted the APEX Radio Apparatus (available locally) and its "mastery over the most advanced radio engineering principles," which "makes distance the obedient slave of your desires and places at your instant command the whole continent of radio enjoyment." As a further point of fact, Springfield's State Teachers College—precursor to Missouri State University—made a significant *scientific* investment some years earlier: "Radio Set to be Installed at State Teachers College," reads a headline in the December 11, 1921 *Springfield Leader and Press*:

> A complete radio station will be installed in the rooms of the science department....
>
> It was in recognition of the imminent possibilities of radio telephony that the Teachers College authorities decided to install the radio equipment. Not only the students of the science department but students of all departments and divisions of the institution will benefit by the modern

---

20. "Radio Fans Hear Foreign Stations," reads a headline in the January 29, 1926 *Springfield Leader and Press*. Foreign reception had become a friendly competition:

> Radio station 2-LO London, England, was heard last night by several local radio fans.... Mrs. Harry Gabriel, 506 East Grand Street, reported she received three or four numbers very distinctively over her radio set.... Radio fans in Marionville were able to reach many foreign stations last night.... A. L. Owens heard a musical program broadcast from Statio OAX, Lima, Peru.

wireless station ....

To disseminate wireless news a Daily Radio News Bulletin will be distributed among students, it is planned. The service of a dozen or more students who are now prepared to "receive" and "send" messages will be used to demonstrate the possibilities and practicabilities of wireless telephony and telegraphy.

The radio telephone receiving station ... will be "tuned" to receive from Washington, Pittsburgh, Chicago, Denver, St. Louis, Kansas City, and points of the lower South....

What a decade will bring to light in the possibilities of radio is a matter of extravagant speculation. Practical scientists and electrical experimenters are confident that the music is now within easy reach of all.... Speculation has it that concerts in distant cities and possibly sermons will be "picked up" in the home.

Apparently, we didn't need to quote the French philosopher on the future of radio transmission. Springfield newspapers had beaten him to the punch.[21] The article ends noting that "the Teachers College in making an outlay for modern radio equipment is keeping abreast of the time." That's what a school is *supposed to do*, and what a school like Missouri State is doing now with the latest imaging, digital, computational, and virtual reality technologies.

In *Commercial and Government Radio Stations of* *the United States* (1921-1923), the U.S. Department of Commerce lists "WQAB Teachers College" among the nation's first "experimental and technical and training school stations" (p. iv). Apparently, the school shared a radio frequency (kHz 833) locally with two smaller commercial stations, "WKAS L. E. Lines Music Co." (10 watts) and "WIAI Heers Store" (20 watts). Another government document, *Amateur Radio Stations of the U.S.* (1923), lists the fairly strong-signaled "9CEG Springfield High School, Benton and Center St." (1,000 watts), along with several privately owned stations of varying broadcast strength: "9CID Kirk T. Pruess, R.R. 11 Park Ave." (750 watts), "9EGI Granville P. Ward, 236 W. State St." (500 watts), "9AUK Charles Birget, 1367 Summit Ave." (250 watts), and "9DOR George F. Lytle, 760 E. Elm" (20 watts). While Springfield High (like Southwest Teachers College) bought state-of-the art equipment, the smaller stations were cobbled together in true amateur style. Beyond the sheer inventive spirit, one feature unites them all: Being "experimental," *none of them survived.*[22]

But it wasn't an amateur's tinkering that piqued entrepreneurial interest. Springfield's Chamber of Commerce was keen on bringing federally licensed, *professional* radio to town, given the technology's outreach in advertising—that is, in boosting the community. In 1927, the Chamber got its wish.[23] "Initial

---

21. If cyborg technology—the hybridizing of human and nonhuman capacities—seems postmodern in sentiment, consider the article, "Telepathy May Be Radio's Big Freak, Experts Believe," printed in the September 6, 1925 issue of the *Springfield Leader*: "Instead of dealing with this phenomenon as a psychic factor, scientists are coming to believe that the mysterious action of one mind on another is accomplished through the transmission of some sort of ether waves...."

Given that the first federal license for public radio broadcast was issued in 1920, we note the speed with which this technology disseminated regionally, nationally and, indeed, globally.

22. Formed in 1927, the Federal Radio Commission effectively killed stations by refusing to renew their licenses. The AM band had become woefully overcrowded (FM would not be introduced until 1941), forcing smaller stations to share frequencies (Goodman). The Radio Act of 1927 "limited radio broadcasting to licensed broadcasters" (Stefon) and eliminated "split frequency" programming, while FRC General Order 32 (1928) effectively removed "experimental status" as a category for licensing. The FRC "cleaned up" the AM airwaves by licensing big commercial stations almost exclusively—which explains the quiet demise of WQAB Teachers College. (After World War II, educational and "public" radio stations would reappear on the FM band.)

23. Local histories of radio typically begin with Ralph D. Foster's KGBX station, which began broadcasting in 1926 in St. Joseph, Missouri and moved to Springfield in 1932. Though Foster put

Program Broadcast of Local Station," reads a headline in the January 3, 1927 *Leader and Press*: "Radio Station WIBM took to the air at the Landers Theater this morning with a test program," with afternoon and evening programs to follow: "... and tonight at 10 o'clock the Chamber of Commerce will broadcast a program at the courtesy of the *Springfield Leader*. Some excellent local artists are scheduled for tonight's program and fans are promised a real treat."

The "test program" proved successful, as reported in the January 19, 1927 *Leader and Press*. Woodruff's own "Kentwood Arms hotel has been selected as the permanent home of WIBM," reads the article. In occupying the main ballroom, "without doubt this arrangement will give WIBM one of the finest studios in the country." The article continues:

> The beauties and virtues of the Ozarks will continue to be made known to the outside world through WIBM and the name of Springfield will be kept constantly before thousands of listeners throughout the middle west.
>
> The conclusion of the enterprise was made possible through the cooperation of some of Springfield's most representative leading men. Certain programs will be given over to nearby towns through the cooperation of their local chamber of commerce bodies.

More than a private commercial enterprise, such a station served the entire business community: Those "representative leading men" who brought WIBM to the Kentwood Arms ballroom were selling an image and reputation—that of a "modernized" Ozarks. They were "letting the world know that Springfield is up and coming—not going or standing still."

The radio, thus, provided the region with its most powerful "boost" yet. As with the technologies of transport, the technologies of radio broadcast conspired with booster policy to create a prosperous, progressive, forward-looking self-image. Global economic depression and war lay in the immediate future; still, Springfield's "leading men" were determined to grow from town to city. And grow it did. As the Ozarks' "Queen City"—the region's largest urban center and rail transportation hub, powered by hydroelectricity, blessed with natural resources, fed by local farms, enjoying a skilled (and largely unionized) labor force, and led by men and women of capital wealth and creative vision—Springfield was poised to lead in the transformation of the Missouri Ozarks.[24] It would achieve this transformation through technologies of communication, transportation, and "citified" culture.

By mid-20th century, the energy, transportation, and communication infrastructure of "modern" Springfield had been laid down and largely completed. By the turn of the new millennium, further changes were in store: Out of the city's increasingly tech-driven service economy, a "postmodern" Springfield would evolve. (Again, it's "technology to the rescue.") And the city's relation to the surrounding region would evolve, as well. What began as a pioneer settlement had at last grown into a metropolitan center encircled by suburban and peri-urban regions and its own vast Ozarks hinterland. *These* subjects, however, deserve an essay in themselves.

Springfield on the broadcast map, clearly radio had arrived before KGBX and Foster's RadiOzark Enterprises. But it didn't stay: The C. L. Carrel Broadcast Company of Chicago operated WIBM under a portable license. On May 4, 1927, following a successful ten-week stint, WIBM returned to the Windy City.

24. Given the masculinist bias in entrepreneurship, it's not inaccurate to speak of the businessmen in Springfield's modernizing; and yet, just south of Springfield in Bonniebrook stood the family home of Rose O'Neill, whose invention (and savvy marketing) of the Kewpie doll had made her "fabulously rich" (McCanse, p. 8)—a millionaire "captain of industry," as she's described in the February 20, 1921 *New York Times* article, "Women Who Lead the Way." Of Ozarkians living in the '20s, few besides Woodruff could rival O'Neill in cultural impact.

Inventing the Ozarks (II):
In Transit to Postmodernity
James S. Baumlin

Many people have an eye on Springfield now.
It's the growingest place in the Midwest.
—Mayor Carl Stillwell, Groundbreaking Ceremonies for the R. T. French Company's Springfield Plant (1971)[25]

[In postwar Springfield,] there was an enormous pent-up demand for everything that could be built in a factory…. Improved radios, for example; and soon there would be the incredible boom in television (who would have believed, in 1945, that Springfield would one day be the site of the world's largest television factory?). Power mowers, once used only on golf courses and vast estates, would soon be a common neighborhood item at a low price (who could have expected Springfield's own Mono Manufacturing to become a leading producer?). Paper cups, once seen only in railroad cars, soon were to find their way into homes and restaurants across the nation (who would have dared bet that the world's largest paper converting plant was shortly to call Springfield its home?).
—Harris E. Dark, *Springfield Missouri: Forty Years of Progress and Growth, 1945-1985* (1984)

Previously, I charted Springfield's meandering path to mid-20th century modernity: By technologies of communication and transport, the Ozarks' "Queen City" overcame its spatiotemporal isolation and, drawing resources from the surrounding region, developed an infrastructure supportive of export-industry and manufacture. If we pass over the Great Depression and World War II and follow Springfield's GIs back home, we'd be entering a vibrant small city ready and able to employ, house, teach, care for, and entertain its citizens, one fully connected to the world beyond. We'd also note that its successes in industrialization and urbanization stood in contrast to its hinterland, where agribusiness was replacing family subsistence-farming, and once-prosperous mining and mill towns were suffering underemployment, economic stress, environmental degradation, and shrinking population.[26]

In transiting from modern to "late modern" or "postmodern" technoculture, *it's the city itself* that evolved, changing demographically as well as economically. For postmodernity marks the triumph, not just of transportation and communication (the subjects of previous discussion), but of urbanization.[27]

---

26. Though I doubt the necessity of doing so, let me make the following disclaimer: Of course one can live the good life in rural regions of the Ozarks. One can find employment, raise a family, and enjoy friendships, pastimes, and the fellowship of neighbors while "staying connected" to the rest of the world through roads that do reach into these communities and cable- and satellite-delivered media that do reach up into the hills and down into the hollers. This undeniable (and pleasant) fact does not negate the general trend regionally, nationally, and globally as urban populations rise while rural populations continue to decline.

27. Though classification by size is simplistic, U.S. Census data list three "city-sized" urban centers in the Ozarks: Springfield (pop. 167,000), Fayetteville (pop. 86,000), and Springdale (pop. 80,000). Large towns are scattered throughout the region: At the northernmost tip lies Jefferson City (pop. 43,000); at the easternmost edge lies Cape Girardeau (pop. 39,000); to the west lies Joplin (pop. 52,000); in the southwest lie Rogers (pop. 68,000) and Bentonville (pop. 52,000). Towns of intermediate size include Nixa (pop. 21,000), Rolla (pop. 20,000), Ozark (pop. 20,000), Poplar Bluff (pop. 17,000), Fort Leonard Wood (pop. 17,000), Republic (pop. 16,000), Lebanon (pop. 15,000), Carthage (pop. 14,000), West Plains (pop. 12,000), and Branson (pop. 11,000).

The Ozarks has hundreds of small towns under 1,000 and

---

25. Quoted in Harris E. Dark, *Springfield, Missouri* (p. 158; emphasis in original).

At the same time, postmodernity marks the region's shift in its structures of economy. By 1960, Springfield had earned its reputation as a Midwest working-class factory town. By 2000, Springfield had become something else. In the city's transit to a postindustrial economy, varied technologies—of information, communication, and media entertainment, among others—would play their roles.

1. '60s Booms and '80s Busts

In *Springfield, Missouri: Invitation to Industry* (1960)—a Chamber of Commerce publication—the opening page reads, in boosterish italics, *"THERE MUST BE A REASON ... why people are moving to Springfield."* A part of that reason lay in "Frisco's west shops," still "the largest railroad shops west of the Mississippi" (*Springfield* p. 6). (Despite the automobile's ascendance, postwar Springfield remained a Frisco town.) A bigger part of that reason lay in the city's diverse industrial/manufacturing base: "Since 1950 these national firms have chosen Springfield, Missouri, as a location: Lily-Tulip Cup Corporation, Royal-McBee, Kraft Foods, Dayton Rubber Co." (*Springfield* p. 69).

Though Kraft had arrived in 1939, it made sense for the company to expand its Springfield operations, given that the Ozarks of the 1950s was "then the most concentrated dairy area in the United States" (Dark, p. 38). Lily-Tulip opened shop in 1952. In 1957, Royal McBee announced that "it would build the world's largest portable typewriter factory in Springfield, creating 1,000 new jobs" (Dark, p. 67), and "another of Springfield's 'world's largest'—a V-belt factory of Dayton Rubber Company" (p. 68)—announced its intentions that same year: By 1960, both factories were fully operational. And both were named among the nation's "10 top plants" for the year, as chosen from 500 by *Factory Magazine* (Dark, p. 69).

There were other grand events: In 1960, "Ozark Air Lines started its first prop-jet service," President Eisenhower "signed the bill creating Wilson's Creek National Battlefield," "Burge Protestant Hospital, ever expanding, was renamed Lester E. Cox Medical Center," St. John's Hospital "added a $1,800,000 annex for psychiatric, medical, and surgical care," and "Paul Mueller, already one of the city's largest plants, was expanding to include the manufacture and sale of stainless steel storage tanks and other equipment for the dairy and food industries" (Dark, pp. 103-05).[28] Besides, Springfield was looking to break the 100,000 mark in population. *Who could deny that it had grown into a city?*

Two decades later, the 1980 *Manufacturers Directory*—another Chamber of Commerce publication—was once again boosting the region's industrial economy. There was a difference, however, in immediate prospects. Whereas the 1950s and '60s were postwar boom-decades, the 1970s and early '80s were busts nation- and worldwide. It took several years for "the Great Recession" (Dark, p. 195) to reach the region.[29] But, "by the end of 1979," as Dark notes, "the Recession had hit Springfield hard," with "the worst part of the Recession" felt "towards the end of 1980" (p. 195). Jobs were lost and businesses closed throughout the

---

28. For discussion of the postwar economy, see Dark's *Springfield, Missouri: Forty Years of Progress and Growth, 1945-1985*. While the Chamber of Commerce was touting the national firms moving in locally, there were local firms—like the Paul Mueller Company—that were growing nationally: "Paralleling quick technological changes in food processing and brewery methods, the company by the 1970s had become the nation's largest supplier of brewery storage and fermenting equipment and one of the biggest processors of stainless steel in the world" (Dark p. 36).

29. In 2019, this term is reserved for the global financial crisis of 2007-2009, caused by the collapse of the U.S. real estate market (which led, in turn, to the U.S. subprime mortgage crisis). But, writing in 1984, Dark describes a different "Great Recession," one belonging to the '70s and early '80s: Marked by the U.S. "savings and loans crisis," it was a time of hyperinflation and "stagflation" leading to high interest rates and high rates of unemployment.

many, indeed, under 100 in population: There's Oxly (pop. 99), for example, and Big Spring (pop. 50), and Peaceful Village (pop. 9).

region; most of these never returned. Family farms were lost, as well.[30]

But, more than a temporary inflationary/financial stumbling, the 1970s recession marked the end of an age—an age, that is, of postwar economic expansion nationwide. Through the 1960s, Springfield rode that wave. But, as that wave receded, it took jobs and industries with it. In 1980, the following companies had payrolls in excess of one hundred (employee totals are given in parentheses), though most *Directory* subscribers were small businesses, employing a dozen or so workers:

Advanced Circuitry Division of Litton Systems on West Kearney (1,000)
Colonial Baking Co. on St. Louis (175)
Dayco Corporation, Springfield Plant on Battle-field and Scenic (1,300)
Fasco Industries of Ozark (700)
Foremost Foods Co. on North Campbell (263)
General Electric Specialty Motor Department on East Sunshine (940)
Holsum Bakers of Springfield on South Grant (150)
Hudson Foods on North Main (375)
International Harvester Pay Line Division on West Maple (230)
Kraft, Inc. on East Bennett (1,000)
Lily Division of Owens-Illinois on North Glenstone (1,250)
MD Pneumatics, Inc. on West Kearney (160)
MFA Milling Co. on Boonville (200)
McBee Looseleaf Binder Products on North Cedarbrook (250)
Mid-America Dairymen, Inc. on West Tampa (249)

Minnesota Mining and Manufacturing Co. on East Trafficway (175)
Paul Mueller Co. on West Phelps (1,200)
R. T. French Co. on East Mustard Way (500)
Reyco Industries on North Prospect (500)
St. Louis-San Francisco "Frisco" Railway Co. on North Lexington (2,000)
Syntex Agribusiness, Inc. on West Bennett (230)
Tindle Mills, Inc. on East Chestnut (135)
Vermillion, Inc. on South Scenic (140)
Zenith Electronics Corporation of Missouri on East Kearney (1,500)

Longtime Springfieldians will remember most of these. Some are still going strong, like Kraft (Levin, *Celebrating* pp. 68-69) and Mueller (p. 202).[31] Some have changed names and owners—most famously the Frisco, which was bought out in 1980 by Burlington Northern. Some have moved a short distance: Vehicle-suspension manufacturer Reyco, for example, took its plant to Mount Vernon. Some have moved out of state; some others out of the country. And some have gone out of business. Companies come and go.

For change, they say, is the only constant. In his chapter, "A Time for Belt-Tightening: 1975-1980" (pp. 164-95), Dark proves a savvy historian and prognosticator, noting that Springfield's "highly diversified economy" (p. 167) had saved it from collapse. But, by the early 1980s, he could read the writing on the wall. The future would be driven not by manufacture primarily, but by "Springfield's unusual ability to develop and sell valuable *services*, especially in the medical,

---

30. Additionally, "much of downtown Springfield, the three-to-four block business area surrounding Park Central Square," had "fall[en] apart, physically and commercially" by the mid-1970s, "unable to compete with the convenience of the many shopping centers that had started up within three to five miles in all directions" (Dark, p. 226).

31. See Rob Levin, *Celebrating Springfield: A Photographic Portrait* (2007). A lavishly illustrated "coffee table book," Levin's is the latest piece of published local boosterism. Note that its list of "Springfield Featured Companies" (pp. 318-25)—the book's underwriters, doubtless—is dominated by banks, schools, law firms, and health clinics—components of the service sector, not manufacture.

insurance, automobile, convention, travel, and tourism fields" (Dark, p. 171; emphasis in original). These are the fields that dominate his final chapter, "Prognosis Affirmative: 1980-1985" (pp. 196-230). They are fields that dominate the regional economy today.

By the turn of the 21st century, Springfield had at last evolved from a town to a city to a metropolis.[32] And, as we move from the machine-age—legacy of the 20th century—into the postindustrial "information age," we find shuttered local factories replaced by a "meds and eds" service economy, whose lifeblood pulses through the city's government offices, medical centers, schools, and such cultural centers as its revitalizing downtown—prime purveyor of arts, entertainments, and urbane lifestyle. Consider the city's largest employers in 2018 ("Doing Business"):[33]

---

32. Actually, the Ozarks' Queen City remains something of a hybrid. In its size, layout, and practices of self-promotion, Springfield most closely resembles the "small city," as described by Mahoney:

> Most small cities are autonomous places with a distinctive urban history. Most have, in spite of some urban sprawl, a clear edge between the city and the country (usually no more than a twenty-minute drive away). Most small cities have relatively few major suburbs surrounding them. Instead they have annexed large areas on their fringe, sometimes absorbing nearby villages and towns. As a result of physical separation from any metropolis, most small cities have a coherent identity as a separate place, and residents view their city as an urban whole containing all the elements of both the small town and the metropolis. This tends to create a divided image which is reflected in policy. Describing one's city as a "big city with a small-town feeling" or exclaiming that the acquisition of a certain institution, facility, or team means no longer being a "small town" are staples of the chambers of commerce and mayors of small cities. A small city aspires to be a metropolis while priding itself on maintaining aspects of the small town. It is both, and, as such, is inevitably described as having "the best of both worlds." (Mahoney 320)

33. It's worth studying the complete list given in the Chamber of Commerce webpage, "Doing Business," particularly in growth areas of information technology and telecommunications. Note that employee totals reflect urban growth generally, as the city's population has increased nearly 58% from 1960 (95,865) to 2018 (est. 167,000).

CoxHealth (11,669): Healthcare
Mercy Hospital Springfield (10,950): Healthcare
Walmart, Inc. (5,372): Retail
Springfield Public Schools (4,100): Education
State of Missouri (4,018): Government
Bass Pro Shops/Tracker Marine
    (3,341): Retail/manufacturing
U.S. Government (3,005): Government services
Missouri State University (2,874): Education
Jack Henry and Associates
    (2,174): Software development
O'Reilly Auto Parts (2,042): Retail
Citizens Memorial Healthcare (1,900): Healthcare
City of Springfield (1,655): Government services
Ozarks Technical Community College
    (1,554): Education
EFCO Corporation (1,550): Manufacturing
SRC Holdings (1,435): Manufacturing

As a culturally and economically diversified metro center heading into the 2020s, Springfield finds itself negotiating further tech-driven changes. The question arises: Where do we go from here?

2. "Cityness" and the Commodification of Culture

The postmodern *habitus*, for good or ill, is the city. But we should add a further warning over vocabulary: Since the late 20th century, we can no longer speak of "urban" and "rural" as clearly defined, stable categories. Paul James points to the problem. While cities globally "have come to dominate landscapes far beyond [their] metropolitan zone," the "changing forms of urbanization" have led to "urban sprawl and the decentralization of non-residential functions[: for example, of] retail parks close to intercity highway junctions, massively increased levels of commuting between urban and rural areas,... and the emergence of polycentric urban configurations" (*Urban Sustaiability*, p. 25).

In today's parlance, *rural* has become "a catch-all category for 'not urban'" (James, p. 26). The regional variety of urban-rural configurations deserves more attention than can be given here; at the least, we need to think in terms of a Springfield metropolitan area—that is, of Springfield-Republic-Nixa-Ozark-Branson conjoined—beyond which lies an expansive hinterland.[34] (In northwest Arkansas, the Ozarks have a second metropolitan area in Bentonville-Rogers-Springdale-Fayetteville). Different regions of the Ozarks do orient themselves toward one or another urban center in a two-way exchange, supplying agricultural and other goods while purchasing consumer goods and entertainments in their turn.

Still, the "settlement categories" as described above belong to the mid- to late 20th century. In the 21st century, we'll learn to think in global-local terms. In this tech-driven shift from urban-rural to global-local, we'll be completing our transit from modern to postmodern. Where one sleeps, and where one does business, can be in different communities on different continents. For instantaneous global communication reconfigures our relationship to physical space—to where we reside and work and play. Think of it in the following terms: In their access to (and command of) communication and information exchange, Springfield, Missouri and Shannon County's Eminence, Missouri are equally interconnected with Beijing, China and Basel, Switzerland. Neither has an absolute advantage in this regard. Both, given access to technology, can do business and "broadcast" worldwide. The difference between Springfield and Eminence lies not in access to technology, but in lifestyle: specifically, in access to "citified" culture with its eclectic architecture, diverse cuisines, museums, and zones of shopping and entertainment.

---

34. As James defines it, the hinterland is a rural area "close enough to a major urban centre for its inhabitants to orient a significant proportion of their activities" to that "dominant urban area of their region" (p. 27). The official Springfield/Branson Combined Statistical Area (CSA) includes Polk, Dallas, Greene, Webster, Christian, Stone, and Taney Counties.

It is silly to try to deny the fact that we Americans are a city people, living in a city economy—and in the process of denying this lose all the true countryside of metropolitan areas too, as we have been steadily losing it at about 3,000 acres a day for the past ten years.

—Jane Jacobs, *The Death and Life of Great American Cities* (1961)

[C]ities that will do well in the next generation have two things in common. First, they have a strong institutional foundation upon which to build a postindustrial economy, particularly the "meds and eds" economy. Second, they have not lost their essential "cityness"—they make room for new kinds of people, who come together in new kinds of ways, and they foster a sense of civic identity that transcends the boundaries of race or class or ethnicity or religion.

—Steven Conn, *Americans Against the City* (2014)

The paradox of a fully developed metropolitan Springfield-Republic-Nixa-Ozark-Branson lies in its current most notable export, which is *not* material or industrially manufactured; nor is it informational or data-based, though it is (as in previous ages) gathered up for "processing" from the surrounding hinterland. This export, rather, is cultural-esthetic. We refer back to Paul Valéry, who predicted radio's capacity "to transform the entire technique of the arts, thereby affecting artistic invention itself." He got that right. In our new global setting, *regional culture* becomes marketable. It's not the railway or the highway but the airwaves that has carried this commodity most notably, broadcasting it locally, soon nationally, and eventually worldwide. We're talking, of course, of the Ozarks' hillbilly, "pioneer" ethos.

Even as the region was building its postwar industrial base, its commodification and export of "countrified" culture had begun. And, when we think of that commodified rural culture, it's the music that comes first to mind. In 1947, KWTO radio-broadcast its *Korn's-a-Krackin'* country-themed variety show from Springfield's Shrine Mosque. Brainchild of local media legend, Ralph D. Foster (1893-1984), KWTO brought the region to the nation: "Keep Watching The Ozarks"—playing off the station's call letters—was its tagline. And Foster took his show "on the road" to Neosho, Joplin, Ava, Marshfield, Jeff City, and elsewhere. "Jefferson City On Nation-Wide Radio Feature," headlines an article in the July 6, 1947 issue of the capital city's *Sunday News and Tribune*:

> Jefferson City will attract nation-wide attention on July 19 when the Korn's-a-Krackin' radio show ... will originate from the stage of the Capitol Theater here and be air-waved through Radio Station KWOS.
>
> It will be the first time that a regular major network show has been broadcast from this capital city and plans are being made to make the most of the occasion....
>
> The Korn's-a-Krackin' show is known as the largest hillbilly variety show on the air and is heard each Saturday night at 9 o'clock over Station KWOS and other Mutual network stations....
>
> During the program time will be provided to tell about Jefferson City.

There's a formula lurking in the article above: Local boosterism, a national audience, radio technology, and Hill Country ethos all conspired to make Foster's RadiOzark Enterprises a success.[35] And this same formula would work for a "newer technology" still: television. In 1953, KTTS-TV—now Springfield's KOLR-10—set up its first studio in a corner room of the old Chamber of Commerce Building. (I'd guess that was an easy decision for the C of C, which relied increasingly on media technology to "spread the word," boosting the city as a place for tourism, settling, and investment.)

In 1953, Foster partnered with another local entrepreneur-legend, Si Siman (1921-1994), to produce the *Ozark Jubilee*—a TV show broadcast locally from the KYTV-TV studios on West Sunshine (Glenn, p. 325). Assembling a team of musical talent from around the Ozarks and Nashville, Siman and Foster pitched the show to network executives at ABC. By 1955, the *Ozark Jubilee*—nationally broadcast live from the Jewell Theater on Jefferson Street—joined ABC's Saturday evening lineup. Much has been written about the *Jubilee* and there's little reason to rehearse its history here.[36] But we cannot ignore its impact on local boosterism. During its ABC premiere, host Red Foley asked, "If you folks want us to come and visit at your house like this every Saturday night, why don't you drop me a line in Springfield, Missouri?" And people did: By the next week, 25,258 cards and letters arrived from 45 of the 48 states (Klise and Payton, p. 301).

For proof that people would "Keep Watching The Ozarks," one might look to the *Jubilee's* fan base: Throughout its six-year network run, the show remained a weekly hit, with millions of TV sets tuning in "from coast to coast." In its February 5, 1956 issue, the *St. Louis Post-Dispatch* declared that "Springfield has become the recognized center of the country music world. In fact, it is generally agreed in television,

---

35. RadiOzark was a transcription service, which recorded short programs, transferred them to disks, and distributed the disks to paying stations for broadcast at a time of their choosing. Foster's business model, thus, was multi-pronged, turning *Korn's-a-Krackin'* into a traveling stage show, local radio broadcast, and nationally syndicated broadcast.

36. See Rita Spears-Stewart, *Remembering the Ozark Jubilee.* See also the essay by Thomas A. Peters, included in this present volume.

recording and radio circles, that Springfield, now a city of 90,000, has shaken Nashville, Tennessee, home of The Grand Ole Opry and long-time mecca of hillbilly musicians, to its very foundations."

For a brief time, other locally produced shows were televised nationwide. "Just think of this," said Si Siman: "Springfield, Missouri, was the third highest origination point for national television—third only to New York and Hollywood. More than Chicago. More than Washington D.C. More than all the other places that originated shows because we were up as high as two and a half hours a week" (Spears-Stewart, p. 7). This TV dominance of country music didn't last long; but it lasted long enough to inspire a town south of Springfield to start its own live "country variety" shows (Payton and Payton, p. 79). The Baldknobbers began in "old town" Branson in 1959; their show has played continuously since. By 1968, the Baldknobbers had built their permanent theater near Presley's Jubilee on Missouri Highway 76. To these shows, add Table Rock Lake; add Marvel Cave; add the old-time theme park, Silver Dollar City; add The Shepherd of the Hills Theater, named after Harold Bell Wright's 1907 novel (and performing a drama-adaptation of the same). Add these together, and "modern Branson" is born.

We cannot overstate the role *Korn's-a-Krackin'* and *Ozark Jubilee* played in constructing the region's postwar self-image—and self-promotion.[37] Through the 1920s to the war years, Springfield continued to project an image of urbanity, progress, and modernity. And now, having at last overcome its spatiotemporal isolation, the region looked nostalgically backwards to an idealized rural past. While the Hill Country ethos found early expression in Wright's novel, it took

postwar media technologies to commodify the Ozarks "countryside," its music, its lakesides and landscapes, and its premodern, pre-industrialized "pioneer" culture. (If Shannon County remains "the authentic" Ozarks, what shall we call Stone or Christian Counties?)

Ozarks historian Brooks Blevins carries this cultural analysis a step further: More than romanticized or idealized, the Hill Country remains a literary-artistic invention. "Like so many Southern subregions that thrive on tourism," writes Blevins, "the Ozarks of *The Shepherd of the Hills* was based on Harold Bell Wright's invented past" (p. 126). In effect, Wright's narrative fiction creates a cultural fiction that tourists cheerfully take as truth:

> The success of the novel ... contributed to the creation and perpetuation of an invented present based on that invented past. The irony is that the arrival of the railroad in southwest Missouri in the early 1900s provided the chief means by which literary tourists and other visitors could find the supposedly isolated, premodern region that Wright memorialized. At the very moment that the era's primary modernizing element brought the outside world and increased marketing opportunities to the rural Ozarkers ... the force and popularity of the region's most famous novel worked to cocoon the area in an image of romantic backwardness. This image in turn drew not only literary tourists but also folklore and folksong collectors, fishermen, hunters, and floaters. Just six years after the release of *The Shepherd of the Hills*, developers built a small dam across the White River below Branson. The dam generated hydroelectricity for Branson and other nearby towns and created Lake Taneycomo, on the shores of which one of the largest resorts in the region sprang up. Eventually automobile traffic into the

---

37. Is it coincidence merely that Springfield's largest half-decade of growth—in population, industry, and general construction—ran concurrently with the *Ozark Jubilee*? The years 1955-1960 were "years of excellence" (Dark, p. 61), when the city "flexed its muscle." Surely the Jubilee did its part in attracting viewers, and visitors, and big-dollar investors.

region would replace rail travel, and post-World War II prosperity would revive the tourism-based fortunes of the Branson area with the creation of an outdoor drama based on *The Shepherd of the Hills*, the development of the Silver Dollar City theme park, and the establishment of a sort of hill-billy music row in the 1960s and '70s, yet another attraction that grew out of and perpetuated the romantic invented past and present of the Ozarks. (p. 126)

While the Hill Country has its ecological and cultural aspects, it's the technologies that most concern us here. A best-selling print novel, a nationally syndicated radio show, and a hit network TV show—technologies all—were the means of mass-produced image-making. More than commodified *culture*, more than media *technology*, "the Ozarks" arose out of their confluence: It is, indeed, a realization of place-as-technoculture.

Again, the "living pioneer" and the "untouched wilderness" are gone: Maybe they're retrievable still in Shannon or Texas Counties, but they're not in Greene or Stone or Christian. Once absorbed into postmodern technoculture, nature ceases to be itself. This is neither a complaint nor a problem to solve: It's a recognition, rather, of our "postmodern condition," in which the "wilderness" is reimagined, reconstructed, managed, and rented out by means of the region's artificial lakes and lakeside resorts, golf-course country clubs, and "nature centers." The premodern "pioneer culture," similarly, is reimagined and reproduced in plays and old-time theme parks whose visitors pay to watch food kettle-cooked on wood-stoked fires, the hand-sewing of quilts, and the blacksmith's anvil-hammerings.

More than any place in the metro region, Branson belongs to postmodernism. And the city-dwelling tourist takes advantage. If you have a car and a credit card and live anywhere nearby, chances are that you've visited the Hill Country, enjoying it scenic vistas and attending some of its shows; if you have a family, you'll make regular trips to the region's theme parks. But our sense of the Ozarks as a one-time wilderness has given way to "the Ozarks" as a nostalgic replica of the same—a stylized, media-produced representation of rural folk culture that stands in apparent antithesis to urbanity, industrialization, and modernity. This commercialized reproduction of wildlife and folk culture, wherein the Ozarks morphs into "the Ozarks"—call it "Shepherd of the Hills Country," if you wish—serves urban technoculture. In its artificial lakes and carefully maintained hiking trails, "the Ozarks" lets us leave the city for a morning or afternoon or weekend of boating, camping, hunting, and fishing, after which we can return to the lodge and its private bath and flush toilet, its fine dining and entertainments, and its full array of electronic devices and amenities (including cable TV and Wi-Fi connections, in-room refrigerators and individually controlled HVAC systems). Having hiked, floated, hunted, and fished, we'll return to the city and the workday world partly refreshed but longing for more.

Blevins is right: The Ozarks Hill Country ethos is "invented." And we need to take that term literally, given that postmodern technoculture transforms all aspects of our lifeworld—material, cultural, aesthetic—into commodifiable, "mediated" experience. Arguably, the single most significant invention that the Ozarks currently markets to the rest of the world is "the Ozarks."[38]

3. Questioning the Future

It's easier to reconstruct a region's "authentic" pre-modern past than to describe its present or predict its

---

38. For discussion of the Morris family (Bass Pro) and the Hershends (Silver Doller City) in "inventing" "the Ozarks," see the essay, "Building Businesses and Building Community, Ozarks-Style," included in this volume.

future. As entry points into a discussion of the Ozarks' place in "the 21st Century Digital World," I pose a series of questions and challenges, inviting readers to do what Ozarkians have always done well: to innovate as best as one can for the common good. But, in point of fact, we cannot speak of possible futures for Springfield and the Ozarks without expanding our purview: We can preserve our community and its folkways—its regional culture and character—*so long as the planet allows.* Surely this segue into ecology comes as no surprise. The global-local nexus presupposes a viable lifeworld; for which reason "our biggest worries," writes Don Ihde, *"ought to be global,* first in the sense of concern for the Earth's environment, and second, in finding [a] means of securing intercultural ... modes of tolerance and cultural pluralism" (p. 115; emphasis in original). If Springfield is to "boost itself" strongly and securely into the 21st century, it must embrace tolerance and pluralism along with an entrepreneurship aimed at global-local solutions to global-local problems.[39]

In previous generations, one might have counted on a town's "representative leading men" and women to lead the way forward. But contemporary technoculture overwhelms us with information, such that no one person can possess the whole. A *surplus* of data is one marker of postmodernity; the *fragmentation* of this data into "expert systems" is another. That is, postmodern technoculture functions by dividing information among "experts," who "specialize" in *components* of larger systems—institutional, disciplinary, commercial, technological—without full

knowledge or understanding of the functioning of the whole.[40]

If we are to thrive as a community within the larger "global village," we need to build a conversation, and a collaboration, among the various "experts" and stakeholders in contemporary technoculture. For the nonce, we'll reduce the conversation to four agents: "the citizen," "the scientist," "the politician," and "the corporation." Out of these four agencies, we can build a viable community. But trust, you'll note, remains a necessary prerequisite: In a decentered, fragmented world, *we rely on others' expertise.*

Such is the postmodern condition: "The citizen" must rely on "the corporation," which must rely on "the politician," which must rely on "the scientist," which must rely on "the citizen," and so on ...

[T]he city of the future, and no very distant future, will have no trolley poles or wires and no horses. All movements will be on rail by silent air motors or by horseless carriages equally silent. All pavements will be asphalt. Unlimited light will be as cheap as unlimited water is today. No coal will be delivered at private houses and no ashes taken

---

39. In an epigraph above, I've quoted Steven Conn (p. 303), who gives two ingredients for successful future economy and culture: 21st century Springfield seems to have transited successfully with respect to its postindustrial economy; as to its cultural transformation into an inclusive, truly cosmopolitan city, this remains a work in progress. Perhaps the boosterism of the 21st century will aim to attract, not just the techno-savvy entrepreneur, but the "global citizen," as well.

40. "As late as the 1870s," Kennedy notes, "families settling on the American prairie would mend their own coffeepots, nail together hog-slaughtering stands, and repair wagon axles" (*Inventology* p. 168). She continues:

"Every active and ingenious farmer should have a good workshop and his own set of tools for repairing implements," wrote a [newspaper] columnist of the time. Back then, a town was not just a collection of houses but also a gathering place for blacksmiths, tinkers, seamstresses, and cobblers who manufactured the accouterments of daily life. Inventors weren't remote experts; they lived next door. (p. 168)

Such pioneer self-reliance belongs to the past: Unless we've expertise in the items following, our SUVs and HVAC systems and iPhones—and even, for goodness' sake, our own coffeepots—lie beyond our mending.

from them. With no horses, no coal, and no ashes, street dust and dirt will be reduced to a minimum. With no factory fires and no kitchen or furnace fires, the air will be as pure in the city as in the country. Trees will have a chance; houses will be warmed and lighted as easily and cheaply as they are now supplied with water.

A city will be a pretty nice place to live in when the first twenty years of the twentieth century are passed.

—"In the Near Future," *Springfield Leader and Press* (March 29, 1897)

The late-twentieth-century "urban renaissance" is surely tentative and, all things considered, small, and it is far too early to predict whether it will last. But the baby boomers, now empty-nesters, who want to give up their driving commute and enjoy the cultural life of cities, and the 30-somethings who are feeding any number of creative urban endeavors, are clearly onto something. And the local food mavens who shop the farmers markets … thereby supporting local farmers are finally linking city and country in mutually supportive ways. Likewise, there are intriguing signs that the antagonism between city and suburb is mellowing at least a bit. As more and more Americans come to reside in metropolitan regions surrounding the central cities, many are beginning to understand that *the futures of both city and region are fundamentally interconnected.* Air and water quality, economic development, transportation networks—all of these and more are problems that crisscross the political boundaries that separate cities from suburbs. Metropolitan regions that succeed in the coming century will be those that recognize this shared destiny and develop a political agenda to foster it.

—Steven Conn, *Americans Against the City* (2014)

Today, we stand at a crossroads. Through the first half of the 20th century, booster-towns across the American Middle West sang paeans to the future. The future *belonged to us*, did it not? And the nation's entrepreneurs were poised to shape it and possess it. In the 21st century, attitudes seem to have downshifted. "The citizen," "the politician," "the corporation," and "the scientist" seem too often to work at cross-purposes. Does "the corporation" *serve* "the citizen"? Do they share the same goals? Surely they share the same community, the same resources, the same planet. While some on the corporate side of technoscience work to protect profits, others work to protect the environment, fending off catastrophe—though which of the following happens first (if any at all) remains anyone's guess: infectious pandemic, the drowning of coastlines through rising sea levels, the release of methane gases with the melting of arctic permafrost, the drying-up of fresh water supply, or warfare leading to "nuclear winter"—not to mention the chance (however small) that an asteroid *will* smash into the Earth, throwing us into another Ice Age. We can't assume that the Ozarks will somehow recover its premodern isolation, remaining unscathed, self-providing, and self-reliant.

"The scientist" prepares models for such possibilities; whether we have the technologies, or even the political will, to meet them remains an unanswered question. Arguably, "the scientist" working today faces greater challenges than "the scientist" working in "machine age" modernism, when polio and tuberculosis, deforestation, economic depression, and military competition among regional powers were the most pressing threats. The stakes were high back then; somehow, they seem higher now, and the time shorter.

Despite our postmodern reliance upon "expert systems," we've fallen into a crisis of trust. Again, "the politician" *must* rely on "the scientist" and "the citizen" *must* rely on "the corporation" if we're to survive, much less thrive, as a community and nation. Urbanized and globally interconnected, our world rests upon technoscience. There's no turning back. Though technoscience made much of the mess that we're in, it's our best chance at cleaning the mess up.[41]

And yet, given the current fierce competition among political-economic narratives, trust in "the scientist" has eroded at a time when this trust is needed most. In the 21st century, the best "booster policy"—for corporations and communities alike—will promote economy and ecology at once, using technological innovation as its means.

 Consider, then, that the materials gathered in TechnOzarks seek to inform public debate over technoscience and its global-local impact. An aim is to make policy—that needful conversation among "the citizen," "the politician," "the corporation," and "the scientist"—more informed, balanced, and intelligible to "the citizen" especially, for whom technology remains largely invisible (until it breaks down). Even as it seeks to inform, TechnOzarks seeks to raise wonder in readers: For science offers ways of seeing (that is,

of expanding human perception) that give their own delight. The images gathered in this anthology show the ease with which science morphs into art, and vice versa. And, in the main, the texts gathered here celebrate as much as commemorate technologies in their local history and contributions to culture.

The questions that follow look beyond 21st century metro Springfield and its Ozarks hinterland. They are social, ethical, and political by implication, and I leave readers to work out their own responses.

Given the modernist conquest of time and space, we have declared ourselves "global citizens." Yet the recent rise of nationalism and "identity politics" expresses anxieties over globalism. The Ozarks is no longer content to adapt technologies to its own terrain; contemporary entrepreneurs aim at global markets—but at what cost culturally, ecologically, politically?

The entrepreneur today may drive to work—out of town or down the street—on behalf of a multinational corporation based in Beijing or Basel and drive home to a lakeside condo or loft apartment downtown. For some, their commute might be by Skype. Can we work globally and live locally at the same time?

We wish to benefit from global markets while maintaining our cultural, political, and economic autonomy and urbane lifestyle. Can we do so in a world where the "American lifestyle" is untenable on a global scale? Can we continue to create technologies that foster lifestyles that consume resources beyond the planet's capacities? Can the Ozarks insulate itself from the rest of the world's problems while enjoying the world's resources? Perhaps more important: Can we continue to produce and consume without caring for the region's ecology—its environmental health?

Further questions arise. Can we set challenges for Ozarks' entrepreneurs today? Can "the citizen" and "the politician" and "the corporation" and "the scientist" agree on priorities in innovation? Can we

---

41. "The Valdez oil spill, Bhopal, the Challenger explosion, Three Mile Island and Chernobyl, and the eco-terrorism of the Gulf War" (Ihde, *Philosophy* p. 120) are six in a long litany of "big-tech" disasters. Others lay closer to home. There's Times Beach (abandoned, its streets contaminated by dioxin); Bridgetown's Westlake landfill (radioactive, a leftover from the Manhattan Project); and Herculaneum (lead-dusted, like other mining towns within the Ozarks' "lead belt").

Currently, the EPA lists thirty-three active "Superfund sites" in Missouri ("Superfund National Priorities List"), including Springfield's Fulbright landfill, a 98-acre site located in the flood plain of the Little Sac River. Owned by the city and operated from 1962 through 1968, the Landfill "accepted industrial and domestic wastes from the Springfield area for disposal, including plating wastes, paint sludge, pesticide residues, waste oil, and wastes containing solvents, metals, acids, and cyanide" ("Fulbright Landfill"). Closed since 2007, the Litton Systems plant on West Kearney contributed to this and other hazardous waste sites, including sites near the Springfield-Branson National Airport. For their current status, see the 2019 report of the Missouri Department of Natural Resources ("Litton Systems"). While contributing to technology, Litton's circuit boards unleashed some powerful pollutants.

work to solve inequalities in global economy and consumption? Can we work to conserve natural resources necessary for "quality of life" in our own urbanized cultural setting? Can we build consensus over "quality of life" issues?

In what ways can contemporary technoscience contribute to "the good life" for all, globally as well as locally?

Why technology in the first place? The answer, anthropologically and philosophically, revolves around humans relating to their environment, whether conceived of as a small territory, or more largely, as contemporarily, to the Earth itself.
—Don Ihde, *Philosophy of Technology* (1993)

Though I've sounded the alarm in paragraphs above, it's toward a hopeful future that I'd cast my glance. Let me end, therefore, with a paean of my own to the future that I believe is already upon us. More than commerce and service, more than the commodification of culture, more than the making of metro Springfield, we are busy reinventing ourselves as a species. Embedded in our lifeworld, the technologies of artificial intelligence (AI) have created cyborg companions for us, responsive to our needs.[42]

Swarm AI may be next to evolve, wherein distributed user networks function as a collective intelligence—a "human swarm." Memory becomes communal, external: When one stores information on a cell phone, the machine "remembers for us." One day, perhaps, memory will expand internally by microchip implant. Our bodies have become engineerable prosthetically, genetically. Increasingly, the boundaries between human and nonhuman dissolve.[43] More than rely on the electric power grid that enervates computer circuitry, *we have become part of that grid*, by virtue of the human-machine interface. Our iPhones and computer tablets are exoskeletal. We are "plugged in."

Here in 2019, I write in the present while gazing across a short horizon to any already-emerging futurity. I take inspiration, however, from a techno-philosopher writing in 1964—more than a half-century ago. A prophet of postmodernity, Marshall McLuhan foresaw our transit from "the mechanical ages" into something wholly new, powered by electrons:

> During the mechanical ages we had extended our bodies in space. Today, after more than a century of electric technology, we have extended our central nervous system itself in a global embrace, abolishing both space and time as far as our planet is concerned. Rapidly, we approach the final phase of the extensions of man—the technological simulation of consciousness, when the creative process of knowing will be collectively and corporately extended to the whole of human society, much as we have already extended our senses and our nerves by the various media. (*Understanding Media*, pp. 3-4)

It was McLuhan who proclaimed the human-machine interface, the fact that "all technologies are extensions of our physical and nervous systems to increase power and speed" (p. 98). And, in this new age—which, again, *is already upon us*, though its implications and applications continue to emerge—our human-social relations change, as well: For "electricity ... decentralizes. It is like the difference between a railway system and an electric grid system: The one requires railheads and big urban centers. Electric power, equally

---

42. Our pop-culture vocabulary has become so saturated by postmodernity that we take terms like "AI technologies" for granted. "VR technologies" (immersive "virtual reality" programs, computer-generated) have become commonplace; we can even write of "cyborg companions" without blushing—the cyborg being a *cyb*ernetic *org*anism, part biology, part machinery.

43. For an intriguing discussion, see the web article by Vivienne Ming, "Why I'm Turning My Son into a Cyborg."

available in the farmhouse and the Executive Suite, permits any place to be a center" (p. 39). Hence, the "center-margin structure" of 20th century geopolitical mapping is now "experiencing an instantaneous reassembling of all its mechanized bits into an organic whole. *This is the new world of the global village*" (p. 101; emphasis added). Surely this reassembling of the "center-margin structure" has implications for metro Springfield and its hinterland.

And, yes, it is to McLuhan that we owe the term "global village." To many in the Ozarks, this remains a foreign phrase, unsettling in its implications. It's a notion, nonetheless, whose time has come for Springfield, the Ozarks, the nation, the world.

Works Cited

1980 Manufacturers Directory of the Springfield Area Chamber of Commerce. Springfield, MO: Chamber of Commerce, 1980.

A Booming City: Springfield, Missouri. Springfield, MO: Lapham and Bro., n.d. (1887). Web transcription retrieved 12 March 2019.

Annual Report of the Commissioner of Patents. Washington, DC: Government Printing Office, 1870-1925.

Blevins, Brooks. *A History of the Ozarks, Volume 1: The Old Ozarks.* Urbana, IL: University of Illinois Press, 2018.

——. *"The Shepherd of the Hills* and Ozarks Tourism." *Living Ozarks: The Ecology and Culture of a Natural Place.* Ed. William B. Edgar, Rachel M. Besara, and James S. Baumlin. Springfield, MO: Ozarks Studies Institute, 2018. 124-27.

Commercial and Government Radio Stations of the United States. Washington, DC: Government Printing Office, 1921-1923.

Conn, Steven. *Americans Against the City: Anti-Urbanism in the Twentieth Century.* Oxford: Oxford University Press, 2014.

Dark, Harris E. *Springfield, Missouri: Forty Years of Progress and Growth, 1945-1985.* Springfield, MO: Dolandark, 1984.

"Doing Business." Springfield Regional Economic Partnership (Chamber of Commerce). Retrieved 23 March 2019. <https://www.springfieldregion.com/doing-business/major-employers/>.

Fairbanks, Jonathan, and Clyde Edwin Tuck, ed. *Past and Present of Greene County, Missouri.* 2 vols. Indianapolis, IN: A. W. Bowen, 1915.

"Fulbright Landfill." Toxic Sites. Retrieved 1 April 2019. <http://www.toxicsites.us/site.php?epa_id=MOD980631139>.

Goodman, Mark. "The Radio Act of 1927 as a Product of Progressivism." *Media History Monographs* 2.2 (1999): 1-20.

Spears-Stewart, Rita. *Remembering the Ozark Jubilee.* Springfield, MO: Stewart, Dillbeck & White, 1993.

Ihde, Don. *Philosophy of Technology: An Introduction.* New York: Paragon, 1993.

Jacobs, Jane. *The Death and Life of Great American Cities.* New York: Knopf Doubleday, 1961.

James, Paul. *Urban Sustainability in Theory and Practice: Circles of Sustainability.* New York: Routledge, 2015.

Kaplan, David M., ed. *Readings in the Philosophy of Technology.* Lanham, MD: Rowman & Littlefield, 2009.

Kennedy, Pagan. *Inventology: How We Dream Up Things That Change the World.* Boston, MA: Houghton Mifflin, 2016.

Klise, Kate, and Crystal Payton. *Insiders' Guide to Branson and the Ozarks Mountains.* Manteo, NC: Insiders' Guides, 1995.

Levin, Rob, ed. *Celebrating Springfield: A Photographic Portrait.* Atlanta, GA: Riverbend Books, 2007.

"Litton Systems, Inc. Site." The Missouri Department of Natural Resources, January 2019. Retrieved 1 April 2019. <https://dnr.mo.gov/env/hwp/sfund/docs/litton-factsheet.pdf>.

Mahoney, Timothy R. "The Small City in American History." *Indiana Magazine of History* 99 (2003): 311-330.

McCanse, Ralph Alan. *Titans and Kewpies: The Life and Art of Rose O'Neill*. New York: Vantage, 1968.

McIntyre, Stephen L., ed. *Springfield's Urban Histories: Essays on the Queen City of the Missouri Ozarks*. Springfield, MO: Moon City Press, 2012.

McLuhan, Marshall. *Understanding Media: The Extensions of Man*. Berkeley, CA: Gingko, 2003.

Ming, Vivienne. "Why I'm Turning My Son into a Cyborg." *Quartz*, 15 July 2019. Retrieved 15 July 2019. <https://qz.com/1650393/transhumanist-parents-are-turning-their-children-into-cyborgs/>.

Morrow, Lynn. "The Auto Could Never Have Been in the Ozarks." Email to James S. Baumlin, 22 March 2019.

Morrow, Lynn, ed. *The Ozarks in Missouri History: Discoveries in an American Region*. Columbia, MO: University of Missouri Press, 2013.

Morrow, Lynn, and Linda Myers-Phinney. *Shepherd of the Hills Country: Tourism Transforms the Ozarks, 1880-1930s*. Fayetteville, AR: University of Arkansas Press, 1999.

Opening of the Atlantic and Pacific Railroad, and Completion of the Southwest Pacific Railroad to Springfield, Mo., May 3, 1870. Springfield, MO: n.p., 1870. 41 pp.

Payton, Leyland, and Crystal Payton. *Branson: Country Themes and Neon Dreams*. Branson, MO: Anderson Publishing, 1993.

Peters, Thomas A. *John T. Woodruff of Springfield, Missouri, in the Ozarks: An Encyclopedic Biography*. Springfield, MO: Pie Supper Press, 2016.

Piehl, Charles K. "The Race of Improvement Springfield Society, 1865–1881." In Morrow pp. 71-100.

Schoolcraft, Henry R. *Journal of a Tour into the Interior of Missouri and Arkansaw, from Potosi, or Mine a Burton, in Missouri Territory, in a South-West Direction, toward the Rocky Mountains, Performed in the Years 1818 and 1819*. London: Richard Phillips and Company, 1821. Web transcription retrieved 1 November 2017.

Springfield Greets You: Dedicated to Our Progressive Citizens of the Ozarks. Springfield, MO: Chamber of Commerce, 1919.

Springfield, Missouri: Invitation to Industry. Springfield, MO: Chamber of Commerce, 1960.

Stefon, Matt. "Fairness Doctrine" (U.S. Communications Policy). *Encyclopaedia Britannica*, 30 December 2018. Retrieved 30 April 2019. <www.britannica.com/topic/Fairness-Doctrine>.

Stevens, Walter Barlow. *Missouri: The Center State, 1821-1915*. Vol 3. Chicago and St. Louis: S. J. Clark Publishing Co., 1915.

"Superfund National Priorities List (NPL)—By State." U.S. Environmental Protection Agency. Retrieved 22 March 2019. <https://www.epa.gov/superfund/national-priorities-list-npl-sites-state#MO>.

Valéry, Paul. "The Conquest of Ubiquity" (1928). Web transcription retrieved 27 February 2019. <https://mtyka.github.io/make/2015/09/12/the-conquest-of-ubiquity.html>.

Wise, J. MacGregor. *Exploring Technology and Social Space*. Thousand Oaks, CA: Sage, 1997.

Woodruff, John T. "John T. Woodruff and Missouri Road Building." From John T. Woodruff's Unpublished Memoirs. *OzarksWatch: The Magazine of the Ozarks* 8.2 (1994): 53-57.

I. Regional History Through the Mid-Twentieth Century

Herding Goats (undated).
Domino Danzero Collection, MSU Special Collections and Archives.

Gasconade Mills:
The Ozarks' First Timber Boom
Lynn Morrow

The Big Piney River courses 105 miles before it reaches the Gasconade River. It lies entrenched in an earlier primeval woodlands where investors developed the Ozarks' first large-scale, commercial lumber industry and its associated rafting and ox-driver occupations. St. Louisans knew the lumber-producing region as "Gasconade Mills." Territorial and early statehood mills were partnerships among small-scale investors, but by the late 1820s, mill sites included absentee landlords in St. Louis. Lumbermen mortgaged or owned a mill business outright and sawmills and rafts were portable. St. Louis consumers took notice that the yellow pine price was half that of white pine imported from the Ohio River.

By the 1830s, lumbering had radically impacted the woodlands with open clearings in the bottoms, making it easier to plant crops and graze cattle. Local governments brought lumbermen into politics, as mill sites became quasi-governmental places where elec-

tions and sales of probate assets took place; they were scenes for court-appointed commissioners to arbitrate financial disagreements, milestones along public roads for travel as county road districts led from mill to mill, and lending places where citizens sought credit from those who had a surplus—the mill owners.

Writers considered the interior Ozarks a "desert," unfit for agriculture. But, water transport represented the centrality of a lumber trade that flowed north to landings in the Missouri River settlements and on to St. Louis. The growing commerce led one businessman to christen his new Mississippi River ferry with the name *Ozark* in 1832. Lumbermen were analogous to itinerant merchants who steered their products over water rather than land. So brisk was the trade that in 1835 Missouri legislators, acting upon complaints from sawmillers statewide, passed a law that enacted "penalties for malicious destruction of rafts, planks, etc., or rafts cut loose and set adrift." The 1849 Missouri

Plank lumber stacked by steam sawmill with men and yoked oxen (1880s).
Courtesy of Lynn Morrow.

General Assembly passed another law that "declared Big Piney Fork of Gasconade River a public highway" to allow suitable passage for commercial rafts—a welcome aid in forest exploitation for farmers. The Ozarks and St. Louis always had a special economic relationship in the 19th century.

Forest products rode upon the rafts. St. Louis merchants negotiated prices for all that could be brought downriver from the hinterlands. This pattern of river trade dated back to colonial generations in the East and came West with migrations into the Ozarks. In practical terms, the Big Piney Valley was simply an outback of trans-Appalachian immigration. Investors brought industrial sawmill parts in their wagons to assemble in the backcountry.

Harrison Landing at the mouth of Little Piney Creek and adjacent to the Gasconade River became a lumber yard until the Civil War. By 1819, Daniel Morgan Boone, his extended family of slaves, and two of his Van Bibber relatives sawed and rafted from there. The landing's namesake, James Harrison, operated a tavern, a profit center for public space during sessions of county court and one of the few places where travelers spent small amounts of cash. However, settlers in the pineries shared a barter economy. The store furnished a minimum of goods, but the staples provided a considerable convenience in the backcountry. Wherever the millers worked, they created an economic nexus with agricultural improvements, merchandising, and blacksmithing.

Carpenters and blacksmiths, known as "mechanics," built and maintained the mills, and most considered themselves "farmers." A farmer in the Ozarks was at least a part-time lumberman who cleared trees for fields and cut wood for fuel. Mechanics fabricated barrels and kegs and set the hoops on them; barrels housed any number of goods for shipment by raft or wagon. Barrels contained salt pork, saltpeter, gunpowder, furs, ginseng, and whiskey, and workmen mixed their products in loads to drop off, just like flatboat pilots did.

In the pineries, hand-labor pitsaws, animal-powered mills, and water-powered mills coexisted. But it was the sash saw—water-powered and vertical—that dominated manufacturing for the rafting trade. Not until 1841 did Henry F. Ormsby import the first steam boiler to provide power for a sawmill, giving rise to Steam Mill Hollow in Texas County. By 1850, more arrived. The size of these machines was smaller than the self-traction engines pictured in late nineteenth-century photographs. Most steam mills produced forty horsepower, but some had only seven. On Big Piney, mills produced nearly nine million board feet of pine plank in 1849 alone (and cord wood for domestic fuel was additional).

The large Pulaski County in 1840 was a commercial leader, not only in the value of pine lumber with twenty-two sawmills, but also in tar (from stumps and roots used as a sealant) and pitch (from the tree for a sealant and to treat wounds). Census-takers counted eleven commercial distilleries in 1839, ranking the county first in southwest Missouri, an indication that locals exported more than they drank. Pulaski led the state in ginseng exports, too. After 1830, settlers who went to Maramec Iron Works took pounds of tar in buckets to trade for store goods or blacksmith services. Maramec Iron was an industrial complement to the millers, as it attracted settlers and created regional markets.

Portability was an advantageous attribute for a sash mill, but portability does not mean that it was cheap or easy to build one. Entrepreneurs looked at topography and decided how the construction and technology would function with spring-fed water. Builders knew their way around carpenters' and blacksmiths' toolboxes. The miller had a working knowledge of woodworking and masonry and was acquainted with mechanical and hydraulic engineering. Planning a mill site and its construction could take a week, and sawing planks required a strong frame to withstand the working of the machinery. Skilled workmen had to be nearby to make repairs—want of blacksmith repair stopped production. And, someone had to shoe the loggers' oxen.

Construction tasks for a sawmill included building the mill dam, which required forty-five to sixty days' labor alone. A carpenter built a covered braced frame structure, 52' x 12' more or less, and fabricated a ladder to move around the project to install flooring, sheeting, shingling, and rafters. He installed plank, sills, and hewed timber at the dam and forebay. The site required a mill gate to control water in the flume. The owner would use either wood or metal to craft and install a carriage, wheels, and head blocks to move logs to the saw blade and then level it all. An un-level carriage was an accident waiting to happen. A blacksmith brought a "set of sawmill irons with a crowbar" and installed cogs and a shaft on a millwheel to drive the blade. He needed special files to sharpen the kerf on thick sawblades and to remove accumulated rust during idleness. Additional items included several wooden buckets on site, and not the least was a barrel of whiskey for the workers.

Portability enabled a variety of business arrangements. Affluent families purchased the machinery and paid for boatmen, teamsters, slaves, and mechanics to transport and install mills. Once the equipment was operative, owners speculated on them, just as one did with real estate, and the capitalist risk spawned partnerships to lessen liability. Partners made annual agreements with sub-contractors to supply and manage the mill and raft plank downriver to middlemen at destination markets. By the 1830s, marketing relationships with buyers in the Missouri River bottoms, St. Charles, and St. Louis were in place. Driving Piney's industrial expansion was the attendant population growth in the Missouri River Valley, but especially in St. Louis. The 1835 city had a population over 8,300 and, ten years later, over 35,000, and urban folks needed dimension lumber.

On Big Piney, the financial arrangements with prominent St. Louis builders and real estate investors represent a unique urban-backcountry association. Joseph Laveille and George Morton's booming construction business was a principal buyer. Laveille was St. Louis' first important architect; the politically active Morton was a major real estate investor and purchaser of lumber from incoming steamboats. Their carpentry, joinery, and construction shop was adjacent to the Catholic Cathedral (which Laveille designed) near the Mississippi River. Laveille and Morton's ventures into Big Piney sawmill production led to contracts where millers and rafters delivered dimension plank and joists to the wharf, a short walk from their shop. Following Laveille's death in 1842, Morton owned three Big Piney mills.

Bourgeoning St. Louis consumed all the pine that millers could send. By 1835, so much was the volume of board feet available that middlemen in the city sent Big Piney lumber to the Deep South. St. Louis's growing population continued to devour dimension lumber up to the devastating cholera epidemic of 1849, the same year that a fire gutted the city. Big Piney millers didn't have to wait long for new orders. The rebuilding of St. Louis during the early 1850s provided the last major stimulus to raft pine plank to the city.

The milling landscape in the Piney bottoms must have been a sight to see. Both steam and water-powered sawmills generated well over a million board feet annually, but water-powered sash mills only occasionally topped a million. A traveler coming upon one of the great springs saw thousands of hillside and ridge stumps, cleared bottoms for oxen and hay, mill frames, reservoirs dammed for power, log and timber frame houses with their gardens, stack after stack of green lumber, and concentrated populations larger than

Ozark postal hamlets. In season, the sound of sawing permeated the valley. These pockets of lumbermen provided agricultural markets for nearby farmers lucky enough to get their patronage.

into rafts. Modern old-house enthusiasts look for the distinctive vertical saw marks on boards to estimate the age of historic buildings.

The widespread introduction of large circular saws after the Civil War gradually replaced the sash saws. The rafting voyagers who left in the middle of the industry avenue between Arthur's Creek, Boiling Spring, and Slabtown traveled approximately 225-250 river miles to St. Louis. Yellow pine stumpage continually receded from mill seats, so labor costs gradually rose.

Sawmills did not run all year; some "sat at rest" for weeks or months. Owners had to balance the labor of woodsmen and ox-drivers cutting logs, along with mill labor and sash sawing, to project a volume for seasonal rafting. As in any business, a disruption in resources, mill repair, or injuries impacted manufacturing. Nevertheless, it took a lot of tree felling, ox hauling, and staging in the river bottom before a miller was ready for export. Lumbermen marshaled hundreds of thousands of board feet in logs and, after sawing, stacked green lumber long before laborers prepared the cribs that made up the rafts.

▲ Sash mill with vertical blade moving up and down in a greased groove (undated). The blade cut on a three-foot downstroke. *Courtesy of Lynn Morrow.*

Ox-drivers with their whips standing among yoked oxen ▼ ready to haul logs to a sawmill in Iron County (1870s). *Courtesy of Lynn Morrow.*

A water-powered, rectangular sash saw, with a wooden waterwheel connected to a pitman crank that drove gears and horsepower to a thick blade, cut more than twenty times that of the legendary manual pit- or whipsaw. Samuel A. Harrison, a lumberman from Hazelton Spring, described the technology: The 7' long blade, secured "in a frame which operated vertically and had a stroke of some 3 feet ... [with] logs fed to the saw by a system of gears ... could cut 1,500 board feet of lumber in 12 hours." To these efforts, add the labor to cut, saw, and transport logs to the mill by oxen, stack and season the lumber, and assemble the cribs

The great majority of the stumpage that oxen brought to the mills came from federal land—why pay taxes when open range timber was free for the taking from the "Ozark desert?" In a real sense, government lumber constituted the Piney rafts. The open woods commons attracted risk-takers with capital or credit lines and a vision toward annual profits. As land was surveyed and profits accumulated, lumbermen purchased their mill seats (springs) and valley land, then invested in upland acreage to expand their farms.

The heart of the commercial pine lay in northwest Texas County. During the 1840s and 1850s, merchants constructed, maintained, and sawed plank at eighteen commercial sawmills along Big Piney River. Missouri builders far downriver installed Ozark plank in thousands of historic structures in the Missouri and mid-Mississippi River Valleys. Piney millers utilized seasonal-to-fulltime labor, used their own slaves, hired more slaves, and ultimately employed white women for cheap labor around the mill site. As elsewhere in Missouri, business families intermarried and maintained common industry goals.

One of the larger slaveholders, St. Louisan Nimrod Snyder, managed a steam mill at his Boiling Spring patent (1841), where raft assembly took place at his lumber yard. Snyder boarded a couple of dozen young men who hired out for work, while most millers boarded several of their own hands. In January 1848, Snyder's St. Louis business partners, Bartholomew and Valentine Rice, acquired legal rights to use George Page's patented, portable, steam-powered circular sawmill in the pineries and became top producers. Virtually all millers hired a young, single woman to do domestic work of cooking and washing at the boarding house.

All sawmillers and loggers needed more than waterpower: They required strength and endurance of oxen, particularly as the edge of the forest receded and logs had to be dragged farther. By 1849, Kentuck-ian James A. Bates—an industry leader—had 60 oxen at Slabtown, where he owned mills at the spring and on Paddy Creek. His brother-in-law, Kentuckian John Burnett Bradford, owned 40 oxen on Paddy Creek, and Henry F. Ormsby managed the same number at Hazelton Spring. Nimrod Synder fed 62 oxen at Boiling Spring. David B. Commons kept 29 for his ox-driving boarders at his mill on Arthur's Creek. Robert W. Rodgers fed 50 oxen on Arthur's Creek and boarded several hands; his dozen slaves topped the count of bonded labor for one master in Texas County. Asa Ellis, namesake for the Texas County town of Ellsworth, worked 56 oxen. Individual loggers kept a half-dozen or more, and most folks who made wagon trips to shop in St. Louis required the same number. Settlers rented oxen for a trip to the city, just as they did for breaking new ground. All oxen pulled stumps. Bates's, Bradford's, and Ormsby's oxen supplied more than one mill simultaneously. No county in Missouri counted nearly as many oxen as did Texas.

No other rural county reported more than three dozen women working at industrial sites. One sawmill counted five, and at least one woman worked at each mill on the 1850 census. Likely, women labored removing the immense amount of sawdust in baskets or carts that accumulated under the mill frame. They probably stacked the slabs that were marketable to steam boilers and for domestic use. On occasion, women could have stacked dimension lumber to season. Whatever their contribution, it is unfortunate that it has escaped preservation in Ozarks folklore.

Each raft had a crew of six to eight men, with one captain for the group; workmen used a sweep and oar locks attached to the raft to steer it in the swift-flowing, multi-channeled Missouri River. An observer noted that the rafters were barefoot, wore a heavily used dingy shirt and ragged trousers, and had a piece of felt for a hat that covered unkempt hair. A pile of blankets lay on the raft. Once delivered, rafters cleaned and washed planks "on the beach at St. Louis." Then, the men purchased a new wardrobe, visited a barber for haircuts and toiletries, and partied at night in the taverns. The "steadier ones among them" purchased "tawdry finery" for their women and kids. Then they "set out on their march" to the pineries, a 150-mile walk, to begin again. The young raftsmen traditionally hailed from "the woods," but St. Louisans journeyed seasonally to Big Piney to team up with Ozarkers for the strenuous outdoor journey back to the Mississippi River.

Young men speculated in taking pine rafts downriver, just like their forebearers did and were still doing on the Alleghany and Ohio Rivers. In 1839, Harvey Woods in Waynesville wrote to his father William in Caledonia about his springtime, six-week "rafting tour" down Big Piney into the Missouri River Valley. He took his rafting wages to finance travel and inspection of real estate north of the Missouri River before returning to Waynesville. Riding rafts for five to seven days and walking long distances home was common in the northern Ozarks.

Sawn lumber rafts varied in size. Rafters stacked 1¼-inch planks six layers deep to make a crib. The term "crib" was common on the frontier to refer to a square or rectangular room with measured dimensions; thus, a "crib of logs" made a room in a house that included a chimney lined with mud and rocks on one end. Piney lumbermen also called them "squares." Sawmillers predetermined the dimensions of the plank cribs, e.g. sixteen, fourteen, or twelve feet in length, according to the market demand. Men attached the cribs to each other to make a linear raft. They secured the cribs, lashed together with leather thongs or held jointly by wooden pins, sapling strips, ropes, or chain links. Cribs, coupled with planks or poles laid lengthwise down the long axis of the raft, were secured by pins in holes bored by an augur. Men called the joined cribs a "string." Cribs had "sand boards" attached to keep river grit from settling among the lumber inside the crib. When entering larger water, such as stops at Spring Creek, Harrison Landing, or the Missouri River, men beached the rafts and coupled cribs together to make longer and wider strings. The result was a rectangular flotilla of planks and joists to dock at the St. Louis levee.

Relics of this sawmill technology are rare, and archaeologists have not excavated a Big Piney mill site. The business is primarily known through court and census records. But, yellow pine boards floated out of the Ozarks into markets and stand yet in buildings along the Missouri River. One can visit Governor Frederick Bates's Thornhill estate at Faust Park in Chesterfield. It overlooks Bonhomme Bottom where rafters docked boards and joists. The house is full of pine and the exposed variable-width boards in the barn represent evidence of the Ozarks' first great lumber industry.

James A. Bates, lumber industry leader, built this house (1855-1857) for his new wife, Sarah Newport, on Slabtown Road, Texas County. *Courtesy of Lynn Morrow.*

"Basket of flowers" trapunto quilt made by Elizabeth Jane Robertson McQuigg (1853). *Courtesy of the History Museum on the Square, Springfield, Missouri.* McQuigg hand-spun the thread used to stitch it and produced her own filling for the quilting.

Homespun:
Textile Production and Technology in Pioneer Greene County
Joan Hampton-Porter

Interior of cabin with spinning wheel belonging to Nancy Ann Simms Price (1926).
Courtesy of the History Museum on the Square, Springfield, Missouri.

Introduction

The idea of connecting technology with a pioneer era may seem incongruous, but one of Merriam-Webster's definitions of technology is "the practical application of knowledge." Arguably, any development of tools whereby the advancement derives from the development of knowledge is, by its very definition, technology. This essay will discuss textile production in pioneer-era Greene County, Missouri, through the words of two such pioneers: William Rountree (1847-1934) and Margaret Gilmore Kelso (1855-1949), whose memoirs are in the possession of the History Museum on the Square.[44]

It will also explore other technological advances, such as improved lighting, which influenced said production.

While many other cities in the American Middle West had ceased to be pioneer many years before, Springfield remained in the pioneer era by virtue of the difficulty of transporting goods and people in and out. There are no major navigable waterways, and the train didn't arrive until 1870. The closest railhead was in Rolla, approximately 110 miles away. There were merchants who procured goods from other areas to be sold in Springfield, but the majority of goods were produced and sold locally.

William Rountree

William Rountree, grandson of Springfield's first schoolmaster, Joseph Rountree (1782-1874), was sent by his father from St. Louis to live with his paternal grandparents in Springfield in 1852 when he was around five, while his father went to seek his fortune in the California gold fields. The remainder of Rountree's childhood was spent with his grandparents. Rountree's memoirs offer an insight into the nearly self-contained world in which he was raised. Textile production was one element of this almost autonomous world. The Rountree farm, which had been settled in 1831, was a small one, a mile and a half south of what is now downtown Springfield (which was then the entire city). Rountree writes,

We were as nearly a self-contained people as it

---

44. The following passages are taken from William Rountree, *Autobiography* (ca. 1932), and Margaret Gilmore Kelso, "Family History" (1938); both typescripts are in the collections of the History Museum on the Square, Springfield, Missouri.

was possible to produce. We had plenty of all of the essentials of living, but it was all produced on the farm by our own efforts. As to money, we had very little. The crops were corn, wheat, flax, cattle, mules, sheep, and hogs. From the wool of the sheep and from the flax was made what clothing we wore. In the fall and winter the women spun and wove the cloth from which our clothing was made for winter. Then the flax was made into linen cloth for our summer clothing. We also raised cotton which went into most of the cloth as warp. In the spring they would commence spinning the wool to make the winter clothing and in the winter were spinning the cotton and flax for the summer wear. Not every place had a loom as there was always someone in the neighborhood that could be hired to do the weaving. The men and boys usually had one suit a year made of cloth purchased at the store and made by the local tailor. This was for Sunday and special occasions. The ladies also bought some dress cloths but made the dresses themselves. All other clothing was made of homespun goods and made at home by hand. There were no sewing machines at this time. The yarns that were spun were dyed at home almost entirely with the bark of native trees. The black walnut bark made a kind of purple. The wild crabapple made a yellow. Combining this yellow with indigo would make a beautiful green. Then there was madder for red and indigo for blue and sumac would make black. These colors were woven into checks and stripes for dresses and into fancy designs for bed-spreads, then called "coverlets." This will give you an idea of the work that went on.

Margaret Gilmore Kelso

Margaret Gilmore was born in a log cabin on Clear Creek in Greene County, Missouri on May 6, 1855 and was married in the same cabin on February 9, 1872 to Jacob Thomas Kelso. Margaret lived to be 93 years old, spending her entire life within four miles of where she was born. She began writing her memoir when she was 83 years old.

Portrait of Margaret Gilmore Kelso (undated). *Courtesy of the History Museum on the Square, Springfield, Missouri.*

Her discerning writing makes it possible to step into her corner of the world. She, too, had a lot to say about textile production, as it was so important and, especially, time-consuming:

I have heard my mother say that when she and father went to housekeeping they moved into an unfinished log cabin, on a dirt floor and built a fire in the wash kettle, until father could make a fireplace and build a stick chimney and daub it with mud. She made her beds on the dirt floor until he could get time to bore holes in the log walls and put in poles, with the bark on, to make a frame for her beds.

When she got her beds made up on these pole frames and spread on her clean covers that she had made by hand in blue and white patterns, woven from wool [and] washed, carded and spun into yarn by herself, she stood off and admired her room and beds and thought they were so pretty. She was so proud of it.

We spun and wove all our own materials, for men's clothing and also for women's clothing. We also made material for our own bedding. We sheared the sheep, scored the wool before it could be carded and spun into thread. At night we had to spin two "cuts" or 144 threads, around the reel, or know a "finger" length on hose or gloves for the family.

We had to wear "linsey-woolsey" dresses for Sunday. Aunt Lou Bradshaw, mother's youngest sister, was invited to a party. She begged grand-mother for a "store-bought" dress. Grandmother wouldn't get it for her. Aunt Lou said, "There won't be anyone else there with a homemade dress on." Grandmother said, "If there isn't it will be because they don't have one, and there won't be anyone there with a dress on that costs as much as yours does."

"Linsey-woolsey" refers to a homespun fabric made of a combination of linen and wool or cotton and wool. What her grandmother meant was, if the value was assessed in terms of the sheer labor required to make the dress, then hers would be the most expensive. All homespun fabric was expensive in terms of labor, but the special, fancy fabrics such as stripes, plaids, and other patterns even more so. Consider the total time spent raising the plant or animal, harvesting the raw material, washing the material, carding it or otherwise preparing to spin it, spinning it, reeling it, obtaining the materials for the dye, preparing the dye, dyeing it, setting up the loom, and weaving the fabric. Then, if it was not a flat item like a coverlet or tablecloth, the pieces would need to be cut out and sewn into a finished product. All of this required time that could have been spent on other necessary aspects of daily life, such as food production and food preservation. Kelso continues:

Grandmother could weave coverlets and tablecloths from a pattern. She called it "double weaving." When she saw a pattern she liked, she would take a strip of paper and put down figures above and below, and take the pattern off. She would put that strip of paper on the loom and follow the pattern in her weaving. She wove white bedspreads and tablecloths. She had a pattern for tablecloths she called "dimity," a pattern of small white dots. She was an expert in dying and weaving all types. She made her own dyes, using cochineal for scarlet, and madder for a dark red or maroon color. The hulls of black walnuts was used for a dye for dark brown. Brown or black was used on their linsey-woolsey clothing, mainly used in men's suits, bedding, etc. They used a soft yellow stone called ochre, which was pounded fine and boiled for dying their yellows.

There is in existence today blocks of quilts that were made in their entirety, thread and all, by Grandmother Polly Julian Edmonson, for the express purpose of covering her growing family of boys. The materials are made in plaids and stripes, with an undyed linsey-woolsey back. The dyes are clear and unfaded today, more than a hundred years later. This was not a work of art, but without doubt it shows much artistic ability. There are also bedspreads of white, handmade in an intricate pattern, and plain white stripes. The block pattern she called her "Sunday Best" spread and the plain stripes her everyday spread.

The conditions these women worked under was often far less than ideal. I say women, because in that time and place it would have been the women producing the cloth, even though both genders may have worked in the fields and with the livestock (and there were male tailors). Many of these women would have been working in log cabins with little light and little ventilation. When better light became available, this would have made the work much easier and more efficient, but they would have still needed to limit ventilation, as this affected the clothmaking process. Warmth was particularly desirable during the spinning process. Again, Kelso provides the salient details:

I think the first remembrance I have of my mother was of waking one night and seeing her sewing with her finger by the light of a grease lamp stuck in a crack in the chimney wall. The stem turned up at the end to hang it by. It held about two tablespoons full of grease. Sometimes we tore a strip of cloth, doubled it back, then twisting it, pushed it down in the grease for a wick. It made a light, but a very poor one.

Our first lights were "grease lamps" which were a saucer of iron with a small lip on the side and a braided rag wick that hung over the lip. The next light was a tallow candle in a candlestick with a small base and a long hook that could be stuck in a crevice between the logs, or between the stones of the chimney. We would melt beef tallow, put[ting] in a little beeswax to harden it. We had a candle mold. First we placed a wick in the mold, then poured the melted tallow and beeswax around it, leaving it to harden. Then they were ready to pack away for winter use. A year's supply was made at one time, and stored for future use. Later on we bought a coal oil lamp. When mother went to fill and light the lamp, she would send all of us children out into the yard. After she had lighted it we would gradually venture nearer, until we were back around the table to see the new light. It was a wonderful sight to see, for us—the brightest light we had ever seen.

Grease lamp (mid-1800s).
Photo by Palmer Johnson, courtesy of the History Museum on the Square, Springfield, Missouri.

During the Civil War, Gilmore's mother put Margaret to work producing materials to sell to the soldiers:

My mother gave me rolls of wool to spin, and let me have thread to knit socks and mittens for the soldiers. She taught me to shape them and let me have the money I sold them for. We knit of night, and kept the spinning wheel going, too. We could spin a nicer, more even thread when the doors were closed and the weather warm.

What do these stories teach us—about the pioneer world and our own? Cloth is such a humble everyday item, people don't think much about it today if they have the necessary blankets to warm their beds and clothes to cover their bodies. Most of us don't think much else about it when we go to buy it at the store—or now, online—other than whether it's attractive and will serve the purpose intended. But cloth on the frontier was a very different commodity, far more precious. Considerable time and labor were spent producing the cloth to have the necessary blankets and clothing. Cloth was so precious, that it was used and used until it couldn't be used anymore. When no longer usable by an adult, that same clothing was cut down to become children's clothing. Coverlets and blankets were cut apart and stitched back together, with the former sides in the middle when the middles were getting worn. When fabric became so worn that it could not be used for its original purposes, it was cut down for smaller uses, such as quilt blocks. Cloth was not a thing to be wasted. In fact, this simple thing called cloth was far from simple to make; it was, as Rountree and Kelso both attest, a valuable homespun commodity that required much knowledge, specialized tools and, above all, labor. Pioneer life depended on it. Spinning wheels rather than drop-spindles and large barn looms rather than smaller simpler looms made the process easier, but it was never easy.

The samples that follow are taken from the collections of the History Museum on the Square.

White-on-white handwoven linen coverlet with handmade lace (mid-1800s). *Photo by Palmer Johnson, courtesy of the History Museum on the Square, Springfield, Missouri.*

Navy and cream handwoven linsey-woolsey coverlet (mid-1800s). *Photo by Palmer Johnson, courtesy of the History Museum on the Square, Springfield, Missouri.*

▲ ▼

Navy, orange, and brown handwoven linsey-woolsey quilt (mid-1800s). *Photo by Palmer Johnson, courtesy of the History Museum on the Square, Springfield, Missouri.*

Mapping Springfield:
The Evolution of Surveying Technologies
Jim Coombs

I. A History in Maps

Early Plat Map (1838)

The first map of Springfield showing the street grid and other detailed features was a General Land Office (GLO) survey plat map, surveyed in 1835 and published in 1838. The GLO was responsible for the surveying, platting, and sale of the public domain lands in the western United States as a source of revenue for the federal government.

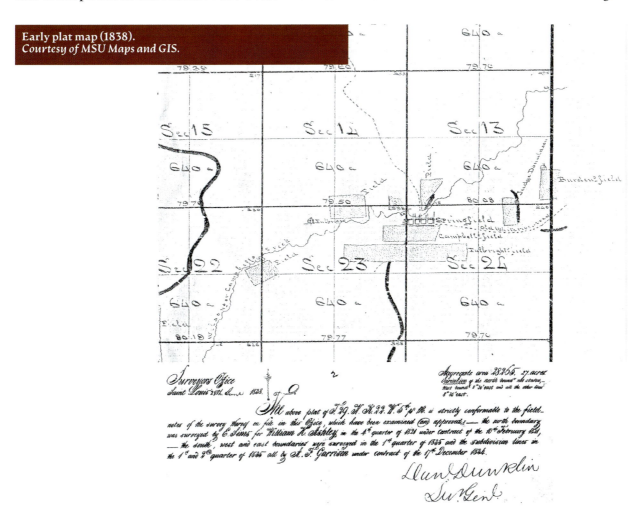

Early plat map (1838).
Courtesy of MSU Maps and GIS.

Panoramic Map (1873)

During the late 19th and early 20th centuries, bird's-eye views were a popular cartographic format used to depict cities and towns in the United States and Canada. Preparation of these maps involved a vast amount of detailed labor. First the artist drew a map of the city showing the pattern of streets in perspective as viewed from above at an oblique angle from an elevation of 2000 to 3000 feet. He then walked the streets of the town, sketching buildings, trees, and other features. The artist would return to a studio and paint a full-scale portrait based on the sketches and notes. Lithograph engravers would then copy and print the finished product. This 1873 bird's-eye view of Springfield was painted by Eli Sheldon Glover, who was artist, printer, and publisher of many views of towns and cities in Missouri, Kansas, and Colorado. These views, known also as panoramic maps, were used as advertisements for a city's commercial potential. This view of Springfield looks toward the southeast.

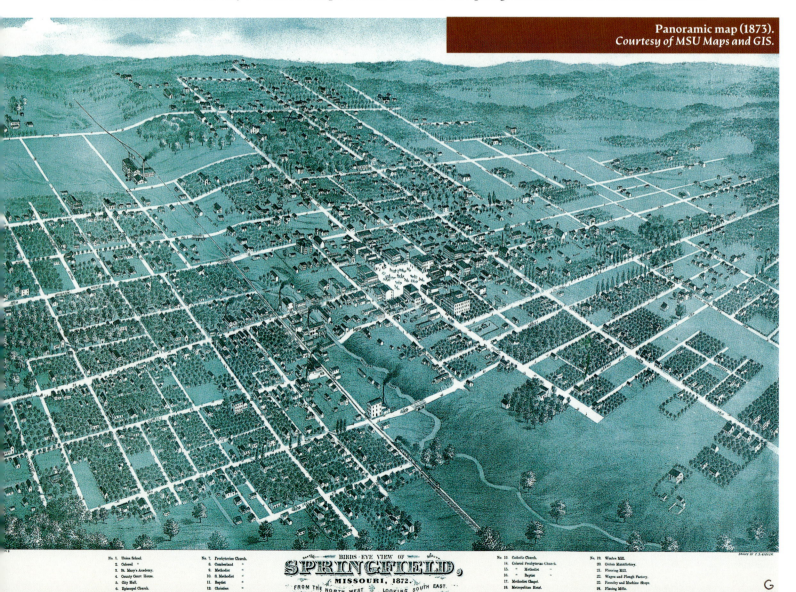

Panoramic map (1873).
Courtesy of MSU Maps and GIS.

18

Topographic Map (1886)

The first topographic map showing Springfield was surveyed in 1884 and published in 1886 by the U.S. Geological Survey at a scale of 1:125,000. It is one of many quadrangles at that scale covering relatively undeveloped western Missouri. USGS created these accurately detailed topographic quadrangles to use as a base for geologic mapping in the field.

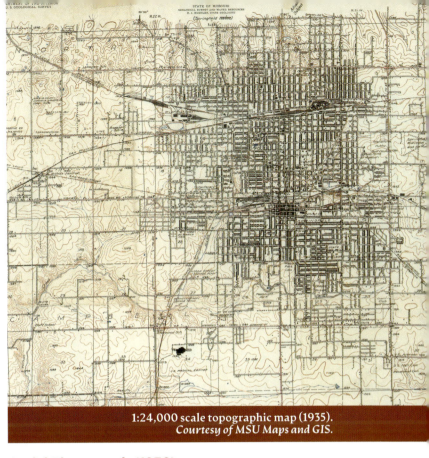

1:24,000 scale topographic map (1935).
Courtesy of MSU Maps and GIS.

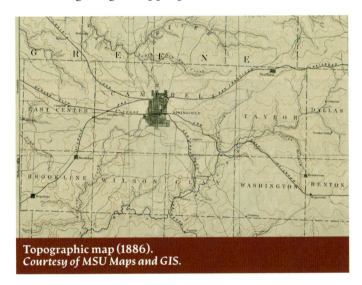

Topographic map (1886).
Courtesy of MSU Maps and GIS.

1:24,000 Scale Topographic Map (1935)

The first 1:24,000 scale topographic map of Springfield was surveyed in 1934 and 1935 by the U.S. Geological Survey. As the country became more fully explored and populated, the Survey began to create topographic quadrangles at the larger 1:24,000 scale as a result of demand for more detail on topographic maps.

Aerial Photograph (1936)

The first aerial photographs of Springfield were taken by the Cultural Adjustment Agency. During the 1930s, one-third of the land surface of the U.S. was photographed. Aerial photography was used to make accurate field measurements to determine compliance with the Agricultural Adjustment Act of 1933, an agricultural conservation program of the New Deal era.

Aerial photograph of Springfield town square (1935).
Courtesy of MSU Maps and GIS.

USGS Topographic Map
from Photogrammetry (1960)

The first USGS topographic map of Springfield using aerial photograph data and photogrammetry was this 1960 edition, using aerial photos taken in 1959:

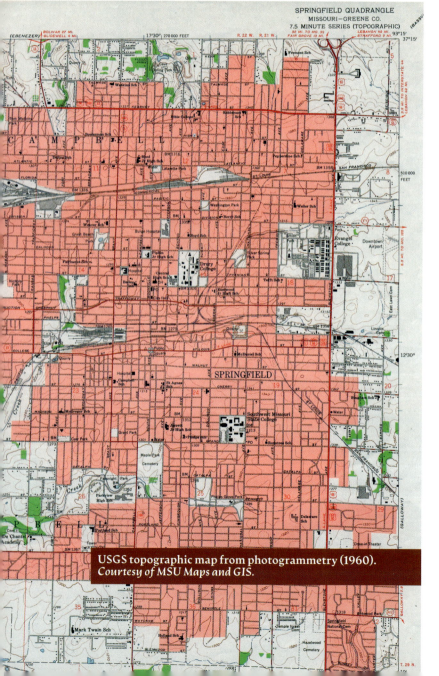

USGS topographic map from photogrammetry (1960).
Courtesy of MSU Maps and GIS.

ArcGIS MyMap (2019)

The development of computer applications may represent the greatest technological innovation in the mapping process. The convergence of geographic information systems (GIS) and the internet has changed the mapping of the world—Springfield, of course, included—forever. Detailed interactive maps are now generated on demand from databases of spatial information and transmitted instantly to our mobile devices. Location services like OpenStreetMap (OSM) offer free, editable images. The following is taken from ArcGIS MyMap:

Lake Springfield and vicinity (2019). © ArcGIS MyMap.
There are also commercial routing and direction services,
like Verizon Media's MapQuest.

ArcGIS Interactive City and County Maps (2019)

The Springfield City and Greene County websites have interactive maps powered by Esri's ArcGIS (described below), showing public notifications. For examples, note the city's EMS incident map, followed by a county map marking flood hazards (see next page).

Interactive city map (2019).
© *Esri, City of Springfield, Missouri.*

Interactive county map, showing sinkhole and flood hazards (2019).
© *Esri, Greene County Assessor.*

II. The Evolving Technologies of Map-Making

Surveying

Created in 1812, the General Land Office (GLO), which made the first survey plat map of Springfield (shown above), was responsible for the surveying, platting, and sale of public lands in the Western United States as a source of revenue for the federal government. The GLO used a rectangular method of describing lands called the Public Land Survey System (PLSS).

The PLSS divided states into 36 square mile township grids, which were further divided into one square mile (640 acre) sections and fractions of sections.

To create these GLO survey plats, a team of surveyors marked and defined the boundaries of each township using the instruments described below. While they were surveying, they made field notes of the measurements they performed. Then they drafted plats of each township to show the distance and bearing between section corners, sometimes including topographic or vegetation information. These maps were the official approved documentation of the survey and were used for land sales. The technology used by the GLO to conduct these surveys included the following.

Surveyor's chain, used to measure distances.

Surveyor's chain.
National Museum of American History, Washington D.C.

The 66-foot long chain, divided into 100 links, was usually marked off into groups of ten by brass rings or tags to simplify intermediate measurement. Each link is 7.92 inches long. A quarter chain, or 25 links, measures 5½ yards, which is a rod. Ten chains measure a furlong, and 80 chains measure a statute mile.

Compass, used to measure the horizontal angle between two points (see next page).

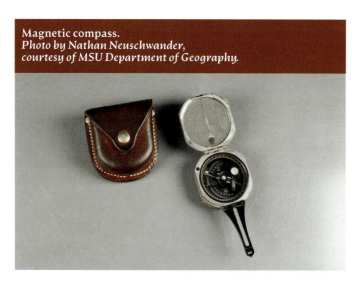

Magnetic compass.
*Photo by Nathan Neuschwander,
courtesy of MSU Department of Geography.*

The mercurial barometer was a delicate instrument about 30 inches long. It was easily (and frequently) broken, but extra glass tubes and a flask of mercury were carried for repairs.

The aneroid barometer, a small watch-like instrument, was more convenient but its measurements were less reliable.

Level staff, used in conjunction with a transit or theodolite to read an elevation up or down from the level of the telescope. From these observations, the surveyor determined differences in elevation of different points.

The magnetic surveyor's compass consisted of a circular box housing a needle pointing to "magnetic north" above 360-degree compass markings.

The solar compass was designed to overcome the difficulties experienced with magnetic declination in areas where iron deposits were prevalent. By making observations on the sun, the latitude of the location could be determined and true north could be determined. An arm extending from the compass included columns at the ends with holes for sighting points. Horizontal angles between two lines were established by sighting two points from the same location.

Barometer, used to determine differences in height by measuring the differences in atmospheric pressure at various elevations.

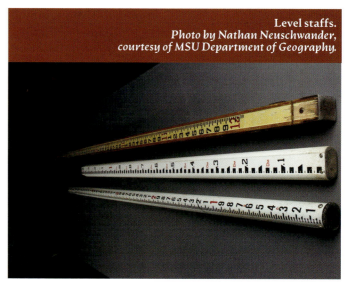

Level staffs.
*Photo by Nathan Neuschwander,
courtesy of MSU Department of Geography.*

Plane table, a 20" x 30" rectangular table mounted on a sturdy tripod, used as a horizontal surface for mounting the compass, transit or theodolite, and for recording measurements (see facing page). The measurements were noted and platted on paper attached to the board.

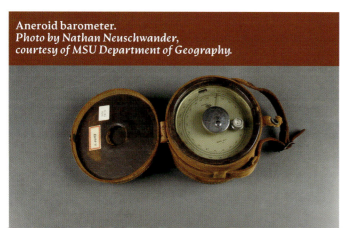

Aneroid barometer.
*Photo by Nathan Neuschwander,
courtesy of MSU Department of Geography.*

Plane table.
U.S. Geological Survey.

USGS Established (1879)

The U.S. Geological Survey (USGS) was created in 1879, charged "with the study of geological structures and economic resources of the public domain." In 1882, a Division of Topographic Mapping was created to produce maps as bases for geologic and economic survey; these illustrated the resources and classification of lands. USGS divided the United States into seven districts. Missouri was in the District of the North Mississippi, along with Ohio, Indiana, Illinois, Michigan, Wisconsin, Minnesota, Dakota, Nebraska, Kansas, and Iowa. USGS mapping in Missouri started in 1884 and included the 1:125,000 Springfield quadrangle (shown earlier).

Steel Tape replaced the surveyor's chain.

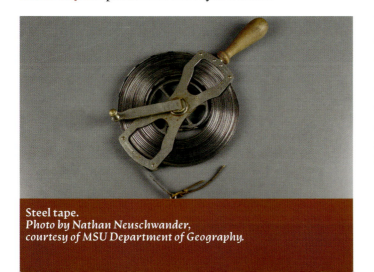

Steel tape.
*Photo by Nathan Neuschwander,
courtesy of MSU Department of Geography.*

By 1887, The surveyors found that measuring base lines with long steel tapes was easier and more accurate than with the surveyor's chain. The 300-foot steel tape also made it practicable to measure much longer base lines.

A team of surveyors and topographers used the GLO plats as base maps to add drainage and other topographic details. The surveyors measured baseline distances with a chain, angles between points with a compass, and elevations with a barometer.

Transits and theodolites gave more accurate measurements.

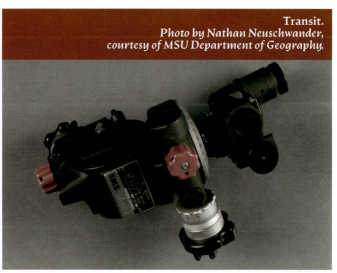

Transit.
*Photo by Nathan Neuschwander,
courtesy of MSU Department of Geography.*

The technology of surveying continued to evolve. Developed in the 19th century, transits and theodolites measured horizontal and vertical angles between points, determined distances, and measured elevations more accurately than measuring with a compass, tape, and barometer. Both instruments consisted of a metal plate marked in degrees and a moveable telescope mounted so that it could rotate around horizontal and vertical axes; by this means, it measured angles between points with great accuracy. The

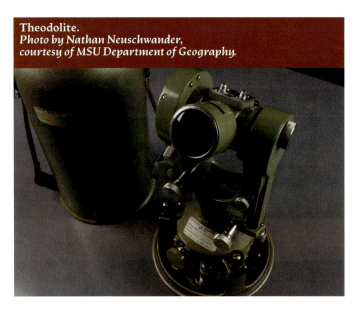

Theodolite.
Photo by Nathan Neuschwander,
courtesy of MSU Department of Geography.

transit had a compass attached to the metal plate, while the theodolite had a level attached to its base and a vertical circle for measuring vertical angles.

Electronic digital theodolite.
Photo by Nathan Neuschwander,
courtesy of MSU Department of Geography.

Improvements in accuracy for transits and theodolites have continued through the 20th and 21st centuries. Today, electronic digital theodolites are the preferred surveying measurement device for surveyors worldwide. Electronic theodolites work quickly and eliminate human error in misreading scales. Infrared measuring devices work in conjunction with complex software to minimize errors and provide more accurate results. Electronic theodolites also record and save the surveying data, eliminating the need for paper and pencil or audio recording devices in the field. The data are later transferred to computers, where complementary software analyzes them for map-making.

The 1:125,000 Springfield quadrangle (published in 1886) and the first 1:24,000 scale topographic map of Springfield (published in 1935, both shown earlier) were created with the surveying and mapping technology just described.

Map Production and Lithographic Printing

The reproduction of USGS maps from the field survey data was done with a three-plate lithographic printing process. The topographic features were engraved on three copper plates. Water features such as streams and canals (hydrography) were engraved on one, for printing in blue; relief (hypsography), shown by contour lines, was engraved on the second, for printing in brown; and towns, cities, roads, railroads, boundaries, projection lines, and lettering were engraved on the third, to be printed in black. Conventional symbols and lettering were established and reduced to the greatest possible simplicity, in order that the maps might be easily understood and be of value to all people.

Copper plate engraving of topographic maps.
U.S. Geological Survey.

Topographic information being transferred from copper plates to a lithographic stone for printing.
U.S. Geological Survey.

required for each color. It had to be very precise; thus, the appearance of the printed map depended largely on the talent and patience of the draftsman.

In pen-and-ink drafting, the line weight depended on the size of the pen point, the fluid qualities of the ink, the surface paper, and the hand-pressure applied on the pen. A new technique of drafting with special tools on coated plastic, known as scribing, rapidly replaced pen-and-ink in the final stages of map production. Since scribing used a metal stylus, it produced a more legible map with neater, sharper, more uniform line work; it did so in a shorter time and at less cost. Scribing output was superior both in quantity and quality to pen-and-ink drafting and made a tenfold increase in map production possible.

Scribing also made it possible to add more detail on the topographic maps. Almost 200 features separated into color groups were on the five color plates used in the printing process. In addition to the three plates described above, a red plate was made for road fills, urban tints, Public Land Survey lines, and other

Pen-and-ink drafting.
U.S. Geological Survey.

Copper-plate engraving was abandoned in pre-World War II years when color-separation drafting was developed. The map detail was reproduced photographically on metal-mounted paper and traced freehand in ink to make the printing plates. A separate drawing was

Scribing method.
U.S. Geological Survey.

were transferred from photographs to paper by the pantograph tracing method, which was tedious but avoided the need for many hours of survey work in swamps, gullies, etc. This method was used until the 1930s, when multiplex machines were invented.

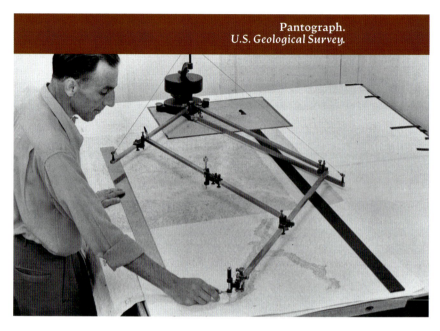

Pantograph.
U.S. Geological Survey.

features, and a green plate was made for woodland tint and other vegetation. The color separations were composited on a five-color lithographic press. The 1960 edition of the Springfield quadrangle (shown above) was made with this process.

Photogrammetry and Pantograph Tracing

The USGS first used aerial photography for producing maps after World War I. Because most of the skilled topographers in the United States were USGS staff, many of them served in the Topographic Corps of the Army, where they used aerial photography for intelligence purposes. They realized the potential of aerial photography for topographic mapping and, after the war, they developed photogrammetry, the art and science of making measurements from photographs. It proved a revolutionary advance in the techniques of mapping: The methods for topographic map-making changed from surveying with instruments in the field to working with aerial photographs and precise plotting instruments in an office. At first, photographic data such as roads, houses, water features, and forests

The multiplex, by direct projection of overlapping photographs printed on glass plates, produced an exact optical model of the terrain to be mapped (see image at end of essay). Horizontal and vertical measurements could be made in the model, and roads, water features, and contours could be drawn. This became the primary method of plotting contour lines on topographic maps from the 1930s to the 1960s. Although projection technologies have continued to advance, they are all based on the apparent change in position of a feature in the two stereo photographs.

Photogrammetry was also used by U.S. Department of Agriculture mapmakers. The Agricultural Adjustment Act of 1933, an agricultural conservation program of the New Deal era designed to boost agricultural prices by reducing surpluses, created a new agency under the USDA, the Agricultural Adjustment Administration, which photographed one-third of the land surface of the U.S. This aerial photography was used to make maps for accurate field measurements to determine compliance with the agricultural conservation program. Taken in 1936, the first aerial photographs of Springfield (including the one shown earlier) were taken by the U.S. Department of Agriculture.

Stereoplotters were the next improvement in photogrammetry technology. The first stereoplotters were projection stereoplotters that used only light rays and optics to adjust the image. The analog stereoplotters came next and used more sophisticated optics to view the image. The analytical stereoplotter is used today. It incorporates a computer that aligns images mathematically. The analytic stereoplotter also stores the data for redrawing at any desired scale.

In the mid to late 20th century, advances in electronic technology led to the next technological revolution in cartography. Computer hardware devices such as analytic stereoplotters, scanners, and printers, along with satellite imagery and database software, democratized and greatly expanded visualization, image processing, and spatial analysis with maps. Then, Geographic Information Systems (GIS) became available. GIS is a modern extension of traditional cartography with one fundamental similarity and two essential differences. They are similar in that both a cartographic document and a GIS contain a base map to which additional data can be added. The differences are that, first, there is no limit to the amount of additional data that can be added to a GIS map; and, second, the GIS uses analysis and statistics to present data, which a cartographic map cannot do.

The U.S. Census Bureau was an early adopter of GIS. The Census Bureau created a national geographic database to support digital data input, error fixing,

and even choropleth mapping for the 1970 census. In the 1980s, the Census Bureau created TIGER GIS data files (Topologically Integrated Geographic Encoding and Referencing) for digital input of the 1990 census. TIGER GIS data became the first freely available digital map of roads, railroads and census boundaries. Those data stimulated technological developments like MapQuest, Yahoo, and Google Maps. In 2005, the data also formed a base for OpenStreetMap in the United States.

In 1986, the Environmental Systems Research Institute (ESRI, now Esri) brought GIS to personal computers when it launched ARC/INFO for PCs. In the 1990s, Esri released ArcView, a desktop application for producing maps via a Windows-based interface. Its software was able to handle both vector and raster data. Gradually, the importance of spatial analysis for decision-making became recognized, and ArcView was adopted by many governments and busi-

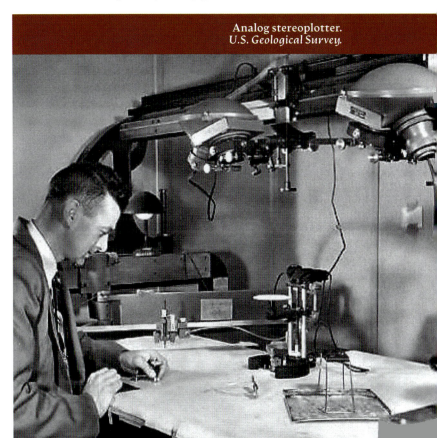

Analog stereoplotter.
U.S. Geological Survey.

nesses due to its ease of use. During the next decade, the internet saw the adoption of GIS technologies at lower and lower levels of government as costs tumbled and the repositories of massive amounts of spatial data, mostly Esri's ArcGIS Online, came within reach of local authorities; this local use is illustrated by the City of Springfield and Greene County interactive maps (shown above). Esri is now the world's leading expert in GIS software development.

As technology improves, cartography and the tools related to it improve as well. Processors are now in gigahertz. GIS data storage is in terabytes. Accuracy has changed from the size of an airport to the size of a small shed. TIGER data, Landsat satellite imagery, and even LiDAR data are accessible for free download. LiDAR (light radar) has complemented and often replaced the stereoplotter for gathering elevation data. LiDAR uses a laser pulse directed at features and detects the amount of time between when the pulse is emitted and when it is detected to determine the difference in elevations.

The globalization of data has contributed greatly to the use of GIS and cartography for more applications every day. With GIS, almost anything can be studied from a geographic point of view. Some technologies that previously were restricted to military uses, like GPS or remote sensing, are now available to everyone. GPS has given users innovative products like car navigation systems and drones.

With technology advancing so fast, it is hard to predict the next step in the development of cartography. Today, the use of cartography and GIS software applications on mobile devices is the latest advance, but tomorrow is a question mark. There is no doubt, though, that computer technology has created a new era in the art and science of map-making.

Resources

Early Plat Map (1838). *Missouri GLO Plats: An index [to] GLO plats in Missouri*. Missouri Department of Natural Resources, Division of Geology and Land Survey. Land Survey Program. Rolla, MO: Missouri Department of Natural Resources, Division of Geology and Land Survey, 2011.

Panoramic Map (1872). *Bird's Eye View of Springfield Missouri, 1872*, drawn by E.S. Glover. Cincinnati, OH: Strobridge & Co. Lith., 1872.

Topographic Map (1886). *Springfield quadrangle, Missouri: 30-minute series (topographic) mapped, edited, and published by the Geological Survey*. Reston, VA: The Survey, 1886.

1:24,000 Scale Topographic Map (1935). *Springfield quadrangle, Missouri: 7.5-minute series (topographic) mapped, edited, and published by the Geological Survey*. Reston, VA: The Survey, 1935.

Aerial Photograph (1936). *Greene County, Missouri, aerial photographs*. Washington, DC: U.S. Agricultural Adjustment Agency, 1936.

USGS Topographic Map from Photogrammetry (1960). *Springfield quadrangle, Missouri: 7.5-minute series (topographic) mapped, edited, and published by the Geological Survey*. Reston, VA: The Survey, 1960.

ArcGIS–MyMap (2019). Image source: http://www.arcgis.com/

Interactive City Map (2019). Image source: http://maps.springfieldmo.gov/fireincidents/

Interactive County Map (2019). Image source: www.greenecountyassessor.org

Lithographic stone, drafting and scribing images, pantograph, multiplex projector, and stereoplotter (2009). Image resource: Evans, Richard T., and Helen M. Frye. *History of the Topographic Branch Division*. Reston, VA: U.S. Geological Survey, 2009.

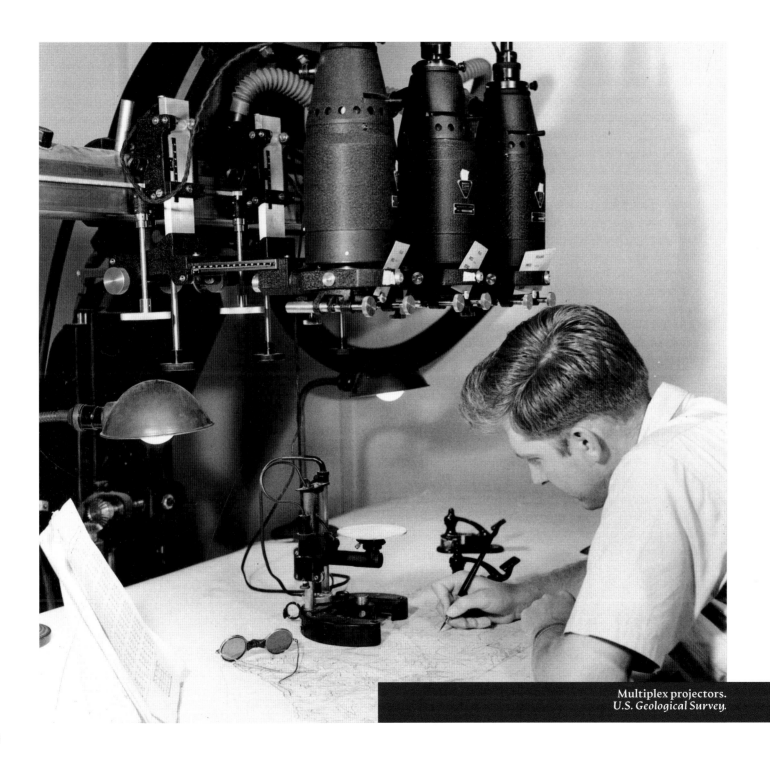

Multiplex projectors.
U.S. Geological Survey.

Walter Majors,

African American Inventor and Entrepreneur
Richard Schur

There was a strange vehicle on the streets yesterday. People gathered about to make a close inspection and see how it was made and horses shied at it. It was no more or less than an automobile, propelled by gasoline and made by a young colored man of Springfield. The trip of this first horseless carriage made in Springfield was not entirely successful but the vehicle moved and could be steered and stopped at will. It did get a rapid move on it and there are some glaring faults in its construction[, but] the young colored man [*sic*] has the right principle and he can perfect the machine so it will carry him on smooth streets at a rapid rate.

—"An Automobile: Made By a Young Springfield Colored Man," *Springfield Leader-Democrat*, Monday April 8, 1901

On Easter Sunday, April 7, 1901, the first automobile built in Springfield, Missouri drove through the city. The *Leader-Democrat* reported that the car scared horses but intrigued residents, who inspected the vehicle, eager to learn more about how it worked.

The car was driven by Walter Majors (*ca.* 1879-1949), a young African American man born and raised in Springfield. According to the newspaper, Majors built the car with the help of his brothers, one of whom worked as a blacksmith at the Springfield Wagon Works. After peeking at the mechanics of a car that came through town, Majors decided to build his own version, with an engine behind the driver's seat and pneumatic bicycle wheels. His car was not perfect, stalling several times and in need of tinkering to com-

Walter Majors at the wheel of his automobile (undated).
Courtesy of MSU Special Collections and Archives.

plete the journey back to his house.[45]

While it did not lead directly to a profitable invention or enterprise, Walter Majors' car served an important symbolic function in Springfield's African American community; it also served as a starting-point for a life of invention and entrepreneurship in Springfield and, later, Saint Louis. This essay explores the life of Walter Majors and how he overcame racial segregation on his path to becoming an inventor and entrepreneur.[46]

Family and Education

Walter Majors was born in Springfield, probably in 1879, to Peyton and Emily Majors. His parents were brought as slaves to the region during the 1850s and stayed after emancipation following the Civil War. Peyton Majors married Emily on April 4, 1871 in Springfield.[47] The Majors had eight children, with seven surviving into adulthood. For most of Walter's youth, the Majors family lived at 822 Washington in the heart of one of the four African American neighborhoods in Springfield. This address stood within two blocks of the Lincoln School, the first African American school in Springfield, which Majors likely attended.

Spanish-American War

Walter Majors first appears in the existing historical records, beginning with his decision to volunteer to serve during the Spanish-American War. In July 1898, Walter Majors, along with his brother Harry and other young African Americans from Springfield, volunteered to serve in Company "L" in the 7th Colored Infantry. Already known for his interest and abilities in technology, Majors was assigned the fairly unusual rank of "wagoner."[48] This rank historically was reserved for the individual who took care of the wagons or transportation for the company. Further, it distinguished him from the other recruits, who were enlisted as privates.

Springfield Entrepreneur

Upon his return to Springfield, Walter Majors married Myrtle Farrier. They owned a home at 628 Phelps Street, which was an integrated street with more white families than black families. In 1899, Majors opened his first business, a bicycle repair shop on North Jefferson, with another young man, George Webb. This business did not last long, but Majors continued developing his technical skills at the Springfield Wagon Works.

In 1904, Walter Majors is listed as proprietor of *The Statesmen*, the second black newspaper in Springfield, Missouri, with offices located on Boonville. The paper printed only a handful of issues. A natural-born entrepreneur, Majors opened a series of other short-lived businesses during the first decade of the 20th cen-

45. Writing in "The Waste Basket" column of the December 26, 1930 *Springfield Leader*, A. J. Hasewell claims to have witnessed Majors' first car drive through Springfield. Describing the events nearly thirty years later, he writes,

> [W]e saw that, which was apparently the sheet iron body of a boy's express wagon of rather larger size than usual and on the axles of this body were four wheels, evidently from some derelict safety bicycles; filling all the space in the wagon box and projecting well beyond the rear was a very small horizontal steam engine; the boiler was an iron cylinder never meant to serve as a boiler but it was secured beneath the wagon bed, and underneath it four old kerosene lamps supplied the heat to produce steam, the power was conveyed from the engine to one rear wheel by a chain and a sprocket wheel.

46. For a discussion of Walter Majors' place in Springfield's African American history, see my essay, "Memories of Walter Majors: Searching for African American History in Springfield," in *Springfield's Urban Histories: Essays on the Queen City of the Missouri Ozarks*, ed. Stephen McIntyre (Springfield, MO: Moon City, 2014), pp. 113-37.

47. The Katherine G. Lederer Ozarks African American History Collection (hereafter Lederer Collection) File 78, Box 20.

48. Lederer Collection File 76, Box 19.

tury: a musical instrument or violin repair business,[49] W. L. Majors Bicycle Repairing and Novelty Works, and the Majors Automobile Company. This last business is notable for several reasons: Not only did he open his car company in 1907, after Springfield's infamous Easter 1906 lynching, but he opened it at 214 N. Jefferson—fairly close to the location of the lynching on the town square.[50] These businesses did not succeed, due in part, perhaps, to how segregation shaped Springfield; still, the local African American community appreciated Majors' entrepreneurial zeal.

St. Louis Business Owner and Inventor

Majors left Springfield,[51] moving to St. Louis in either late 1907 or early 1908. He eventually settled in the historically black district known as the Ville. Unlike his effort to live and work in integrated portions of Springfield, Majors found a measure of success in this segregated part of the city, where there was a sufficient consumer-base of African Americans to support his businesses. Also, white racism, while always an element of the social and economic structure, was

a bit further removed from him and his enterprises. Initially, he worked as a mechanic. Soon, however, his career took a potentially surprising turn. Between 1912 and 1914, Majors worked at Annie Turnbo Malone's Poro Beauty College. Poro sold beauty and hair-care products for African Americans, and its owner, Annie Malone, become one of the richest African American women during this time. Not simply a financial success, Malone and Poro symbolized African American beauty, intelligence, and business acumen. At Poro, Annie Malone took Majors under her wing, teaching him about marketing and sales.

The experience at Poro allowed him to set up his own beauty school, the Oxford College of Hair and Beauty, between 1914 and 1916.[52] Majors toured the Midwest, seeking out students and building up the reputation and reach of his business.[53] Working at Poro and his own beauty college afforded him time to work on his own inventions.[54] As part of his efforts to promote beauty and beauty products for African Americans, Majors patented a kind of hot comb and an early hair drier.

Later, Majors opened his own garage. The garage allowed him the opportunity to own a business, earn money by repairing cars and probably taxicabs, and have a space sufficient for him to work on his inventions. In his garage, Majors worked on inventions (see list below) that spanned a wide range of interests, from hair care to taxi maintenance to home heating. Here are the patents held by Walter Majors:

---

49. Interview with Harold McPherson (November 15, 2008).

50. In April 1906, a sizable crowd of white Springfieldians, estimated to be as large as 4,000 people, lynched three African Americans, Fred Coker, Horace Duncan, and Will Allen. A local woman claimed an African American man had raped her and these three men were arrested, although it is doubtful these men were in any way connected to the supposed crime. The crowd surrounded the jail and killed the three men following a mock trial in the town square. There was a federal investigation of the incident, in which eighteen members of the mob were indicted but never convicted. The grand jury concluded that Coker and Duncan had nothing to do with the rape. Local African Americans tell stories of hiding for up to a week to escape the mob. For a detailed account of the events, see Kimberly Harper, *White Man's Heaven: The Lynching and Exclusion of Blacks in the Southern Ozarks, 1894-1909* (Fayetteville, AR: University of Arkansas Press, 2010).

51. Many have argued and speculated that Majors left Springfield after the 1906 lynching (see footnote above), because the state of race relations made it difficult, if not impossible, for an African American to succeed in business at this time.

52. Its relation to Malone's Poro Beauty Colleges is unclear based on existing records. It appears to have been an extension of her network, rather than a direct competitor. See "Majors' Oxford College of Hair Culture, St. Louis," *The Freeman, An Illustrated Colored Newspaper* (April 25, 1914), 3. See also Lederer Collection File 17, Box 82 and File 76, Box 19.

53. "Duck Majors Here on 4,000 Mile Tour," *Springfield News-Leader* (September 24, 1916), p. 1.

54. "W. L. Majors Dies at Home," *St. Louis American* (December 15, 1949), p. 1.

Coin-controlled taxicab controller (1913),
     U.S. Patent 1,069,558
Heater for water coolers in cars (1914),
     U.S. Patent 1,121,266
Motor controlling device for taxis (1915),
     U.S. Patent 1,123,906
Hair drier (1915), U.S. Patent 1,124,235
Oil stove (1920), U.S. Patent 1,331,162
Anti-skid device for a car (1922),
     U.S. Patent 1,422,285
Mimeographic attachment (1922),
     U.S. Patent 1,422,286
Hair and scalp treatment (1923),
     U.S. Patent 1,466,629
Carburetor auxiliary substitute (1926),
     U.S. Patent 1,596,885
Heating apparatus (1930), U.S. Patent 1,783,576
Oil burner (1932), U.S. Patent 1,862,691

From the vantage point of the 21st century, the automobile and beauty products may seem ordinary and unremarkable. However, during the life of Walter Majors, cars and beauty products challenged racial stereotypes. Whites saw African Americans as rural and uninterested in technology. Mastering the automobile as a driver, mechanic, and inventor exploded that myth. Similarly, the nascent black beauty industry sought to contest racist assumptions about beauty.

Visiting Springfield
Though Majors made his living in St. Louis, he regularly returned to Springfield to visit friends and promote his business interests—which included the latest makes and models of automobiles.[55] The newspapers were happy to note his visits, reporting them as triumphant returns. The papers also used these occasions to retell the stories about his first car, sometimes adding (or possibly inventing) new details not mentioned in earlier versions. In these stories during the 1910s and 1920s, Majors became treated as a returning hero whose opinions about roads (including the challenges faced by Ozark drivers during this period), technology, and the greater world were quoted and viewed as significant.

Symbolic Value of Walter Majors
Majors was, and remains, a symbol of racial pride and progress in Springfield. He was remembered, for example, in the 1933-1934 *City Directory* in Louis Smith's article, "The Part The Negro Played From the Oxcart to the Airplane in Springfield."[56] In 1984, Abraham Clark started the Walter Majors Classic Car Club and several local leaders developed plans to construct a Walter Majors Cultural Institute of the Ozarks, but this was never realized.[57] Majors is also listed on a plaque in Founder's Park in downtown Springfield and depicted in the large mural there.[58]

For all the jobs he held and the businesses he owned, Majors' death certificate listed his occupation as "inventor-retired." His obituary also emphasized the patents he received. His life and work should not be simply viewed in terms of his wealth or the importance of his inventions. Rather, Majors' importance lies in his daring to break racial stereotypes during the era of segregation by crafting new inventions. Even if his automobile or other inventions did not create exceptional financial or business success, he was, and remains, an inspiration for African Americans in the Ozarks.

---

55. For examples, see "Negroes Arrested on Charge of Speeding," *Springfield News-Leader* (September 12, 1909); "Duck Majors Here on 4,000 Mile Tour," *Springfield News-Leader* (September 24, 1916); and "Woes of First Auto Recalled," *Springfield Leader and Press* (June 12, 1928).

56. 1933-1934 City Directory, in Lederer Collection Box 20, File 160.

57. "Black Families of the Ozarks, Volume 4C," *Greene County Archive Bulletin* No. 68, p. 356 (there are three pictures in this source, including a business card), and "Walter Majors Cultural Institute of the Ozarks," Lederer Collection Box 19, File 169.

58. The plaque was printed with the wrong date for the first drive in Majors' car.

W.H. Goss Mill
at Timmersville Mo 1912

Son Goss

Bill Goss

Big Machines from the Past Century:

Photos from MSU Special Collections and the History Museum on the Square
Selected by Tracie Gieselman-Holthaus

▲ Panoramic photograph of Ozark Land and Lumber Co.,
Shannon County, Missouri (1903).
*Ozark Land and Lumber Company Collection,
MSU Special Collections and Archives.* **Taken by the *American
Lumberman* photographer.**

◀ W. T. Gates, "Mill at Summersville, Mo." (1912).
Bill A. Gates Collection, MSU Special Collections and Archives.

▼ W. T. Gates, "Threshing on McCaskills Farm,"
Summersville, Missouri (1912).
Bill A. Gates Collection, MSU Special Collections and Archives.

Road grading machinery (ca. 1915). ▲
*The Streets Collection, History Museum on the Square,
Springfield, Missouri.*

▲ Horse-drawn boiler in Eminence, Shannon County, Missouri (undated).
*Bill A. Gates Collection, MSU Special
Collections and Archives.*

"Using locomotion hoisting machinery in mill," Ozark Land and Lumber Co.,Shannon County, ▲
Missouri (undated). *Ozark Land and Lumber Company Collection,
MSU Special Collections and Archives.*

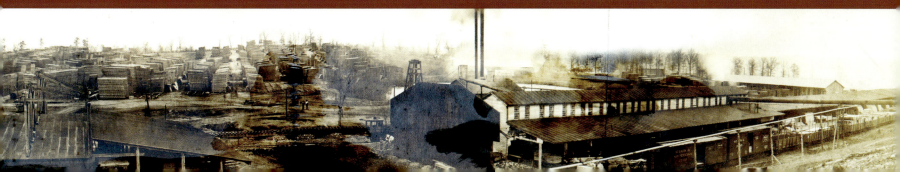

James L. Dalton:
From "Barefooted Boy in the Ozarks" to American Inventor, Entrepreneur, and Industrialist
Brooks R. Blevins

James L. Dalton stared out at readers of *Hearst's Magazine* in the autumn of 1920.[59] Below the portrait was the byline B. C. Forbes, the young Scottish-born chronicler of American wealth and power who had only three years earlier launched his own magazine. If ever there was a picture of the titan of industry, Dalton's was it. Piercing eyes, aquiline nose, gray-flecked Van Dyke, white collar, and black tie—a middle-aged man at the height of success, atop a multi-million-dollar corporation with sales offices around the globe. Though only half a dozen years removed from living in a small town on the edge of the Missouri Ozarks, here was Dalton—his life story featured in one of America's glitziest magazines alongside a series of Forbes sketches of the business world's most powerful movers and shakers. Here he was in the company of men whose names resonate to this day—Rockefeller, du Pont, Swift, Eastman. Here was James L. Dalton, American inventor, entrepreneur, and industrialist—"A Barefooted Boy in the Ozarks."[60]

Dalton's wasn't the only titan story lacking old money, prep schools, and the Ivy League, but his was the only biography in which region became a central character. The Ozarks was just beginning to emerge in the public consciousness as an identifiable cultural region—the kind of place that should produce fiddlers, illiterate farmers, and moonshiners, not millionaire industrialists. And B. C. Forbes was not one to overlook the Horatio Alger echoes of any success

Portrait of James L. Dalton (1920). *Hearst's Magazine.*

---

59. The following is excerpted from Blevins's *A History of the Ozarks, Volume 3: The Ozarkers* (forthcoming 2021 by the University of Illinois Press).

60. B. C. Forbes, "A Barefooted Boy in the Ozarks: James L. Dalton," *Hearst's Magazine* (October 1920), p. 26.

story that traced its origins to such an unlikely place. Neither was sensationalist publishing tycoon William Randolph Hearst, who, oddly enough, claimed roots in the same geographic region that produced James L. Dalton (1866-1926). Hearst's parents—one a wealthy goldmine owner and former U.S. Senator and the other a great cultural benefactor and first woman regent of the University of California—were both born and raised on antebellum farms in the Missouri Ozarks. For Forbes, Dalton's rise was nothing less than "Lincolnesque," his childhood in a log cabin on a "scraggy farm" more an obstacle to be overcome than a training ground for hard work and determination. In Forbes's rendition, the rolling hills of Dalton's boyhood became isolating mountains, the seventeen miles to the nearest railroad a formidable forty-mile journey. Like the remarkable souls who occasionally sprang from inferior native bloodlines in the local color novels of the Ozarks, Dalton's intelligence and skills could only have been the gifts of a gracious god, or the result of natural accident. "Although he was never in his life taught to 'parse' a sentence, he is now a brilliant speaker and writer, the recipient of requests from leading chambers of commerce and other bodies throughout the country to address their conventions and banquets."[61]

But Forbes need not have resorted to exaggeration and regional stereotype to stress the unlikely life journey of James Lewis Dalton. Born in rural Ripley County, Missouri, a year and a half after his Confederate father returned home from the war, Dalton moved with his parents a few years later to a farm in the Eleven Point River valley of neighboring Randolph County, Arkansas, where he grew up in a community named for the family. A precocious and mechanically inclined child, Dalton spent long hours experimenting with machinery and crafted a working wooden model of his mother's sewing machine. When his father's death left the family in straitened circumstances, James received a rudimentary education in local one-room schoolhouses and later trekked more than sixty miles through the hills to attend a respected academy

in Izard County, where he may even have parsed a few Latin sentences during his stay.

At the age of eighteen, Dalton took the sixty dollars he had cleared from a small cotton crop and rode the rails northward, finding work at a dry goods house in St. Louis. After a brief apprenticeship he returned to the Ozarks and found work as a store clerk in Doniphan, Missouri, soon mastered the business, and married the sister of his employers, and at age twenty relocated to Poplar Bluff to run a branch of his employers' enterprise. Buying out his brothers-in-law, Dalton capitalized on Poplar Bluff's location on the St. Louis, Iron Mountain, and Southern Railway and its proximity to both the hills and the flatlands to build up southeastern Missouri's largest department store by the turn of the century.[62]

It was around the dawn of the new century that the former mechanical wunderkind took an interest in developing a smaller, simpler adding machine. Collaborating with St. Louis inventor Hubert Hopkins, who received a patent for a ten-key device in 1904, James L. Dalton launched the Dalton Adding Machine Company in a little shed in Poplar Bluff. Leaving the store's operation to his eldest son a few years later, Dalton poured all his energy into growing the new company, receiving three patents for improvements on his machine and even serving as chief travelling salesman. In 1914, with the rapidly expanding Dalton Adding Machine Company outgrowing its facilities and marketing capacity in little Poplar Bluff, a town of only 7,000 people, Dalton moved his operations to a Cincinnati suburb. By 1920, the company was the second largest of its kind on the planet, with annual sales of $12 million, and various models of Dalton adding machines could be found in offices around the world. Following his untimely death due to a burst appendix in 1926, James L. Dalton's successors sold the company,

61. Ibid.

62. Ibid.; Lawrence Dalton, *The History of Randolph County* (Little Rock, AR: Democrat Printing and Lithographing Co., 1946).

which was merged with Remington Typewriter and Rand Kardex Bureau to form Remington Rand in 1927. The new corporation continued to manufacture Dalton adding machines into the 1950s.[63]

The story of James L. Dalton and his adding machine is obviously an extraordinary one, but his ingenuity and drive were not as exceptional as one might be tempted to believe. The Ozarks suffered no shortage of inventors and entrepreneurs in the generations following the Civil War. Almost none of them reaped the material rewards enjoyed by Dalton, and some of them received almost no monetary compensation whatsoever. Yet, the inventor's spirit was alive and well in the great age of mechanical technology. In Springfield—home to the region's leading value-added factory, the Springfield Wagon Company—no tinkerer better exemplified this spirit than blacksmith and mechanic Walter "Ducky" Majors. The African American inventor built the city's first horseless carriage, a steam-driven vehicle on which he toured the town square in 1901. Ahead of his time as an entrepreneur of failed repair shops for bicycles and automobiles, Majors eventually made his way to St. Louis, where he obtained patents on inventions ranging from an electric hairdryer to a coin-operated taxicab controller.[64]

Given the region's strong dependence on agriculture, it is no surprise that many of the Ozarks' inventions came from farmers. Webster County, Missouri,

farmer Jonathan D. Whittenburg patented a design for "flood fence" in 1885. The fence was designed to collapse and not wash away during a flood. Far to the south in Stone County, Arkansas, mechanically inclined farmer James H. Younger received a patent for a hay press in 1915. According to family memory Younger later sold the patent to the International Harvester corporation, but an ancient, rusting prototype still sits in an overgrown field behind a descendant's house today. The dairy industry's significant impact on southwestern Missouri influenced at least three patents from Ozarks inventors trying to improve the butter churn. The last of those, patented by Cedar County, Missouri's Virgil Pyle in 1930, was a simple screw-top agitator that could be attached to any Mason jar. Pyle and his brothers-in-law had high hopes for their Buttercup Churn Corporation, but a similar, cheaper version introduced by Sears, Roebuck and Co. a few years later ended their business venture.[65]

The story of Virgil Pyle, James L. Dalton, and other Ozarks inventors and entrepreneurs encapsulates key developments and themes in Ozarks life between the Civil War era and the Great Depression. Though the region and anyone connected to it rarely escaped the stereotypes that tended to reduce the Ozarks to a model of premodern, backward life, the people of the Ozarks followed the same variety of pursuits and chased the same basic dreams that motivated human

63. Forbes, "Barefooted Boy," pp. 26, 53; Peggy Aldrich Kidwell, "Calculating Machine," in Robert Bud and Deborah Jean Warner, eds., *Instruments of Science: An Historical Encyclopedia* (New York and London: The Science Museum of London and the National Museum of American History, Smithsonian Institution, in association with Garland Publishing, 1998), p. 77; *The Cincinnatian* (June 1914), p. 1.

64. Richard Schur, "Memories of Walter Majors: Searching for African American History in Springfield," in Stephen L. McIntyre, *Springfield's Urban Histories: Essays on the Queen City of the Missouri Ozarks* (Springfield, MO: Moon City Press, 2012), pp. 119-125.

65. Jonathan D. Whittenburg (1885), Flood-Fence, U.S. Patent 326,931 (filed June 26, 1885 and issued September 22, 1885); *History of Laclede, Camden, Dallas, Webster, Wright, Texas, Pulaski, Phelps and Dent Counties, Missouri* (Chicago: Goodspeed Publishing, 1889), p. 888; Jonathan Johnson and James H. Younger (1915), Hay-Press, U.S. Patent 1,159,948 (filed January 4, 1915 and issued November 9, 1915); Abner B. Johnson (1900), Churn, U.S. Patent 655,871 (filed June 4, 1900 and issued August 14, 1900); Dee Ayers and Hal Dee Fossett (1922), Churn, U.S. Patent 1,450,238 (filed May 29, 1922, and issued April 3, 1923); Virgil Matthew Pyle (1930), Agitator, U.S. Patent 1,762,421 (filed November 16, 1929 and issued June 10, 1930); Leland Fox, *Tall Tales from the Sage of Cane Hill* (Greenfield, MO: Vedette Publishing, 1971), p. 23.

activity in other parts of the United States. Industry and commerce provided links with national and international markets. Like the country at large, the Ozarks was no monolith. Some people prospered, while others struggled to survive. Some Ozarkers thrived amid a new world of railroads and steam engines and cash payments, while others lived lives essentially no different from those of their parents and grandparents. Despite the region's integration into broader markets and cultural currents, however, the Ozarks remained a peripheral region before the Great Depression, a place more likely to exploit and export its natural resources than to add value to them through manufacturing, a place unable to absorb its excess labor supply. As James L. Dalton discovered, it was still a place that more often than not exported its most able and ambitious people, along with their ideas and their entrepreneurial creations.

Water and Power:

Powersite Dam and Lake Taneycomo on the White River
Thomas A. Peters

Powersite dam in operation, Taney County, Missouri (1913).
Courtesy of Lynn Morrow.

Running water and electricity are the basic tenets of a civilized society. Without the two, the progression of mankind ceases to move forward. The practical application of electricity during the early 20th Century in the United States propelled American society forward, surpassing other countries in the world.

> —Brad Belk, *Celebrating a Century* (2009)

Today, Powersite Dam seems rather small and insignificant. The reservoir it created, Lake Taneycomo, seems small, too, barely wider than the original banks of the White River it impounded.

When it was built, however, Powersite was the first hydroelectric dam in Missouri and the largest west of the Mississippi River. Designed in 1911 by Nils F. Ambursen, a Norwegian-American engineer who built many concrete slab-and-buttress dams primarily in New England, Powersite began producing electricity in 1913. And it was the first major dam built in the Ozarks, predating Bagnell, Bull Shoals, and Table Rock.

By the time Powersite Dam was completed, major technologies (e.g., railroads) and cultural forces (e.g., the novel, *Shepherd of the Hills*) were already having major effects on life in the White River valley. The dam, and especially the resulting Lake Taneycomo, acted as a stimulus to tourism and general economic development in the region. From 1913 into the early 1950s, Lake Taneycomo was a regional tourist attraction. In the Ozarks as elsewhere, technology and culture are inextricably intertwined. It all relies on water and power. And, in the case of Powersite, it all goes back to the White River:

> From transportation corridor to early wood-based industries to mussel shell button factories

to major infrastructure projects to float trips to tourists to resort and golf communities to retirees to music shows to convention centers, the upper White is the thread and binding agent from then to now, and to tomorrow. (Parnell 2006, p. 15)

Although the Powersite/Taneycomo complex fueled the development of flatwater tourism in the Missouri Ozarks, it was built primarily to provide electric power for the tri-state lead and zinc mining district of southwest Missouri, southeast Kansas, and northeast Oklahoma, centered around Joplin. As the mining operations transitioned from human, animal, and steam power to electricity, they needed a reliable, inexpensive source of power. The Empire District Electric Company created Powersite Dam to meet the growing need for electricity.[66] In sum, Powersite was built to support and advance the mining activities in this region.

Hydroelectric power promised to fuel the growth of manufacture, as well. Some people saw the building of this hydroelectric dam as a necessary step in developing the manufacturing potential of the White River region. As declared in the *Springfield News Leader* of September 27, 1911, "This dam marks the area of water-power development on the White River and its tributaries which in a few years will make the White River a manufacturing center of no mean importance."

The $4 million dam on the White River was perceived by Missouri economists as the beginning of the development and exploitation of a great latent natural resource, waterpower (*St. Louis Post-Dispatch*, December 14, 1913, p.78). Just as today we see the Great Plains as one of the best wind-power regions in the world, one hundred years ago, civil engineers saw the Ozarks as a region perfect for hydroelectric power. So notes

the *St. Louis Post-Dispatch* of December 14, 1913:

> Few states are so fortunate topographically in possibilities for the development of waterpower as Missouri. Its numerous rivers, particularly those south of the Missouri, have the necessary rapid current for waterpower and the necessary high banks for dam construction. T. O. Kennedy, one of the engineers who built the Branson dam, said recently that there is room for a 50-foot dam every 25 miles on the White. (p.78)

Other rivers in the Ozarks were envisioned as good sources of hydroelectric power: "What is being done on the White could be done equally as well on the Meramec, the Osage, the Gasconade, the St. Francis, the Black, the Current, the James and many of their tributaries, [Minard L. Holman] said. By utilizing the waterpower site and supplementing hydro-electric plants with steam plants he believed the Ozark territory could be made a beehive of industry" (*St. Louis Post-Dispatch*, December 14, 1913, p.79).

Municipal uses (e.g., streetlights) and domestic uses (e.g., kitchen appliances) for the electricity came later. In fact, the city of Joplin did not begin using electricity generated by Powersite Dam until late in 1918, even though the Empire District Electric Company had been making proposals to the city for several years (Belk 2009, p. 21). Branson did not receive any electricity until 1916, and Forsyth, the town closest to the dam, was not electrified until 1919 (Albers and Stacey 2001, p. 40).

Flood control was not a major reason for building Powersite Dam. Any environmental concerns that were expressed about the dam and the reservoir focused on the movement of fish upstream to spawn.

In the fertile entrepreneurial early years of the 20th century, other methods for harnessing waterpower without dams and reservoirs were actively

---

66. The Empire District Electric Company always referred to it as the Ozark Beach Dam, although it is popularly known as Powersite Dam (*Springfield Leader and Press*, March 10, 1963, p. 39).

explored by some. For instance, in 1907 William Henry Standish, a colorful character who (being a Civil War veteran) preferred to be called General Standish, proposed digging large tunnels connecting two bends in the James River. The resulting sluice, he argued, would power turbines without impounding the river (*Daily Arkansas Gazette*, December 25, 1907).

The U.S. Congress actually approved Standish's idea, but President Teddy Roosevelt vetoed the plan, perhaps fearing that the Standish Sluice would have a virtual monopoly over the roaring waters of the James. General Standish eddied and frothed for a few years. When William Howard Taft succeeded Roosevelt as president, Standish again petitioned Congress and again his plan was approved. Although the subterranean sluice was never built on the James River, one was built on the Niangua River ("Ghost," 2018), producing electricity sufficient to power Camdenton, the newly incorporated Camden County seat.

Powersite Dam began producing power on September 1, 1913. When operating at full capacity, it generated 28,000 horsepower per day (*St. Louis Post-Dispatch*, December 14, 1913, p.78).

Financing the Dam: Captains of Industry

Scarcely less interesting than the story of the dam itself is the story of how it was financed. A firm of St. Louis promoters first conceived the project and employed Holman & Laird, consulting engineers of St. Louis, to select a possible location and draw up the plans. Holman & Laird explored the White and hit upon the Branson site.... Every effort was made to enlist St. Louis capital, but without result. It is said on credible authority that the promoters spent about $50,000 before they finally sold to New York interests.

The firm of Henry L. Doherty & Co. was not hard to convince of the amazing potentialities of the Ozark country. When they took hold of the thing they went even further than Holman & Laird had dared dream, made a bigger dam and converted the whole thing into a much bigger project. With the necessary money, the necessary courage and the necessary confidence they lost little time getting to the actual work.
—"Foretells New and Greater Industrial Era for the State," *St. Louis Post-Dispatch* (December 14, 1913)

As Brad Belk (2009) notes, financing for the construction of the dam originally was supposed to come from Saint Louis: "In 1909, a group of St. Louis businessmen organized the Ozark Power and Water Company to build a hydro plant on White River near Forsyth, Missouri. Unable to carry out its program, the Ozark Power and Water Company was taken over by the newly formed Cities Service Company" (p. 16)—a New York firm, which supplied the necessary capital. As reported in the December 14, 1913 issue of *St. Louis Post-Dispatch*, "the firm of Henry L. Doherty & Company was not hard to convince of the amazing potentialities of the Ozark Country" (p. 79). Doherty & Company "began purchasing power companies in southwest Missouri and southeast Kansas" and, after consolidating several smaller companies, "named the new company The Empire District Electric Company,"[67] referring to the financial support received from the New York financier ("History," 2019).

Doherty—the financier-hero of this story—had been involved in utilities almost his entire life, always in white collar office positions. "Henry Latham Doherty," writes Brad Belk, "was born on May 15, 1870, in Columbus, Ohio." Belk (2009) continues:

---

67. The Empire District—part of the name of the company that eventually built and operated the dam—refers to New York's nickname as "the Empire State," not to the Ozarks Empire, a multi-state region in the center of the nation.

At the age of twelve he began his career in the utility business as an office boy for the Columbus Gas Company. By the age of twenty, he had worked his way up the ladder to become manager of the Columbus Gas Company. At the age of 35, he created his own company—Henry L. Doherty & Company. The year was 1905 and his company was engaged in the reorganization, management and financing of public utility concerns. Doherty's interests became so great that he created a separate holding company to handle its interests—the Cities Service Company. (p. 34)[68]

The Empire District Electric headquarters has always been in Joplin, which was the *de facto* capital of the tri-state mining region (Belk 2009, p. 14).

Raising capital for Powersite did not end when the initial construction was complete. In April of 1922, a group of 80 bankers from New York, Philadelphia, Boston, Pittsburgh, and Chicago came to Branson on a special train to inspect the dam in which they collectively had invested $2 million in bonds issued by the Doherty Company of New York (*Winfield [Kan.] Daily News*, April 3, 1922). Again, it was financing from far beyond the Ozarks that bankrolled Powersite Dam.[69]

---

68. In recounting the history of Empire District Electric, Belk (2009) writes of Doherty's role in the financing and development of utilities nationwide:

> The Henry L. Doherty & Company was created in 1905 to act as a fiscal agent for numerous utility companies. Five years later, the Cities Service Company began acting as a holding company for both utility and non-utility Doherty interests. Established on September 2, 1910, the Cities Service Company was a New York based holding company deriving income from dividends generated by stock held in subsidiary corporations. Cities Service was established to acquire public utility companies, such as natural gas, electric and transportation firms, but later their interests shifted to the burgeoning petroleum industry. (p. 13)

69. Although Powersite was a private project, the federal government did grant a permit to build and operate the dam. Back in the

Construction at Camp Ozark, Hall Photo Co. (*ca.* 1912). *Courtesy of Lynn Morrow.*

Camp Ozark: The Construction Camp

Construction of Powersite Dam lasted well over a year, and the workforce and payroll for the project were prodigious for the sparsely populated Ozarks of that time. The December 14, 1913 issue of the *St. Louis Post-Dispatch* made note of this feat: "Before the Branson work began, a site for an encampment was laid out near the scene. This became a small city—in fact, at one time it contained more inhabitants than any town in Taney County. It was properly policed, kept in sanitary condition and schools were provided for the children of the small army of workers" (p.78).

Humans—wielding picks and shovels and pushing wheelbarrows—and work animals made up the labor force for this massive project. Wooden forms, made with timber from the surrounding region, were used to hold the concrete. It took over a year and a half to construct the dam, beginning in 1911 and ending in 1913.

---

early 20th century, the U.S. Federal Power Commission (FPC) typically issued a hydroelectric dam permit for a period of 50 years. In August of 1967 the FPC announced that it would not seek to recapture (i.e., nationalize) the Powersite hydroelectric dam when its license to operate expired on August 31, 1968 (*Springfield News-Leader*, August 7, 1967, p. 11).

"Working on the White River Dam," Hall Photo Co. (*ca.* 1912). Courtesy of Lynn Morrow.

The construction camp, named Camp Ozark, was filled with hardworking, hard-living men and their families, and the men were making good wages:

> During the dam's construction phase, the workers' camp numbered over 1,000 laborers…. For nearly two years, the payroll of the camp for just the day laborers approximated $1,500 a day. Sanitary conditions were good and the camp's water supply was carefully guarded at all times. (Belk 2009, p, 16)

A young, quiet, serious bachelor by the name of Albert Parnell knew a commercial opportunity when he saw one and, along with his partner Ven Yandell, opened up a general store right in the camp (Parnell 2006, p. 25). Both Ozarkers and immigrants worked to build the dam: "While many workmen from the region around Taney County were used in the construction, a large number of immigrants were employed" as well (*Springfield Leader and Press*, March 10, 1963, D3).

Construction of Powersite Dam

As local journalist Lucile Morris Upton noted, the Branson Commercial Club came up with the name, Lake Taneycomo, building a regional portmanteau word out of *Taney Co*[unty] *Mo.*, with an added subtle allusion to romantic Lake Como in Italy (*Springfield Leader and Press*, March 10, 1963, D3).[70] Preparing the site for the dam involved a lot of hard work. As reported in the December 14, 1913 issue of the *St. Louis Post-Dispatch*, "the excavators removed about 70,000 cubic yards of earth and 20,000 of rock" (p. 78).

Getting construction materials to the dam site was a challenge in itself. The railroad had come to Branson and Hollister in 1906, so bringing materials down the main Frisco line to Springfield, then south to Branson, was the transportation corridor used for most of the construction materials. Waterborne construction materials were loaded onto barges and floated down the river from Branson to the site: "Contractors for the dam relied almost entirely on water transportation between Branson and Powersite. Barges and boats hauled down heavy equipment and supplies" (*Springfield Leader and Press*, March 10, 1963, p. 39). Wagons hauled materials over land, too.

Transportation to the dam and reservoir area needed to be improved. As reported in the *Springfield News-Leader* of March 3, 1912, a pike was being constructed from Hollister to Kirbyville (p. 7). An electric railway also was proposed from Hollister to Powersite, then on up to Forsyth (p. 7), though this latter project never materialized.

T. O. Kennedy was the overall manager of this massive construction project. Keeping an account in the Hollister bank, he and the bank's owner, William

---

70. Upton writes in commemoration of the fiftieth anniversary of the dam and its impoundment.

H. Johnson—a key developer of Hollister—became friends. As Upton notes, "Frequently Johnson was invited by Kennedy to make the roundtrip on a boat which hauled special supplies. He recalls there was a kerosene (coal oil in those days) stove on the boat and they made egg sandwiches and fried good old country ham on it. It took six hours to make the trip back against the current" (*Springfield Leader and Press*, March 10, 1963, p. 39). Johnson was an enthusiastic supporter of the dam project and may have been one of the first Taney Countians to grasp the tourism potential in Lake Taneycomo.

Remarkably, for all the person-hours used to build this dam, coupled with the size of the project and the heavy reliance on manual labor, there was only one fatality: Carl Kressig, "who drowned when thrown from a barge making a landing at the dam" (Anonymous, 1963).

After the dam was completed in 1913 and the reservoir filled, repairs and improvements to the dam continued throughout the decades. In 1922, only ten years after the dam was built, $400,000 was spent to repair it (*Douglas County [Kan.] Herald*, May 25, 1922, p. 1). The June 23, 1922 issue of the *Springfield Daily Leader* gives more detail: "Work of strengthening the dam of the Ozark Power & Water company across White river at Powersite has been resumed under the direction of Engineer Fee of the New York offices of the Doherty company. The Ozark company, which supplies hydro-electric power to the Springfield Gas & Electric company, is one of the numerous subsidiaries of the Doherty company" (p. 7). The article continues:

> The work of repairing the big concrete barrier across the river was started last year, and for several weeks a force of 250 laborers was worked. Thousands of barrels of cement were used in filling huge [h]oles that had been washed out at the base of the dam on the lower side by the rush of

water over the spillway. (*Springfield Daily Leader*, June 23, 1922, p. 7)

Roy Linton, who operated a planing mill just west of Yellville, Arkansas, provided some of the lumber used for the forms for that project (*Mountain Echo [Yellville, Ar.]*, September 7, 1922, p. 2).

The basic design challenge was that most of the water flowed over the top of the dam. The renovation work also included adding five feet to the height of the dam, as well as adding two more power-generating turbines, bringing the total number to seven (*Springfield Daily Leader*, June 23, 1922, p. 7). Belk (2009) gives a summary:

> Work began on re-enforcing the foundation of the spillway during the fall of 1921 at the Powersite Dam. A total of thirteen thousand cubic yards of reinforced concrete were used in the construction. The work crews applied an average of 500 cubic yards of concrete per day. Most of the reinforcing was done underwater. The cement used in the construction project was shipped to Branson by train. From Branson the cement was loaded to a fleet of barrages and shipped seventeen miles down the river to the dam.... In addition, 200 tons of steel were used in the construction project. (p. 25)

Fast-forward two decades to another round of repairs: As reported in the *Macon (Mo.) Chronicle-Herald* of April 24, 1941, Empire Electric announced that the pool level on Taneycomo would again be lowered during the prime tourist season so that extensive repairs could be made to the dam (p. 6). During floods and other periods of high water, the water flowed over the top of the dam. Because the White River watershed is prone to flooding, these frequent large overflows took their toll on the dam.

▲ Barges at the construction site, Hall Photo Co. (*ca.* 1912). *Courtesy of Lynn Morrow.*

"Hauling Cement Rods to the White River Dam from ▼ Branson," Hall Photo Co. (1912). *Courtesy of Lynn Morrow.*

▲ Dam construction, Hall Photo Co. (*ca.* 1912). ▶
Courtesy of Lynn Morrow.

Dam construction, Hall Photo Co. (*ca.* 1912).
Courtesy of Lynn Morrow.

WHITE RIVER DAM
20 HALL PHOTO CO

Operations, Floods, and the General Life of the Dam

Under the supervision of Sam E. Gard,[71] the dam's smooth, safe operation seemed assured: "This immense plant is so perfect mechanically that relatively few men are required to operate it. The working staff is divided into day and night shifts of six men each" (*St. Louis Post-Dispatch*, December 14, 1913, p.79).

71. Sam E. Gard, "who had previously run the Riverton plant, was transferred, becoming the first plant superintendent. He worked in that capacity from 1913 until his death in August 1931" (Belk 2009, p. 18).

And yet, flooding occurred immediately. In late March of 1913, torrential rains threatened the new dam, and the filling of Lake Taneycomo happened much more quickly than planned—only a day and a half. The heavy rains threatened to destroy the new dam and inundate the town of Forsyth, two miles downstream, as well as the surrounding farmland (*Daily Arkansas Gazette* [*Little Rock*], March 26, 1913, p. 1).

In March of 1920, severe weather again focused national attention on this resort region. A tornado killed thirteen people, including ten people in the Turkey Creek Valley south of Hollister in Taney County, and torrential rains caused many streams and rivers in southwest Missouri to overflow, threatening Powersite Dam again (*Miami [Fla.] News*, March 12, 1920, p. 1). In April of 1927, torrential rains caused flooding again in the Taneycomo district. All the residents of Forsyth, downstream from Powersite, were warned to head for the hills (*Springfield Leader and Press*, April 15, 1927, p. 1). There were reports that "several large boats of the Sammy Lane Boat company have broken from their moorings and have floated down the lake" (*Springfield Leader and Press*, April 15, 1927, p. 1). The height of the water flowing over Powersite Dam was measured at nineteen feet, ten inches, but officials of the Empire District Electric Company said there was no danger of the dam breaking or washing away (*St. Louis Post-Dispatch*, April 16, 1927, p. 1). The previous highwater mark over Powersite had been fourteen feet, one inch. Another period of heavy rain in May 1943 caused water to flow over the dam to the height of 11.6 feet (*St. Louis Post-Dispatch*, May 20, 1943, p. 3). The flooding two years later set a new record: 20.15 feet over the dam (*Moberly [Mo.] Monitor-Index*, April 16, 1945, p. 1).

Droughts diminished and interrupted the hydroelectric operations, too. In April of 1954, the challenge in the White River watershed was not flooding, but drought: "Residents said [Bull Shoals Reservoir] was 48 feet below high water. A boat dock owned by Jim Owen, former mayor of Branson and a local promoter of float trips and sport fishing, has been left on mud and dust flats. Owen referred us to John Morris farther downstream. We couldn't find Morris. He had moved downstream to escape being stranded by the receding water" (*Chicago Tribune*, April 5, 1954, p. 44).

Accidents plagued the dam and lake. Over the years and decades, numerous boats capsized or sunk, and many people drowned. Frank ("Cicero") Weaver

of the Weaver Brothers and Elviry vaudeville troupe once hit something submerged in Lake Taneycomo as he raced his speedboat down the lake. The boat began to sink, but Frank made it to shore unscathed. In July of 1933, three men from Forsyth drowned when the boat they were fishing in from below the dam capsized: An automatic gate in Powersite Dam opened abruptly and the onrush of water overwhelmed them (*Sedalia [Mo.] Democrat*, July 9, 1933, p. 1).

On August 8, 1954, Charles ("the Old Chuckaroo") Norman, a St. Louis disc-jockey and radio time salesman, unwittingly piloted his speed boat over the brink of Powersite Dam. After a months-long convalescence in Barnes-Jewish Hospital in St. Louis, he recovered from his physical injuries and gathered his wits. In early 1956, he sued the Empire District Electric Company for $100,000. "Frankly," he said, "I didn't know there was a dam there until I woke up in the hospital!" (*Springfield Leader and Press*, April 18, 1956, p. 13).[72]

Transmission and Use of the Electricity

At first, the hydroelectric energy flowed primarily to Joplin and Springfield. As Belk (2009) writes, "the transmission system from the dam extended 150 miles on 66,000-volt lines. A series of substations transported the electricity along the line. They were located at Ozark, Springfield, Monett, Aurora, Diamond and Joplin. The principal service markets included the Springfield and the Joplin-Picher District" (p. 17).[73]

---

72. "The Old Chuckaroo" Charles Norman was known as a dashing, witty, chain-smoking confirmed bachelor-playboy, fond of highballs, who lived with his mother until she died in 1960. At one point he owned a pink Cadillac, which apparently he never crashed (Hinman, 2005).

73. The Joplin newspaper elaborates: "This plant [at Powersite] is connected to the Empire system at Joplin by a 66,000-volt transmission line 150 miles long, which serves Forsyth, Branson, Ozark, Springfield, Aurora, Monett, Peirce [sic] City, Neosho, Diamond, and Granby (*Joplin Globe*, July 9, 1924, p. 3).

The effects on industry were felt immediately: "By supplying ample power and lighting service to the lead and zinc fields, to the coal mines, quarries and the varied industries, the Empire District Electric Company has been an important factor in the life and development of the territory" (*Joplin Globe*, July 9, 1924, p. 3).

Hard as this is to imagine today, much of the rest of the region lived without electricity. Plans were in place to change that. The resulting hydroelectric power was going to provide electricity to a large area of the surrounding region: "It [the dam] is to develop 32,500 horsepower of electricity which will be enough when transmitted by cable to supply all the light and power for Springfield, Joplin, Carthage and all the other towns within a radius of one hundred and fifty miles of the dam. The dam is expected to furnish power for the zinc and lead districts of Missouri, Kansas and Oklahoma, which are within its territory" (*Springfield News-Leader*, April 6, 1913, p. 13).

By the mid-1920s, approximately 75% of the ore mined in the Joplin district was mined with power provided by the Empire District Electric Company (*Joplin Globe*, July 9, 1924, p. 3). But, as more dams were constructed and the region joined ever-larger grid systems, the power from Powersite became a smaller and smaller contributor to the overall power grid in the region. By 1992, Powersite was producing only 3% of the total electricity produced by Empire District Electric Company (Belk 2009, p. 145).

Lake Taneycomo and Ozarks Tourism

Ozark Beach Dam, the completed structure, was notable for two things: it was at the time the largest hydroelectric dam west of the Mississippi River, and it created the Midwest's largest recreational impoundment. The Lake Taneycomo area became so popular with tourists that it was called "The Playground of the Middle West" in promotion and the "Taneycomo District" locally.
—Robert Gilmore, *An Overview and Survey of Lake Taneycomo Beach Towns and Resorts* (1990)

From the mid-1910s through the mid-1950s, the Lake Taneycomo District thrived as a resort area, though with a bit of a dip during the Great Depression and when Lake of the Ozarks became a competing Ozarks lake attraction. Lake Taneycomo offered the right experience in the right place at the right time: It provided "Arcadian travelers recreation in a natural, beautiful riverine setting which was touted as inspirational and wholesome while at the same time offering the amenities and conveniences of comfortable resorts and well-equipped outfitters" (Morrow and Myers-Phinney 1999, p. 36).

Development of communities and tourism around the dam and reservoir began almost immediately. William H. Johnson was selling lots for the proposed town of Powersite on the bluff just to the east of the dam (*Springfield News-Leader*, March 3, 1912, p. 7). Advertisements noted, "You cannot fail to make a 'killing.' Powersite will grow like 'sixty.' A few dollars spent today will return big profits tomorrow.... Great things are happening in the White River Country and the new town of Powersite will be the center of activity" (*Springfield News-Leader*, March 3, 1912, p. 7). However, there were several problems with the location of the village of Powersite: It was on the other side of the river from Forsyth, the closest town, and it was hundreds of feet above the lakeshore itself.

Today, Powersite Village is a sleepy little town, with little or no tourism activity. That was not always the case, especially in the early years. In August of 1915, William H. Hamby, educated at Springfield's Drury College and at Mizzou, friend of Bonniebrook's Rose O'Neill and well-known magazine writer, spent several days at Taneycomo scouting a location for the fall 1915 meeting of the Missouri Writers' Guild. The

Guild had recently organized at the University of Missouri in Columbia during Journalism Week (*Springfield News-Leader*, March 7, 1915, p. 4). The Cliff Hotel at Powersite was chosen as the location for the October gathering of the group, being "ideally located for such a meeting, commanding as it does one of the most magnificent views of lake, river and hills imaginable and offering for the pleasure of the visitor the sports of all three" (*Springfield News-Leader*, March 7, 1915, p. 4).

In August of 1915, the Cliff Hotel was in the process of being enlarged and converted to a private club. Articles of incorporation were filed by the Cliff House Club Company with the Missouri Secretary of State. According to the promoter and manager R. H. Wilson, one thousand shares had been sold at $100 each, with an initial capitalization of $100,000. This was a common way to finance a new venture at this time and place. The proposed officers of the new corporation included R. W. and C. H. Wilson of Powersite, A. W. Cooper of Mountain Grove, Mrs. S. R. Lee of Indianapolis, Indiana, and J. H. Mason, an attorney in Springfield (*Springfield News-Leader*, August 24, 1915, p. 1). "The paths leading to the hotel ascend 250 feet," declared the local newspaper, "but this will be remedied with the construction of a tramway or electric elevator for the transporting of passengers from the boat landing to the club house and grounds" (*Springfield News-Leader*, August 24, 1915, p. 1).

From its creation in 1913 until Lake of the Ozarks was formed in the early '30s, Lake Taneycomo was the only lake, natural or artificial, in the Ozarks Region. In 1919, when the Ozark Playgrounds Association was formed in Joplin, Lake Taneycomo was still the only lake in the entire area (*Neosho Daily News*, April 8, 1964, p. 13).

The Taneycomo recreation area and the Roaring Twenties fit hand in glove. In early 1920, the board of directors of the H. L. Doherty group in New York City—the group that had purchased the surrounding land and financed the construction of Powersite Dam—announced their intentions to invest a half-million dollars in the emerging Lake Taneycomo resort area: "The Doherty company is a $200,000,000 corporation of New York, owning a majority of the stock in numerous public utility projects in all parts of the United States. They are known locally as the financial backers of the Powersite project as well as the owners of the street railway system at Sedalia" (*Springfield News-Leader*, February 15, 1920, p. 1). Initial plans were for the construction of three hotels: one at Hollister, one at Forsyth, and one at the mouth of Bull Creek.

By the summer of 1920, due to increased tourism and traffic, there were plans to build a highway bridge over the White River at Forsyth, below the dam (*Springfield News-Leader*, July 13, 1920, p. 5). The basic problem was that, at Forsyth, the White River usually was too deep to ford and too shallow to ferry across.

In November of 1921, during the "Land of a Thousand Smiles" conference of the Ozarks Playground Association held in Joplin, R. C. "Colonel" Ford of Forsyth reported that approximately 50,000 people had visited the Taneycomo area during the tourist season just ending (*Joplin Globe*, November 23, 1921, p. 2). And tourist traffic continued to increase.

William H. Johnson, a key developer of the region, described Lake Taneycomo in the mid-1930s: "Lake Taneycomo, a widening of White River caused by a 54-foot dam above Forsyth, is 25 miles in length, extending six miles above Hollister. It is lined with innumerable resorts and recreation camps and navigated by hundreds of pleasure craft, and site of annual boat racing" (*Baxter Bulletin* [*Mountain Home, Ar.*], June 26, 1936, p. 18). Hollister, Missouri was platted and developed before Powersite Dam was built, but the formation of Lake Taneycomo certainly helped attract tourists to the region. In the same article, Johnson waxed poetic, describing Hollister as "a bower of flowers," adding that "Hollister has no factories or payrolls of any kind and invites none. The community is planned and designed for recreation, pleasure and homes" (*Baxter Bulletin*, June 26, 1936, p. 18).[74]

It would keep this reputation through ensuing decades. In May 1954, heading into the summer vacation season, the St. Louis newspaper described the White River recreation area as "an old and progressive vacation land":

> The Taneycomo area, created more than a quarter of a century ago when Powersite dam was built, and since developed into an area boasting every conceivable recreation, now blends into Bull Shoals lake below the dam, which was completed last year…. On the northwestern edge of this man-made paradise is the famous Taneycomo lake area, comprised of Forsyth, Branson, Rockaway Beach and Ozark Beach. Here can be found all of the recreational facilities created by an old and progressive vacation land. (*St. Louis Post-Dispatch*, May 2, 1954, J 6)

In January of 1957, prior to the filling of Table Rock Lake, a group of people took one last float trip down the upper White River. "It took a four-horse team to pull our boat around the Powersite Dam," said one adventurer: "The teamster set us down in shallow water below the dam" (*Springfield News-Leader*, January 15, 1957, p. 10).

The opening of Table Rock in the late '50s, with its tall dam and outflow from low in the lake rather than over the top, as at Powersite, had a chilling effect on the Lake Taneycomo tourist region—literally. The cold water coming from Table Rock Lake made Lake Taneycomo good for trout fishing, but not for swimming and water skiing.

Visitors to Taneycomo never completely died out, however. On June 18, 1961, the Pierce City Catfish Club held its twelfth annual outing at Powersite Dam. The largest catfish caught that day weighed 38 pounds. Mr. and Mrs. C. C. Williford were special guests at the event (*Springfield Leader and Press*, June 19, 1961, p. 14). He was the longstanding and well-loved weather forecaster on KWTO-AM 560 radio in Springfield.

During the first weekend of June in 1963, the Chambers of Commerce of Forsyth, Branson, Hollister, and Rockaway Beach invited Governor and Mrs. John M. Dalton down to celebrate the fiftieth anniversary of Powersite Dam and Lake Taneycomo. In the morning, the entourage started at the School of the Ozarks, paraded through Branson to Rockaway Beach (Lebanon High School Band paraded and played along with them), then embarked on the water to head down to the dam. At noon, at Electric Park near the dam, a barbequed chicken lunch was served to all. Don McKee, chairman of the board of the Empire District Electric Company, was one of the speakers. Ted Kennedy, who had served as the chief engineer in charge of construction of the dam fifty years earlier, was unable to attend due to poor health (*Springfield Leader and Press*, May 26, 1963, p. 42).

Precedence and Impact

In 1922, Ralph C. Ott, a Springfield artist, produced a huge oval painting sixteen feet wide and nine feet high, depicting the Powersite dam and power plant. Titled "The Power from the Hills," it celebrated the general theme of waterpower (*Springfield Republican*, July 11, 1922, p. 8). The painting was hung in the Natural Resources Museum inside the state capitol building in Jefferson City (*Shelby County [Mo.] Herald*, August 29, 1923, p. 2). But, more than celebrate waterpower, Ott's painting expressed a people's hope in the future

---

74. For further discussion, see Paul Johns (2016), "MOzark Moments: Hollister and Its Exceptional Promoters," and Donald R. Holliday (1996), "Hollister-on-the-White-River."

of technology—and of the role electricity would play in that future.

During the 1910s and '20s, the power of electricity to transform the region, especially in transportation, industry, and domestic life, was widely believed. "Truly," declared the *Shelby County Herald,* "The Power from the Hills is almost limitless. By the time these new dams in the heart of the Ozarks are in operation R. R. trains as well as factories will be using electric motor power generally, and in a few years electric power will be supreme" (*Shelby County Herald,* August 29, 1923, p. 2).

While the negative consequences of dams and reservoirs were at least vaguely known by many Ozarkers in the first half of the 20th century, most people seemed to think that, on balance, dams and reservoirs were good for the region. One problem with Powersite Dam was that it restricted the flow, not only of the water, but also of the life within the water, especially fish. For instance, prior to the construction of Table Rock Dam, a tagged channel catfish was recorded to have swum miles upstream. As reported in the July 31, 1953 issue of the *Washington (Mo.) Citizen,* "One [channel catfish] was tag[ged] near Edgewater Beach in Lake Taneycomo May 21, 1951, and recovered at the mouth of Aunt Creek in the James river last month. It had been free 804 days, during which it had moved through Taneycomo, up the White river and into the James, a distance of 47 miles" (p. 10). That's an average daily upstream migration of 309 feet!

The fishing, especially for bass, was particularly good below the actual dam site. This, of course, was long before the development of Bull Shoals Dam and Reservoir. One of the early environmental concerns about the dam centered on impairments of migrating bass and other species of fish that swim up the White River to spawn. Great schools of fish "have congregated below the dam and in their helpless state are being slaughtered by the thousands" (*St. Louis*

Post-Dispatch, May 3, 1914, p. 20). Fishermen who exploited the situation were called fish pigs, as noted in the *Post-Dispatch* of May 3, 1914:

> W. B. Burks, a Springfield wholesale hat dealer, was fishing below the dam a few days ago and caught three large red horse with hooks that were not baited. Frequently a bass will snap a minnow off the hook and make his getaway. In jerking his hook out of the water, Burks on three different occasions snagged a red horse in the side and landed him on the bank. The water is simply seething with the fish. (p. 20)

By 1921, discussions had begun to build a larger dam downstream on the White River. One early proposal was to build the dam near Cotter, Arkansas, and call it Moark, because the resulting reservoir would straddle Missouri and Arkansas (*Arkansas Democrat [Little Rock],* April 30, 1921, p. 2). In October of 1922, a group of four hydroelectric and geological engineers led by E. N. Mayer of Detroit, Michigan, floated down the White River from Forsyth to the mouth of the Buffalo River, looking for dam sites (*Southern Standard [Arkadelphia, Ar.],* October 19, 1922, p. 6). One newspaper account suggested that the team of engineers was representing Henry Ford, the automobile magnate (*La Plata [Mo.] Republican,* October 27, 1922, p. 2).[75]

Also included in the 1922 planning and discussions was the idea of building another dam above Branson. That second dam would not include power-generating equipment, but the second impoundment would provide enough water year-round to keep the existing seven turbines at Powersite Dam turning: "It is the belief of engineers that the two dams in White river will impound sufficient water to permit of the operation of seven turbines every month in the

---

75. In the early 1950s, Bull Shoals Dam was built upstream from Cotter, Arkansas, as a federal project.

year, even in periods of drouth. It is not possible to operate the five turbines now available when there is a prolonged drouth in the White river country" (*Springfield Daily Leader*, June 23, 1922, p. 7).

Better fish ladders and even locks for boats were discussed from time to time. In May of 1926, J. H. Hand, secretary of the White River Chamber of Commerce, wrote an editorial bemoaning a move by Springfield, Missouri, asking the War Department "to adopt a policy designed to make White river navigable from its mouth to a point in Barry County Missouri, by requiring the installation of locks in proposed power dams at Table Rock above Hollister and at Wild Cat Shoal, above Cotter" (*The Mountain Echo* [*Yellville, Ar.*], May 13, 1926, p. 1).

A decade later, in late December of 1936, the federal power commission gave conditional approval for a 102,000-horsepower hydroelectric dam project to be built near Wildcat Shoals on the White River. The reservoir would back up 120 miles all the way to Forsyth. As reported in the *Thayer (Mo.) News*, If the proposed dam and reservoir at Table Rock upstream would be built, the resulting chain of dams (Table Rock, Powersite, and Wildcat Shoals) "would become the largest chain of artificial lakes in the world" (*Thayer News*, January 1, 1937, p. 2). Whether or not these remain "the largest chain," that indeed is what has happened, to an even greater extent than was imagined in the 1930s.

Conclusion

Hydroelectric power is a renewable extractive industry. Lead and zinc mining are not renewable extractive industries, and forestry is at least marginally renewable. The hydroelectric industry is not without negative outcomes, however. It flooded some of the best farmland in the Ozarks. It effectively killed the local float trip industry. And it radically changed the tourism and fishing industries. By electrifying and remaking the landscape, it could be argued that the building

of Powersite Dam in the early years of the 20th century created the possibility of a "modern Missouri Ozarks."

Was Powersite Dam a financial success? The costs of maintaining and repairing it over the decades have been substantial, and its ability to provide consistently large amounts of electric power did not completely materialize. Technologies can fail in spectacular, tragic ways, but they also can suffer from basic assumptional and design flaws and shortcomings that prevent them from becoming complete successes. That seems to have been the case with Powersite, and it was duly noted within a decade of its construction: "The dam at Powersite, which generates 20,000 horsepower, never has been profitable because it has not a large storage reservoir. The creation of the dam at Table Rock will cure this defect, engineers say" (*St. Louis Star and Times*, October 21, 1929, p. 20).

Now, in the 21st century, Powersite dam is considered small, with diminished overall significance. The reservoir created by Bull Shoals Dam downstream backs almost all the way up to Powersite. When Table Rock Dam was built upstream in the '50s, Powersite Dam and Lake Taneycomo became quaint, aging, throwback technologies to larger post-World War II federal hydroelectric projects. When people in the Springfield/Branson combined metropolitan area and beyond go "to the lake" in the summer months, most likely they are headed to Table Rock Lake, not to Lake of the Ozarks, Bull Shoals, or Lake Taneycomo. Nevertheless, Powersite Dam and Lake Taneycomo started it all, over one hundred years ago.

References

Albers, Jo Stacey, and Dorothy Stacey. 2001. *Hometown Branson: Early History*. Branson, MO: Loafers Glory Publications.

Anonymous. 1963. "Lake Taneycomo and Powersite Dam: Fiftieth Anniversary." *White River Valley Historical Quarterly*, vol. 1 no. 7, pp. 2-10. Retrieved

30 March 2019.`<https://thelibrary.org/lochist/periodicals/wrv/V1/N7/Sp63b.htm>.

Belk, Brad. 2009. *Celebrating a Century of Service.* Battleground, WA: Pediment Publishing.

Gilmore, Robert. 1990. *An Overview and Survey of Lake Taneycomo Beach Towns and Resorts: Phase 1.* A report for the Historic Preservation Program of the Missouri Department of Natural Resources. Retrieved 30 March 2019. <https://dnr.mo.gov/shpo/survey/TAAS011-R.pdf>.

Holliday, Donald R. 1996. "Hollister-on-the-White-River: Its Golden Years." *OzarksWatch: The Magazine of the Ozarks* vol. 9 no. 1, pp. 6-10.

Hinman, Kristen. 2005. "The Old Chuckaroo." *Riverfront Times*, April 27. Retrieved 30 March 2019. <https://www.riverfronttimes.com/stlouis/the-old-chuckaroo/Content?oid=2483559>.

"History." Liberty Utilities, Empire District. Retrieved 30 March 2019. <https://www.empiredistrict.com/About/History>.

"The Ghost of Virgin Bluff." James River Basin Partnership. Retrieved 30 March 2019. <https://www.jamesriverbasin.com/river-ramblings/2018/10/31/the-ghost-of-virgins-bluff>.

Johns, Paul. 2016. "MOzarks Moments: Hollister and Its Exceptional Promoters." *Christian County Headliner News* (June 16). Retrieved 30 March 2019. <http://ccheadliner.com/opinion/mozark-moments-hollister-and-its-exceptional-promoters/article_73f88964-3308-11e6-bcca-fb9b77eb0e93.html>.

Morrow, Lynn, and Linda Myers-Phinney. 1999. *Shepherd of the Hills Country: Tourism Transforms the Ozarks, 1880s-1930s.* Fayetteville, AR: University of Arkansas Press.

Parnell, Todd. 2006. *Postcards from Branson: A Century of Family Reminiscence.* Springfield, MO: PFLP Publishing.

"White River Dam," Hall Photo Co. (1913). *Courtesy of Lynn Morrow.*

"Vitaphone Victims" of the Jazz Age:
Movies, Economy, and Cultural Backlash in 1920s-1930s Springfield
Shannon Mawhiney

The shift from silent movies to sound-on-film marked a shift in cultural image-making as well as in technology: Call it an evolution in technoculture—in the convergence of film technologies, economic stressors, and social/demographic upheavals, both regional and national. The history given below proceeds within dichotomies: of local vs. national trends in entertainment; of "live" vs. recorded music; of theater owners vs. union labor; of the economy (local and national) before, during, and after the Depression; of rural vs. urban ethos; and, preeminently, of "traditional" vs. "modern" values, sentiments, and lifestyles. In Springfield and the Ozarks of the late 1920s, the "technologies of leisure" were implicated in the economy and ethos in ways that created considerable local backlash: Such is the thesis of this essay. We begin with considerations of economy.

Despite the difficult financial situation of the Great Depression, movies and movie theaters continued to be attractions in the 1930s,[76] nationally as well as locally in the Ozarks. But for a subset of employees in the motion picture industry, the Depression compounded hardships already begun in the 1920s. Prohibition had already cost union musicians gigs at bars, clubs, and dance halls that once served alcohol legally.[77] Still, the arrival of recorded sound in movies tolled the economic death knell for those who played "live" music at otherwise silent motion pictures. In fact, the technologies of radio, phonograph recording, and "talking" movies were changing habits of entertainment—and changing the ways people spent money on entertainment—effectively putting musicians out of work. These organists and theater musicians were "Vitaphone Victims," as the local newspaper notes, of the Jazz Age.[78]

Springfield's Gem Theatre Orchestra (*ca.* 1910). *Courtesy of the History Museum on the Square, Springfield, Missouri.*

Though synchronized film-and-sound was developed as early as 1884, the technology proliferated nationwide in the late 1920s, with over 200 different systems competing for sales in 1929.[79] The new

---

76. Garff B. Wilson, *A History of American Acting* (Bloomington, IN: Indiana University Press, 1966), pp. 249-250.

77. George Seltzer, *Music Matters: The Performer and the American Federation of Musicians* (Metuchen, NJ: Scarecrow Press, 1989), p. 23.

78. "Superfluous Music," *Springfield (Mo.) Leader* (January 15, 1930).

79. Kenneth MacGowan, "The Coming of Sound to the Screen," *The Quarterly of Film Radio and Television* vol. 10, no. 2 (Winter 1955), p. 145.

technology first hit Springfield's musicians on June 22, 1928, when the Landers Theatre installed sound equipment and played their first movie with synchronized sound—*Old San Francisco*, with music and sound effects—along with short "speaking photoplays" to a preview audience. These played to a public audience the next day.[80] The Landers played its first "all talking picture" on November 1, 1928, with a showing of *The Home Towners*.[81] *The Jazz Singer* followed on November 11, 1928,[82] slightly more than a year after the famously successful film debuted in New York, with recorded crowd noise interrupted by the voice of Al Jolson. These "talkies" and movies with synchronized, pre-recorded music were an immediate godsend for an industry hit hard by the popularity of radio—one of film's competing "technologies of leisure,"[83] which the population largely preferred.[84]

Recording Technology Vs. the Union Musician
Regrettably for union members, the American Federation of Musicians did not take the threat of pre-recorded music seriously at first, seeing it as a passing fad that could not compete with the allure of live, professional musicianship.[85] To be fair, the theaters themselves had not yet given up on live music; Springfield's Landers Theatre had its organ repaired in 1928, at the same time that it added both Vitaphone and Movietone equipment.[86] And musicians still played during newsreels and advertisements.[87] But all this was before the Great Depression. By 1930—even before the full effects of the Depression had been felt regionally—the Springfield Musicians Union Local 150 was suffering from unemployment and underemployment; so much so, that it held two fundraising concerts for its members and for the local itself, whose treasury had been depleted.[88] These turned into scandal when Eddie Girard, president of Local 150 and organ player at the Gillioz, absconded with the monies raised (approximately $150), reputedly heading to Chicago. According to his wife, the union owed him $45 in back pay.[89]

By 1930, professional vaudeville was as good as dead in Springfield: Put simply, it cost less to rent a feature film and Vitaphone vaudeville "shorts" than to pay for live acts. Unfortunately for local musicians, 1930 also saw Springfield clubs and dance halls turning them away, partly due to the loss of clientele, but also due to the introduction of "automatic phonographs,"[90] coin-operated precursors to jukeboxes.[91] In dance halls as well as in movie theaters, recording technology was replacing the "live" musician. The technology was, in a word, cheaper. And whether it

80. Sara Sarett, "Victims' Wails Heard in Movie as Walls Fall," *Springfield Leader*, June 23, 1928. This was not the Landers' first "first." A fire that had gutted the theater in the previous decade led to renovations and an installation of projection equipment that transformed the old vaudeville theater into a "combination theater," featuring both "live" acts and (silent) film.

81. Landers Theatre advertisement, *Springfield Leader* (November 1, 1928).

82. Landers Theatre advertisement, *Springfield Leader* (November 11, 1928).

83. *The Gillioz "Theatre Beautiful": Remembering Springfield's Theatre History, 1926-2006*, ed. James S. Baumlin (Springfield, MO: Moon City, 2006), p. 34.

84. Wilson, *American Acting*, p. 249. Whereas people went to the theater for film, the radio came to the people, broadcasting directly into homes. The radio, thus, was a form of *domestic* entertainment, shared by all family members and requiring no purchase of tickets. For its ubiquity, cost, and convenience, the radio was king.

85. *The New York Times* (October 5, 1944); Seltzer, *Music*, pp. 24-25.

86. "Talking Movie to Be Tested," *Springfield Leader* (June 13, 1928).

87. "Superfluous," *Springfield Leader* (January 15, 1930).

88. "Head of Musicians Union Missing," *Sedalia (Mo.) Democrat* (February 6, 1930).

89. "Leader Gone; Money Sought by Musicians," *Springfield Leader* (February 5, 1930).

90. Allen Oliver, "Springfield Slants," *Springfield Leader* (January 15, 1930).

91. Ibid.

GILLIOZ
THEATRE BEAUTIFUL

Now—
First "All Talking" Picture

Lights of New York

Every Character Speaks
Every Word

VITAPHONE

VAUDEVILLE

Sissle & Blake and
Vincent Lopez Band

Continuous 1 to 11   Usual Prices

---

4 Days Starting Sunday
First "Movietone"
Talking Picture

Not A War Picture

"The Air Circus"

with Sue Carol - Arthur Lake

Hear and See Them
TALK---SING---DANCE

Vitaphone Vaudeville

Con. 1 to 11.

Usual prices

delayed the inevitable or hastened it, Springfield's Musicians Union did not allow its members to play for wages lower than the union scale, which kept professional musicians from performing at venues that could no longer afford them.[92] And venues began taking advantage. Theaters like the Gillioz continued to include "live" shows, though these featured local amateurs—talent shows, beauty pageants, public readings, etc. In 1930, when a local theater held a parade with amateur (that is, non-union) musicians, there was no strong union pushback.[93] Emboldened, the dance hall at Cherry Crest Park hired a non-union band a year later,[94] with no obvious consequences from union backlash.

After 20,000 theater musicians nationwide lost their jobs in the late 1920s,[95] the American Federation of Musicians sent letters to all union locals, urging them to raise public awareness of the "threatened rape of the musical art."[96] In 1930, ads placed in the *Springfield Leader and Press* decried the ills of "mechanical music," asking, "Are YOU getting YOUR money's worth in the theatre?" By filling out a form (included in these ads), readers were asked to join the Music Defense League, which "opposed ... the elimination of Living Music from the Theatre."[97] The Federation gave out three million of these free memberships nation-

---

92. Ibid.

93. "About Town," *Springfield Leader* (June 2, 1930).

94. Correspondence from E. F. Lloyd, June 24, 1931, LA 14, Box 1, Folder 3, Greater Springfield [Missouri] Central Labor Council, Special Collections and Archives, Missouri State University Libraries.

95. "History," American Federation of Musicians, retrieved January 26, 2019 (https://www.afm.org/about/history-2).

96. Maurice Mermey, "The Vanishing Fiddler," *The North American Review* 227, no. 3 (March 1929), p. 306.

97. American Federation of Musicians, "VOTE TODAY!!!" *Springfield Leader* (January 27, 1930).

wide,[98] but their pleas otherwise fell on deaf ears. By 1930, three of the town's major theaters—the Landers, Jefferson, and Gillioz—each employed an organist. And that was the extent of their professional "live" musical offering[99]—quite a falling-off from the thirty-piece orchestra that accompanied the Landers Theatre's first silent film, the racist epic *Birth of a Nation*, in 1916.[100] In 1937, Fred Dawes, president of the Springfield Association of Musicians, was blaming amateur musicians (specifically high schoolers) for accepting low-wage jobs at "amusement places" around Springfield, thereby deflating the wages offered to his local union's meager membership of thirty.[101] But the problem was larger than a depressed economy: The entertainment industry was itself changing, pressured by technology.

Technology Vs. Economy Vs. the Union Projectionist

Another occupation affected by films with synchronized sound and music was that of the movie theater projectionist, who had to learn to operate the new sound equipment. At its biennial convention in 1930, the International Alliance of Theatrical Stage Employees and Moving Picture Machine Operators (of which Springfield Local 447 was a member union) looked to technology as a hedge against the mounting economic depression: As noted in its "President's Report," the advent of sound "has given tremendous stimulus to motion picture theatres, and the members of projectionists' local unions have had steady employment and increased compensation for the exacting nature of

the duties they have been called upon to perform."[102] But by the next convention in 1932, the reports were dire. As the president of the International Alliance remarked,

> At the conclusion of the [1930] convention, our organization was enjoying a most enviable position, in so far as the employment, wages and working conditions of our members were concerned. As you all well know, at the present time, we are faced with the inevitable result of the tremendous toll exacted, emanating from the deplorable outcome of the prevailing unemployment situation. Of necessity, we were forced to accept graduated reductions, which I am only too pleased to report by no means equal, or even come close to, the loss suffered by those active in numerous other fields of industry.[103]

By the next convention in 1934, even technology could not save the projectionists. In the struggle between specialized entertainment technology and global economy, the economy "won"—and the union projectionists lost. As the "President's Report" declared, it "would be both difficult and unwise to attempt to paint a glamorous picture of deeds and accomplishments of the past two years, due to the economic strife that was prevalent throughout the world."[104]

98. Robert D. Leiter, *The Musicians and Petrillo* (New York: Bookman Associates, 1953), p. 60.

99. "Leader Gone," *Springfield Leader* (February 5, 1930).

100. Robert C. Glazier, "Condensed History of the Queen City of the Ozarks (Part XVIII)," *Springfield!* 20, no. 5 (October 1998), p.

101. Harry Browning (?), Works Progress Administration union histories (*ca.* 1936-1937). Springfield Association of Musicians, Vertical Files, MSU Special Collections and Archives.

102. "President's Report," *Proceedings of the 30th Convention of the International Alliance of Theatrical Stage Employes and Moving Picture Machine Operators of the United States and Canada* (June 1930), p. 76.

103. "President's Report," *Proceedings of the 31st Convention of the International Alliance of Theatrical Stage Employes and Moving Picture Machine Operators of the United States and Canada* (June 1932), p. 119.

104. "President's Report," *Proceedings of the 32nd Convention of the International Alliance of Theatrical Stage Employes and Moving Picture Machine Operators of the United States and Canada* (June 1934), p. 165.

"Max Hollingsworth, hard at work doing projectionists duties," I.A.T.S.E. Local 447 (undated). *Ozarks Labor Union Photographs Collection, Missouri State University Special Collections and Archives.*

In 1932, Springfield's projectionists union was locked in a contract battle—one being fought nationwide—over theater managers' decision to employ one projectionist for each movie, rather than the standard two. The union argued that the newer equipment was too unsafe to be run by one person only, given the still-high risk of projection room fires (amplified by the flammable silver nitrate film, which remained in use until the 1950s). Management claimed that the technology was safe enough. Besides, the theaters could not afford to pay so many employees: They had put their money into the technology, spending upwards of $15,000 on the required sound equipment.[105] And, like

the economy generally, theater attendance remained depressed.[106] Though the union countered with an offer to accept reduced wages, the owners would not budge: The Landers closed for the summer of 1932, while the Fox-Plaza pondered a similar fate.[107] It was a time of flux, as some of the town's theaters sold out to new owners.[108] Others closed permanently.

"Urban Life" Vs. "Rural Sentiments" Vs. the "Technologies of Leisure"

Clearly, advances in film/recording technology impacted the lives of local musicians and projectionists, whose daily bread depended on theater work primarily. It's time to consider the impact of this technology upon the larger Springfield community. Within our own media-addicted age, one might suppose that the new technologies of entertainment were fully and cheerfully embraced—"the greatest thing since sliced bread." But this was far from the case. At the same time that "live" music was losing its stronghold in the entertainment industry, movies and movie theaters were seen by some as threats to the region's traditionally rural, "conservative morality."

For more was changing than technology: The demographics of the American Middle West was changing, as well. For the first time in history, the majority of Americans were living in cities. People may

105. Leiter, *Musicians*, p. 57. Compare this to Springfield's Jefferson Theatre adding a pipe organ for movie accompaniment in 1923

for the same amount, on top of having to pay musicians to play it: "Pipe Organ Installed in Jefferson Theatre," *Springfield Republican* (August 17, 1923).

106. Paul Whitington, "How the Great Depression Inspired Hollywood's Golden Age." *Independent.ie* (October 4, 2008), retrieved September 24, 2018 (https://www.independent.ie/entertainment/movies/how-the-great-depression-inspired-hollywoods-golden-age-26481978.html).

107. *Springfield Leader*, (June 11, 1932). In 1930, the Electric Theatre was sold and reopened as the Paramount; in 1934, it was sold to the Fox chain (Baumlin, *Gillioz* p. 46).

108. John Sellars, "The Long History of the Fox Theater," *Springfield News-Leader* (August 19, 2017).

have left the countryside for the city, but they brought many of their rural traditions and values with them.[109] And these, being socially conservative in the main, looked with deep disapproval upon the moral "laxity" of the city and its ways—its entertainments especially, which included jazz and "sexualized" dance, drinking in an age of Prohibition, and, yes, the movies, where young unmarried men and women practiced that new-fangled notion of "dating" beyond the watchful eyes of parents, seated side-by-side in a dimly lit theater as images of lust, intrigue, and illicit pleasure played across the silver screen. The Roaring Twenties were a mixture of moral codes and contradictions: of Prohibition vs. speakeasies, of "youth culture" vs. traditional (parental) authority, and of the old-fashioned "Gibson Girl" vs. "Thoroughly Modern Millie," with her bobbed haircut, short skirt, and right to vote.

The seeming moral crisis besetting 1920s Springfield can be stated in a phrase: More than a reaction against immorality, it was a reaction against urban-modernist "technologies of leisure." Identified with the older generation—straitlaced, hardworking, temperate, self-denying—a socially conservative Springfield sought to rein in the excesses of "modern lifestyle," thereby preserving the city's youth. In 1921, a Springfield "blue law" ordinance banned movies on Sundays, passing with a 66% majority and attributed largely to the number of women voting,[110] with about half of the voting population casting ballots.[111] It remained in place for five years, until the owner of the Grand Theatre, W. W. Smith, openly challenged

the ordinance by playing a movie on a Sunday in 1926. Seven minutes after the show began, he was arrested and fined $100; but three months later, because of the questionable legality of the ordinance in the first place,[112] the ban was lifted.[113] Smith's victory was seen widely as a defeat for morals: "Practically every movie theater is a school," wrote evangelist John Carrara in 1922, "and the movie star the teacher of crime and immorality."[114]

Leading the local reaction against "amusements" was Springfield's Assemblies of God, whose church members were supposed to give up drinking, dancing—and moviegoing. The Gospel Publishing House of the Assemblies of God made its strongest case in a tract titled *"The Movies": The Greatest Religious Menace* (*ca.* 1929). Among many other faults, the movies were "a persistent enemy of the Christian observance of the Lord's Day." They mocked the sanctity of marriage, normalizing divorce and encouraging "a standard of life that will ultimately destroy the home." They also mocked the clergy: "Whenever a Protestant minister figures in a drama it is always a caricature."[115] Films of the late 1920s and '30s did rely on caricature, often at the expense of traditional cultural mores; in this respect, they reversed trends of earlier silent films, which idealized and celebrated rural life.[116]

Other preachers from other denominations joined in the battle against "amusements." In 1927, Reverend Innis Harris of Springfield's Grace Methodist Episcopal Church wrote a newspaper editorial excoriating the evil in modern movies. As he saw it, the first great

---

109. Hal S. Barron, "Rural America on the Silent Screen," *Agricultural History* 80, no. 4 (Autumn 2006), p. 384.

110. "Sunday Closing Ordinance Gets Big Majority," *Springfield Republican* (December 12, 1921). Given that women had just won the right to vote—the 19th Amendment having become law in 1920—the large turnout ought not to surprise. As to their majority support of the "blue law," one might consider the historic strength of the local Women's Temperance movement.

111. Ibid.

112. "Sunday Closing Act to Be Tested When Smith Arraigned," *Springfield Leader* (December 12, 1926).

113. "Picture Shows to Operate Sunday," *Springfield Leader* (December 16, 1926).

114. John Carrara, *Enemies of Youth* (Grand Rapids, MI: Zondervan, 1922), pp. 81-82.

115. Baumlin, *Gillioz*, p. 31.

116. Barron, "Rural," p. 385.

assault against conventional morality came with the Great War, whose "wholesale slaughter cheapened life and left the world brutal and criminal minded." But the movies came in a close second. If the technologies of war "cheapened life," the technologies of entertainment corrupted youth, teaching viciousness and criminality:

> There are [movies] which are entertaining, educational, but many of them are absolutely vicious, and running through them are evil suggestions and a stream of slime. Brawls, crimes and the licentious are vividly and predomin[antly] pictured. The bully, burglar and blasphemer are oftimes the heroes and lauded. And young lives go out all fired up to repeat in actual life what they saw on the screen. They excite wrong desire rather than do they incite to pure thinking and noble endeavor.[117]

Though these tempted "the plastic minds of American youth,"[118] the new movies proved equally seductive in their urban settings, themes, and styles.

In effect, the silent film's "rural sentiments" yielded to the sound-on-film's newly found fascination with the city and its characters. "False and superficial to downright dishonest,"[119] the urban ethos gave the Reverend Harris sufficient grist for his mill. He could have pointed to *Sunrise: A Song of Two Humans* (1927). Winner at the 1st Academy Awards (1929) and considered an early film masterpiece, the movie shows the vacationing "Woman from the City"—a *femme fatale*—luring "the Man" into an illicit affair. Bored with family and farm life, "the Man" conspires to murder "the Wife" by drowning, though his awakening conscience saves her life, their marriage, and the farm. Let's add up the themes: lakeside vacationing—think Taneycomo—predatory city woman, bored farmer scratching out a living on the land, and the allure of "the City." To the moviegoer, it's an award winner; to the moralist, it's mischief afoot.

Some Springfieldians avoided the theater altogether, along with the tavern and dance hall, on religious grounds; but, more than preaching, it was money (or the lack thereof) that kept theater attendance down throughout the Depression years. In their habits of entertainment, many fell under the spell of big-city life. Concomitantly, some locals may have welcomed the Motion Picture Production Code (or Hays Code), which censored the more scandalous themes in films nationwide when it began to be seriously enforced in 1934.[120] Besides, city folk were being depicted with pretty much the same virtues as their country cousins.[121] Consider *Gold Diggers of 1933*, the top-grossing film for that year. Filmed in the Depression (and making continuous reference to it), the plot revolves around a stage show threatened with closure for unpaid bills. When Brad—the show's songwriter: mysterious, young, handsome, and well-heeled—shows up with cash, starlets Polly, Trixie, Carol, and Faye suspect his wealth. (Organized crime, perhaps?) When Brad saves the production, his brother is next to show up—to save Brad from seduction, presumably, by one of the show's "gold diggers." It turns out that Brad is a millionaire, not a criminal, and that Brad's stuffy old brother is wrong about the girls (and about showbusiness generally). The movie's cultural mes-

---

117. Innis D. Harris, "Lest Thou Forget," *Springfield Leader* (August 22, 1927). On this point, the Rev. Innis agrees with the Gospel House tract, which declared that "the Picture Show is a full graded course of schooling in the technique of crime…. Safe-cracking, pick-pocketing, abduction, murder, white slaving—every sort of crime is enacted in the most skillful manner possible, so that all the student need to do is to do as he is taught" (Baumlin, *Gillioz* p. 31).

118. Baumlin, *Gillioz*, p. 31.

119. Barron, "Rural," p. 394.

120. Ina Rae Hark, *American Cinema of the 1930s: Themes and Variations* (New Brunswick, NJ: Rutgers University Press, 2007), p. xii.

121. Barron, "Rural," p. 407.

sage is obvious: *There were good people in the city, too,* and moralizing can be mere prejudice—hypocrisy in disguise. (This, by the way, is the movie that gave us the Depression-era anthem, "We're in the Money.")

If American movies of the late 1920s and '30s celebrated the big-city ethos—that is, the world and lifestyles and character-types of Bostonians, Chicagoans, New Yorkers above all—then where would Springfield, a middling town in the American Middle West, fit into that big, fast-paced, morally ambiguous film-generated fantasy? Even today, Springfield (and its larger metro region) finds itself caught between two sets of images, one urban and progressive, one rural and traditional.[122] A century ago, this same divided self-image played out in the city's theaters, pulpits, and newspapers. It was a mid-sized Midwestern city struggling to fashion an identity out of polarities in ethos and lifestyle, set on becoming modernized and urbanized while projecting a higher moral character than the big-city stereotype portrayed in movies of the 1920s and '30s.[123]

The October 2, 1928, issue of the *Springfield Leader and Press* illustrates this paradox on its front page, which features the sensationalized story of Mae West's arrest in New York City for an "indecent performance" of her play, *Pleasure Man*. The same front page reports on an upcoming Springfield performance of a German acting troupe's *Passion Play*—a "spectacular event" with "fine performances" of the trial and death of Jesus Christ.[124] The front page gives two more headline stories side by side. The first declares the federal government's estimate of Springfield's population as inaccurately low (closer to 61,000 versus the federal estimate of 51,700, with extremely high growth rates in the eight years since the last U.S. Census). The sec-

ond—in larger typeface than the Springfield article—reports on the declining population of Manhattan (a loss of 200,000 in three years). The paper appears to show what its readers wanted to see: that morally upright Springfield was a rapidly growing urban center while Manhattan, being of higher population and lower morals, was losing its appeal.

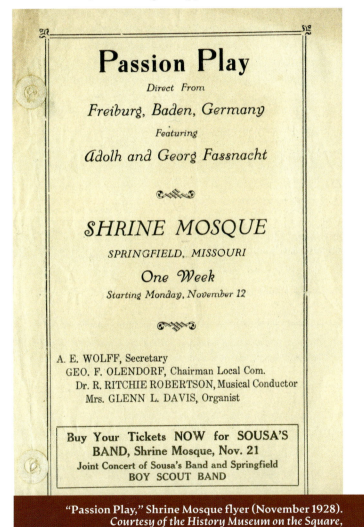

"Passion Play," Shrine Mosque flyer (November 1928). *Courtesy of the History Museum on the Square, Springfield, Missouri.*

122. Jonathan Groves, "The Ozarks hits the crossroads of urban and rural," *Springfield News-Leader* (February 26, 2006).

123. Barron, "Rural," p. 394.

124. *Springfield Leader* (October 2, 1928).

The Depression brought with it refugees from larger cities, many of whom were relatives of Ozarkers and had left for industrial work during previous decades.[125] These refugees, accustomed to big-city life, brought back with them a taste for the types of movies that were more abundantly available to them before returning to the Ozarks. And if the local theaters wanted to keep the money coming in, they needed to appeal to as broad an audience as they could. This change may have occurred regardless of the influence of former city dwellers, however. The motion picture industry as a whole consolidated and conglomerated during the Depression in order to stay afloat.[126] The types of movies shown in any particular location became more dependent on cost than on taste, and it was much cheaper for a studio to share many copies of one film. And they had many more customers in high-population cities,[127] superseding any rural population's desire for a particular type of movie.[128] To maximize sales, studios eventually settled on mixing genres to ensure that their films appealed to differing tastes (such as musicals within comedies),[129] while tailoring their advertising to each particular audience.[130]

By the mid-1930s, the conflict seemed to be waning some. Rural themes continued but focused on a nostalgic past, while urban themes represented the new and modern, irrespective of locale.[131]

The percentage of movies set in rural locations dropped to 5%, compared to 20% in 1915.[132] And when Prohibition was repealed in 1933, the musicians that had been displaced by recorded music recovered some of their lost employment by playing once again in nightclubs and bars[133]—giving them a slight cushion when movie theaters (among other industries) felt the full financial effects of the Depression. With the arrival of World War II came a dramatic increase in employment overall and a new reason to go to the movies: twice-weekly newsreels with updates about the war.[134]

The perennial complaint of an older generation chastising the new for its lack of morality—a pattern far older than the Jazz Age—would of course repeat itself through ensuing decades. And the nostalgia of each generation for a certain kind of music, a certain type of movie, a certain level of technology, and a certain set of values retreated into the past, "fated to pass into the limbo of discarded customs"[135]—like live musical accompaniment to the old silent films.

125. Robert Flanders, "When the Big Change Came in Here," *OzarksWatch* vol. 7 no. 4 (Spring 1994), p. 1.

126. Ronny Regev, "Hollywood Works: How Creativity Became Labor in the Studio System," *Enterprise & Society* 17, no. 3 (September 2016), retrieved March 25, 2019 (https://doi.org/10.1017/eso.2015.89).

127. Charles W. Eagles, "Urban-Rural Conflict in the 1920s: A Historiographical Assessment," *The Historian* 49, no. 1 (November 1986), p. 27.

128. Donald Crafton, *The Talkies: American Cinema's Transition to Sound, 1926-1931*, ed. Charles Harpole (Berkeley, CA: University of California Press, 1997), pp. 17-18.

129. Henry Jenkins, III, "'Shall We Make It for New York or for Distribution?': Eddie Cantor, 'Whoopee,' and Regional Resistance to the Talkies," *Cinema Journal* 29, no. 3 (Spring 1990), p. 38.

130. Jenkins, "Shall We," pp. 46-47.

131. Barron, "Rural," p. 404.

132. Barron, "Rural," p. 410.

133. Leiter, *Musicians*, p. 64.

134. Wilson, *American Acting*, p. 250.

135. Preston J. Hubbard, "Synchronized Sound and Movie-House Musicians, 1926-29," *American Music* vol. 3 no. 4 (Winter 1985), p. 431.

SPRINGFIELD
ELECTRIC

ELECTRIC
THEATRE

THEATRE OF THE STARS
STAGE - MAY KENNEDY MC-CORDS "OZARK FOLK LORE"
SCREEN - JAN KIEPURA "MY HEART IS CALLING"

Springfield's Electric Theatre (ca. 1935).
Courtesy of the History Museum on the Square,
Springfield, Missouri.

Transporting the Ozarks (I):
Photos from MSU Special Collections and the
History Museum on the Square
Selected by Tracie Gieselman-Holthaus

Springfield Public Square, with electric tower and trolley tracks (*ca.* 1896). *Public Square Collection, History Museum on the Square, Springfield, Missouri.* In 1896, the Springfield Traction Co. helped pay for the Gottfried Tower, which supplied electricity to its streetcars.

▲ Rural mail delivery by mule team (*ca.* 1910). *Oscar W. Carter Collection, Missouri State University Special Collections and Archives.* The mail carrier, Minnie, was photograpapher Oscar Carter's sister.

▼ Streetcar on the Town Square (1919). *Transportation Collection, History Museum on the Square, Springfield, Missouri.* The Springfield Traction Co. operated from 1879 to 1937, evolving from horse and mule to steam and electricity. In 1937, the city's public transportation system was "modernized," with buses replacing electric streetcars.

Highway clearing machine (1927). ▲
Transportation Collection, History Museum on the Square, Springfield, Missouri.

"Accident on Mo. Super Highway" (1925). *Domino Danzero* ▼
Collection, Missouri State University Special Collections and Archives. The "Super Highway" depicted in this traffic jam is a gravel road (note the orchard to the right).

Cars parked on the Town Square (1924). *Public Square Collection, History Museum on the Square, Springfield, Missouri.* The automobile was already facing an endemic urban problem: lack of parking space.

"The Frisco Firefly" (*ca.* 1940). *C. C. Roberts Collection, Missouri State University Special Collections and Archives.* Rarely are technology and aesthetics so firmly wedded as in this steam-powered streamliner, which ran on the St. Louis-Tulsa-Oklahoma City Line (via Springfield) from 1939 through 1960.

▲ Airplanes at McClure Flying Field, East Division Street, Springfield (1930s). *Transportation Collection, History Museum on the Square, Springfield, Missouri.* The field continues operation as the privately owned Springfield Downtown Airport.

Frisco business car (1948). *C. C. Roberts Collection, MSU Special Collections and Archives.* In 1967, the last passenger train would leave the Springfield Frisco depot; the depot itself would be demolished a decade later. ▲

▼ Colonial Bread trucks in front of Colonial Baking Company, St. Louis Street (1935). *Business Collection, History Museum on the Square, Springfield, Missouri.* Ready to make their morning deliveries …

Campbell "66" Express trucks and driver, Norval Latham (undated). *Transportation Collection, History Museum on the Square, Springfield, Missouri.* With its Snortin' Norton camel mascot-image, "Humpin to Please" was the Campbell "66" motto. The company operated from 1933 to 1997. ▼

Passengers boarding Ozark Airlines, Springfield Municipal Airport (1960). *Betty Love Collection, History Museum on the Square, Springfield, Missouri*. Ozark Airlines operated from 1950 to 1986. To the city's northwest, the Municipal Airport would grow into today's Springfield-Branson National Airport.

"Red Gold" of the Ozarks:
Tomato Canneries in the Ozarks, 1890-1960
Tom Dicke

For nearly three generations, canneries, especially tomato canneries, were an integral part of Ozarks life. Every autumn for over half a century, thousands of small canneries scattered across the Ozarks came alive and, in the space a few months, packed millions of pounds of tomatoes. The first commercial cannery in the Ozarks opened in Springdale, Arkansas in 1885 and, by 1900, Ozarks canners accounted for just over 20% of all tomatoes canned in the United States. Tomatoes remained the "Red Gold of the Ozarks" until the early 1950s when, within a few years, the canneries all but vanished. The reasons why have much more to do with technological "fit" than with economics. Commercial canning flourished because it was a cheap, simple, small-scale technology that was exceptionally well suited to the needs and abilities of small, cash-poor Ozarks farmers. When it faded, the technology for picking and processing tomatoes was still fundamentally the same and canneries were still profitable; but the Ozarks had changed in ways that made the part-time canner an anachronism—out of place in the economy and out of step with the needs and lives of 1950s Ozarkers.[136]

Commercial canning began in 1810 when Nicolas Appert, a French winemaker, chef, and inventor, published a book detailing his methods for preserving food in hermetically sealed containers. Appert began his work in hopes of winning a 12,000 franc prize offered by the Emperor Napoleon to anyone who could come up with a new, safe, and reliable method of preserving food for prolonged periods. After over a decade of trial and error, Appert found a solution that relied on exposing airtight containers to high heat; no one knew why this worked but, if done properly, Appert's method unarguably produced safe, edible foods that could keep for years. Napoleon offered the prize in hopes of finding a way to feed France's far-flung armies reliably and, for the next several decades, the main market for canned food was the military, pioneers, or others without access to other food alternatives.[137]

Commercial canning came to the United States in 1819 but remained a novelty until the Civil War. As late as 1860, canned food was expensive when compared to fresh, and many Americans were suspicious about its safety. In the short term, the war stimulated production, from five million cans in 1861 to thirty million in 1865; it sped technological development and gave canners the resources and expertise to move into the civilian market after the war. Perhaps most important, the war introduced hundreds of thousands who might not otherwise have been exposed to the benefits of canned food as a safe and fairly tasty supplement or substitute to fresh food.[138]

The American canning industry continued to expand quickly in the years after the war. On the supply side, it was an easy business to begin. The

---

136. For a more detailed treatment the Ozarks canning industry, see Tom Dicke, "Red Gold of the Ozarks: The Rise and Decline of Tomato Canning, 1885-1955," *Agricultural History* 79 (2005), pp. 1-26. http://www.jstor.org/stable/3744875.

137. Descriptions of Appert's methods are remarkably similar to those of small "shade tree" canners in the late 19th and early 20th century Ozarks. See, for example, Mrs. Glen King, "Memories of Tomato Canning Days," *Webster County Historical Society Journal* (December 1976), p. 12.

138. See Mark W. Wilde, "Industrialization of Food Processing in the United States, 1860-1960" (Ph.D. diss., University of Delaware, 1988).

technology was simple to understand, operate, and maintain. Farmers who wanted to supplement their income by packing their own crop could get into the business for almost nothing. Cans could be filled by hand, sealed with simple tools and sterilized using one or more large caldrons. Larger outfits, those packing many thousands of cans per season, were partially mechanized with conveyors to move the cans and steam retorts to sterilize them; the machinery could cost several thousand dollars although, with luck, the investment would pay for itself in a year or two. On the demand side, canned food was generally much more palatable than salted or dried food (the other main alternatives) and found a ready market, especially among urbanites who lacked the space for more than vestigial gardens.[139]

Canned food found a place at the table at roughly the same time the expansion of the railroad was tying the Ozarks more tightly to the national economy. Although nearly anything could be canned, beans, corn, peas, and tomatoes were most popular in the late 19th century. And of these, tomatoes were the overwhelming favorite, sometimes making up more than half of the total vegetable pack in the U.S. Tomatoes fit comfortably into the Ozarks economy. The small farmers who dominated the region generally practiced mixed agriculture and were open to trying new crops that did not disrupt their existing mix. For many, tomatoes were an ideal choice. They grew well in the often-indifferent Ozarks soil, and the work of planting and picking could be done using family labor. Once canned, the market for tomatoes was usually strong, since local wholesalers found they fit well into their already-established marketing channels. Given the potential profits and low cost of entering the market,

for many it was a short step from growing to canning. The end result was that, by 1900, canneries were as common as country stores throughout the Ozarks.[140]

Two types of canner flourished in the Ozarks. Both were part-timers. The more economically important were the commercial packers who processed tomatoes and, perhaps, berries or apples. Canning and associated tasks were their main business, though they packed three months or so per year. The more numerous were farmer-canners, for whom canning was a low-risk and potentially profitable sideline. This group got into the business by adding a few acres of tomatoes to their existing mix of crops and then spent a week to a month packing these (and perhaps those of a neighbor or two). In the early days of the industry, these "shade-tree canners" often set up anywhere close to their crop with access to good water. Their "canneries" were typically crude, with dirt floors and open sides. They often sterilized their pack outside in large caldrons. By the early 20th century, the technology had become a little more complex, but canning was still an easy business to get into. Used equipment was cheap, family labor almost free, and brokers would provide cans and other supplies with payment due after the pack sold. Prices and profits varied considerably season to season but could be substantial. In 1933, for example, an Ozarks farmer could have sold his tomatoes to a local canner for eight to twelve dollars a ton or can them himself and sell his pack to a local broker for roughly seventy-five dollars a ton. Even with the additional expense and effort, the profits were enough to compensate for the few weeks of labor required to can their own crop. Starting in the mid-1930s this group began to decline as more stringent food safety laws forced many out of business.[141]

139. For a solid overview of the early canning industry, see Arthur L. Hunt, "Canning and Preserving," in *Twelfth Census, Manufacturers, Part III Special Reports on Selected Industries* (Washington, DC: GPO, 1902), pp. 463-80.

140. Harvey A. Levenstein, *Revolution at the Table: The Transformation of the American Diet* (New York: Oxford University Press, 1988), Chapters 2 and 3; Dicke, "Red Gold."

141. Neil Sagerser, "The Ozarks Tomato: A Cash Crop for Farmers

▲ Canning factory in Billings, Missouri (1920). *Courtesy of the State Historical Society of Missouri.*

Interior, canning factory in Billings, Missouri (1920). ▼ *Courtesy of the State Historical Society of Missouri.*

Commercial canners were larger and more mechanized but used the same basic technology. Like the farmer-canner, they might be found anywhere with access to plentiful fresh water and a road, though most clustered on the Springfield Plain where, as early as 1900, canned tomatoes were the leading agricultural export of many counties. Packers were drawn by the large number of small farms, which provided a steady source of supply. Raw tomatoes did not travel well and needed to be as fresh as possible when canned to preserve their color, texture, and flavor. As late as 1950, when roads were significantly better than a generation before, most Ozark tomatoes were grown within three miles of where they were packed.[142]

A commercial canner's work began before the plants were in the ground. In the weeks or months before the growing season started, a canner would contract with farmers for delivery in the fall. Because tomatoes were just one of several crops grown by local farmers, fields were small. Contracts for five for or six acres were fairly typical. The packer often provided tomato seed or seedlings and sometimes fertilizer and insecticide free of charge. In exchange, the farmer agreed to sell his crop to the packer at a set price.[143]

Farmers typically began tomato plants in a seedbed in early spring. After the plants reached six to eight inches in height, they were transplanted to a prepared field. Once established, the plants required little maintenance other than periodic cultivation to reduce weeds. After the fruit appeared, tomato worms could be a problem and needed to be picked off by hand or sprayed with insecticide. Harvesting generally began in August or September. Farmers were responsible for delivering the fruit to the factory so a farmer's children did most of the picking. Early on, yields of three to five tons per acre were common and improved significantly over time, due to better varieties. Pickers packed tomatoes for transport in stackable crates holding a bit less than a bushel.[144]

Once tomatoes reached the factory, they were weighed and placed in storage until processing. Typically the canner kept a tally of deliveries and settled with the farmer after the pack sold. Ideally, canners packed the tomatoes the day they came in; but if stored in vats of water that was changed regularly to keep it cool and clean, tomatoes could be kept in good condition for up to a week.

Canners typically hired local boys to unload the crop, wash it, and discard any obviously damaged or otherwise unusable tomatoes. Next, they loaded the tomatoes into wire baskets and scalded them to loosen the skins. Lastly, they loaded the steaming fruit into buckets and carried it to girls and young women who cored the tomatoes and squeezed or peeled them out of their skins. Peelers dropped the fruit into fresh buckets, which were then carried to packers. These women generally hand-filled cans by forcing the fruit through holes in a table or trough positioned above the can line. After filling, they placed the cans in a heat bath to expel excess air from the headspace. Then they used a crimper to seal the tops closed and loaded the cans into large wire baskets, which were lowered into boiling water long enough to sterilize and preserve their contents.

in the 1920s and 1930s," *Douglas County Historical Society Journal* (Summer 2002), pp.16-19. Court records also give detailed price information: see, for example, *Ozark Mountain Canning Company v. The Consolidated Companies Incorporated*, Case no. 1,276 (September 1934), Circuit Court of Greene County, Missouri.

142. Irene A. Moke, "Canning In Northwestern Arkansas: Springdale, Arkansas," *Economic Geography* 28 (April 1952), pp. 151-59.

143. Examples of growers' contracts can be found in court records of disputes between canners and growers. For a sample contract, see *Ozark Mountain Canning Company v. A. H. Sweeney*, Case no. 2,566 (May 1935), Circuit Court of Greene County, Missouri.

144. Robert McGill, "Red Gold Ozark Tomatoes," *OzarksWatch* vol. 9 no. 1 (1996), p. 23; Margaret Lucas, "Ozarks Canners and Freezers Progress Report" (1963), Webster County Historical Society; King, "Memories," p. 12.

After sterilization, the cans were brought outside to cool. Usually they were boxed quickly, but if there were concerns about spoilage, cans would be left for three days or more to check for bulging. A bulged can indicated that the contents were spoiling and the can needed to be discarded. Suspect cans could be tapped with a metal rod: A spoiling can sounded hollow. After processing, the canner either labeled the cans with the packer's brand or sold them to a broker who provided labels. These were typically hand-glued, and a fast worker could label one thousand cans an hour. Once labeled, the cans were packed in cardboard cases and stored until picked up by the broker.[145]

The workday ran from dawn to dusk. Most cannery workers moved into shacks or tent cities that grew up around the cannery during the pack. Like the tomatoes they packed, workers came from the area. Between planting, picking, and processing, the factory provided short-term employment for most, if not all, working-age members of many families. Local wages were low: Peelers and fillers, for example, were generally paid five cents per bucket for most of this period. In the early decades of the industry, this was only slightly below the national average. Beginning in the 1930s, the gap increased noticeably and, by the early 1940s, Ozarkers earned slightly less than half the wage paid to California cannery workers.[146]

From the late 19th century through the end of World War II, canneries were as common as country stores throughout the Ozarks. Individual canneries

145. Thelma Keithley Bilyeu, "My Life Story," *Ozarks Watch* vol. 8 no. 2 (1995), pp. 4-21. As for local labels, there was AIR LINE BRAND Tomatoes, packaged by Roy Nelson Canning Co. of Crane; BRID-WELL'S OLD SOL Hand Packed Tomatoes, packaged by Bridwell's Canning Co. of Marshfield; SHELL KNOB BASIN BRAND Hand Packed Standard Tomatoes, packaged by Epperly Canning Co. of Shell Knob; OZARK CHIEF BRAND Tomatoes, packed for Marshfield Supply Co.; and GREERS SPECIAL BRAND Tomatoes, packaged by Greer Bros. and McKnight, again of Marshfield. These were a few of the literally hundreds of brands packed regionally.

Image of canning label (above) courtesy of the Barry County History Museum, Cassville, Missouri.

146. C. E. Campbell, "Organization and Management of Tomato Canning Factories in Arkansas," *Arkansas Agricultural Experiment Station Bulletin* (June 1929), p. 10; Missouri Bureau of Labor Statistics, 30th *Annual Report* (Jefferson City, MO: Office of the Bureau of Labor Statistics and Inspection, 1908); Bertha M. Nienburg, "Hours and Earnings in Canning and Preserving Industries, 1937 to 1939," *Monthly Labor Review* (February 1941), p. 435.

changed hands or shut down, and individual farmers quit growing or canning but, overall, the industry remained deeply entrenched throughout the region. For a variety of reasons, that changed quickly in the years after World War II. Changes in canning technology made it possible for larger, more efficient canners to process a full line of goods, which pushed Ozarks canners to the margins of the industry. Changes in consumer tastes, which swung strongly in favor of national brands, exacerbated the problem. This left small, single-line packers—like those found in the Ozarks—concentrated in less profitable areas like store brands, institutional supplies, and other niche markets, which gave potential canners no incentive to enter the business.

This would have been a manageable problem if Ozarks life had not changed in ways that were fundamentally incompatible with small-scale packing. The most important change here was the sharp decline in the number of small farmers. They were the ones who provided both the raw material and the labor to process it. As late as 1940, small farmers were the backbone of the Ozarks economy; by 1955, that was no longer the case. In 1930, for example, there were 11,172 general farms in the eight-county area surrounding Springfield, Missouri. By 1954, that number had declined to under 700. Many farms disappeared altogether as increased mechanization drove a wave of consolidation. Of those that remained, most specialized in a single crop or in dairying; and, while the land could support large-scale tomato cultivation, it was an unlikely choice for Ozarks farmers. Too many other crops were easier and more profitable to produce.

Likewise, for those still interested in living on the land and working off the farm, there were better opportunities than canning. There was no incentive to take a few months of part-time work canning tomatoes when year-round work in town was now possible due to better roads and cheap, reliable transportation.

For the part-time farmer who commuted to work in town and stayed on the farm as a preference of lifestyle, cattle ranching became the most popular part-time option by far, since the work was not too physically demanding and could be done without disrupting a full-time job off the farm. As it became more difficult for canners to find labor, to find adequate supplies to pack, and to get a good price for what they could produce, canner after canner left the business. It was an easy decision, since small-scale canning was always a part-time business, even for commercial canners. Small canners in the Ozarks did not generally fail; instead, one season, they simply failed to reopen.[147]

By 1960, the only significant concentration of canners remained in the area around Springdale, Arkansas. This area survived as an important canning center, not because of any significant natural advantage, but because the previous seventy years had left a critical mass of infrastructure and expertise. This provided local packers with an important competitive advantage that allowed, and continues to allow, them to compete successfully with areas better suited naturally to the canning business. Though canning remains an important industry in the Arkansas Ozarks, its relationship to the area has fundamentally changed. Instead of an industry thoroughly enmeshed in and reliant on the local economy and community for its success, contemporary canners have become largely independent of their local environment.[148]

---

147. Milton Rafferty, *The Ozarks: Land and Life*, 2nd ed. (Fayetteville, AR: University of Arkansas Press, 2001), pp. 158-174; Milton Rafferty, "Agricultural Change in the Western Ozarks," *Missouri Historical Review* 69 (April 1965), pp. 299-322; Milton Rafferty, "Population and Settlement Changes in Two Ozarks Localities," *Rural Sociology* 38 (Spring 1973), pp. 46-56.

148. For a brief look at a contemporary canner, see the Allens corporation webpage, "Authentic Southern Veggies" (https://www.allens.com); for a broader view, see "Food Science Engineering," University of Arkansas Division of Agriculture Research and Extension (https://institute-of-food-science-and-engineering.uark.edu/default.aspx).

"Taking Care of a Creek":
A Profile of George Langenberg of Rosebud, Missouri
Alex T. Primm

Focusing on how a German-American farmer takes care of his creek, this article helped create my career as an oral historian. This is from my initial fieldwork, the Ozark Rivers Oral History Project, sponsored by the Meramec Regional Planning Commission based in Rolla. Funding came from the Missouri Humanities Council and the James Foundation, which manages Maramec Spring Park. There's continuing concern about water quality in this large spring and a large population of ornery Rainbow Trout.

While this article mainly involves managing land, Mr. Langenberg's interview offers insights into Ozark farming. We're dealing with the northern Ozarks here, Gasconade County, settled largely by German Catholic families following the failed progressive revolutions in the old country of 1848. My wife's family is from this area, just west of Vienna on Little Tavern Creek. The rolling hills of this Missouri Rhineland region tend to have better soils than other parts of the Ozarks. Crops and families benefit.

In part because of this article, I joined other oral history projects that the U.S. Geological Survey carried out in the 1980s and '90s from its water resource division in Rolla. Oral history proved to be a useful methodology for this agency, because farmers often do not keep records of field management over time. But they often have clear memories of land-use changes during their lifetimes. Anyone interested in fishing, agriculture, or running rivers should enjoy Mr. Langenberg and his observations on the care and maintenance of streams.

The following is edited from my manuscript, Rivers and Rainbows: Oral Histories from the Middle of America. *A slightly longer version ran in the April 21, 1982 issue of the* Gasconade County Republican *from Owensville.*

"If you have a stream you've got a lot of work"

South of the town of Rosebud, the Red Oak Creek makes a gentle bend by a field that seems as broad as any in Kansas.

George A. Langenberg, 87, who lives in town (about 5 miles to the north), knows the story of this field, all 189 acres of it. He has spent over 40 years working this land.

He has also studied the Red Oak Creek. Knowing how and why and where the stream flows is part of the secret to keeping the bottomland productive, Langenberg says.

Most times the Red Oak is such a small creek that it is usually not a problem. Fishermen rarely bother to even try it, Langenberg said.

But when a flood comes, the Red Oak becomes a torrent. Last year the floods were the worst ever. Neither George nor his son Arvil, who is now part owner and main operator of the place, were able to get any kind of crop planted.

"It was wet all the time," Langenberg said late last fall when he took time to talk about his field and the creek.

"It was frustrating not to be able to plant the crop. I like to turn the dirt," he said. "I like to see how good I can get it to come up. That's what interests me."

The condition of his farm buildings, rusty parts under an oak, an old tractor that still works, all have stories and recall challenges for Langenberg.

"It was July 10 or 11, 1947, I forget the exact day," he began when showing me the barn. "We had an awful rain up on Soap Creek and I'd been cultivating corn down there on the bottom. When it come up I worked as long as I could, but I seen I was going to get caught. Oh, I knew it was going to get up big, and I'd left my fenders from the cultivator down on a stump there. I got down there and I see it's coming up good through that slough. Oh, it's a-going with one wheat shock after another floating down the middle of the Red Oak double time.

"Oh, then I got scared so I turned around and I couldn't get up this incline. The problem was my back-cultivator gang couldn't get up and it was slick there. The cultivator kept getting caught in these weeds. As foolish as I was, I went up to get a couple of log chains to put around each wheel. I shouldn't have done that," he said.

"The slough was already this deep and halfway up the magneto and over the axles. So I tried to back up but I seen I couldn't make it. I went off to the side of this forked sycamore and got up in it till it went down. All you could see of that tractor was the radiator cap and the steering wheel."

"If you had done something stupid, you could have been washed downstream," I said.

"I would have, if I tried to cross that slough," Langenberg said. "Big bundles of wheat were going down right past me. Quite an experience. It shows a man's mind isn't always thinking right. I could have hooked those back gangs up. Six inches would have done it."

Much of the Red Oak would have been permanently tamed by an impoundment, which had been proposed by the U.S. Army Corps of Engineers in the 1950s, he said, but the farmers didn't want it.

"We went and saw the state senator, Don Owens it was then, and told him it's no good for the farmers and he never was for it. But now, I don't know, there's not so many farmers. The Corps of Engineers might get what they want sometime in the future because the people don't understand these little creeks anymore."

After a quick visit to some Indian mounds, we drove to the bottom along the Red Oak. "If you have a stream, you've got a lot of work," he said. "Let me show you how I control the creek."

Two hundred yards down from the Highway T bridge, we left the truck next to a row of trees. They were sawed-off river maples that paralleled the road and formed a line perpendicular to the creek for several hundred feet out into the field.

"How'd the trees get that way?"

"I cut them off," Langenberg said. "You don't want big trees out here in the field. It'll shade out the crops a little, but worse, it'll catch all the logs and debris the flood brings down. I don't want that, I keep this line of short trees to slow down the current of the flood and let the big timber float on by. The high water's not bad. I don't mind the floods so much. See how I plow down here, against the slope. That makes it harder for that floodwater to take my soil. I get it from up above."

Once a nearby farmer complimented him for the good soil the bottom had, George said. "You ought to know how good it is, I says to him, and he just looks at me and smiles." He knows lots of it comes from his place.

"I've seen farms lose six inches or more off their fields in a flood. Or it'll change course and take an acre or ten if you're not careful. This bottom when I got it was all growed up with sycamore and sloughs. It was a jungle along the bottom, the best land. I had to start from the beginning, just like my father taught me. It was sure a lot of work."

We walked over to the edge of the field, which was maybe ten feet above the surface of the creek and just about that far back from it. As we walked along the field's hard mud bank, George explained and showed the steps he takes to protect it and make the maximum amount of land available for the crops.

The basic principle to hold the bank is keeping the creek's flow even across its channel, George said. Any curve along the creek must be smooth and no gravel bars or root piles should be allowed to develop and deflect the current.

"If you let it hit one bank, then it'll hit it again down below," he explained. An obstruction-free channel causes less friction, so that any floodwater will be more likely to flow evenly by and not gouge out craters along the creek or take out even bigger bites.

After a couple of minutes we came to a line of river maples along the bank whose outermost branches had been sawed three-quarters through and left hanging.

"Along the bank you want to leave some trees," he said. "All those roots hold the soil. You can't plow right down to the edge anyway, so the trees help you. The branches when they hang out over the river are a problem, though. They catch the trees and timber coming down with the flood and hold them, making a woodpile that's going to cause that flood to go over your ankle and take some ground with it. That's why I leave them hanging in the creek."

At some ledges across the creek, we stopped. "I had to get a cable and pull some of those rocks out of the creek. I put them in a hole on this side to keep the bank here. Now it flows good there, pretty good except just at the end of them, you see," he pointed across the creek. "There's a little gravel bar there. Just a few of those rocks falling off the little bluff did it. I'll have to go over and clean that up, get that gravel bar out of there because it could start something big."

The line of gravel along the bluff was barely six feet long, hardly even a foot into the channel and not big enough to support more than a few tiny weeds.

"It's really that big of a problem?"

"If you let it go, it is. See this," he pointed to a telephone pole that was wired to a pair of foot-thick maple stumps. "This is my newest project this fall."

The telephone pole had come from the electric

co-op, which replaced a line crossing the farm. George had pointed it out earlier during our walk, saying he allowed them to do the work provided the crew cleaned up once the work was done.

"Well, by each of those holes, I found a handful of old copper staples. A cow would eat those and they'd kill her, but I guess those kids don't know that."

A whole two-foot thick burr oak tree had washed down in the last flood, hung up by some soft maples along the creek, and forced up against the bank. George explained the necessity for the 30-foot pole. It angled upstream and held the top of the oak into the shore and a pile of branches into the hole dug by flood. "Eventually that will all fill in with soil," George said.

As he spoke, a light German accent gave a musical cadence to his voice. His grandparents had emigrated from Germany, and he was born January 1, 1895, the first of four boys, and was sent out by his family to help another farm family as his brothers grew able to do his chores. After serving with the Pioneer Engineers in France during World War I, he came back to Rosebud and married the former Laura Rosa Buehrer in 1923.

His service station became the first Chevrolet dealership in the area and a major truck repair garage along U.S. 50 when it was paved. The Depression wasn't bad for him, George said, but it meant long hours often late into the night repairing the big trucks, and the hard times did "blow up" the bank in town, the doctor left, the hardware store closed up, and one of the two mail routes eventually was lost.

"In 1937 I bought the first farm, 130 acres. My wife said okay, as long as we don't have to move out there. In 1942, I got the second one, 151 acres right next to it."

At the far end of the field, George asked if I was ready to go back. "Sure," I said. "Well, let me show you this first," he insisted.

I followed him through a thicket, which ended in twenty feet at a county road crossing the Red Oak on a solid bed of gravel.

"Look upstream." The creek shone blue against the deep shade under the few limbs near the forest, but upstream was a broad channel fit for swimming, though Langenberg said it was half-polluted from the upstream Owensville treatment plant and only a few inches deep.

"Now look down the other way." Several huge fallen trees crossed the creek, but the gravel seemed so thick that what water remained seemed stagnant pools.

"My neighbor asks me what to do, but there's so much to it," George said. "It sure doesn't hurt to have the county come down and take some of that gravel for the road. When it floods, it just goes all over his field. It's sad."

Walking back, we stopped to look over a sycamore cut on the far bank. "The neighbor over there won't mind, but you got to be careful when you cut timber on someone else. You see there's two sycamores there, but I just cut the one. That was enough. The creek was running too fast there. Putting that one tree in was just enough.

"There's one other man down this creek who works his creek bottom like I do. I've seen him walk along and study the creek, just watching it like it's an animal, something living. He learned the same as me, from his father. You've got to take time to see what the river's doing. I'll see where he's dropped one tree up along the bank. He could have cut others, but one's enough. He'll just take what he needs. You've got to study it, be interested in it. That's the secret to this whole thing."

The farmer who loves his creek—and knows how to care for it ...

Since completing the Ozark Rivers Oral History Project, I became fascinated with the Native American heritage of the Ozarks. There's much that the Native Americans can teach us about proper care for the land and its streams. They have understood that human life is bound to the region's grasses, trees, and creatures of land, water, and air—and that the region's soils and streams are part of that unified, living ecology. Langenberg understood this, too. The roots of the trees that line a creek bed preserve the soils that grow the crops that feed the farmer's family, etc. The Native Americans have long practiced an ecotechnology whose tools may have been simple but whose knowledge runs deep. That's the essence of any successful technology: It's not a tool alone but a practice—an understanding of how to use a tool when and where and why.

Did you notice where Mr. Langenberg points out several small mounds at the edge of his big bottomland field? Though tangential to the subject of managing the creek, I could tell they fascinated him, because he hadn't bothered them or dug into them. Mr. Langenberg's respect of the mounds has inspired my own respect for native sites, which may involve different cultures over time.

The main point George makes is that streams are alive and require deep study to understand. You can tell it's work that George Langenberg enjoyed. Maybe this doesn't grab you? To me it's fascinating, in part because the management ideas for a stream came from his father and grandfather. Langenberg brings his German heritage to Ozarks farming, but that heritage shares a little something with the Native American: Both traditions rest not in exploiting, but in caring for the land and its streams. He took the knowledge passed down to him through generations, developed his own keen eye, and, with a tool as "simple" as an axe or chainsaw, helped his stream take care of its own floodwater problems.

I tried to explain his basic ideas, and several hydrologists at the U.S. Geological Survey office in Rolla could see that I had taken care to let Mr. Langenberg tell his story. It's a story—indeed, a living tradition of successful ecotechnology—that the "modern" hydrologist could stand to relearn.

Transporting the Ozarks (II):
"The Frisco" Shops and Depot from MSU Special
Collections and the History Museum on the Square
Selected by Tracie Gieselman-Holthaus and Christopher Bono

Frisco Depot, Springfield, Missouri (1940s). *Domino Danzero Collection, MSU Special Collections and Archives.*

Machining operation on large gears for Tulsa wheel lathe, photo taken at West Shops (December 18, 1937). *Frisco Lab Photos Collection, MSU Special Collections and Archives.*

Engine room of West Shops' power plant (January 9, 1938). *Frisco Lab Photos Collection, MSU Special Collections and Archives.*

Aerial photo of Frisco shops and railyard (undated). *Business Collection, History Museum on the Square, Springfield, Missouri.*

Frisco Depot in disrepair (February 5, 1977). *Business Collection, History Museum on the Square, Springfield, Missouri.* ▲

Frisco Depot the morning after demolition (March 6, 1977). *Business Collection, History Museum on the Square, Springfield, Missouri.* ▶

Technology, Water, and Consequence:
Our Upstream Choices and Their Downstream Effects
Michael R. Kromrey and Loring Bullard

A pure and abundant supply of water is one of the necessities of every progressive community....

The water from Valley Mill spring sinks again into the earth, filters through clay and sandstone a distance of four miles, and reappears as Fulbright Spring. Artificial filters costing the water company $100,000 have been installed, so that it now has one of the finest filtration plants in the country. This, and other expenditures which are being made, came as a result of great faith in Springfield's future. The water supply, filtration system and pumping plant are adequate for a city of from sixty to seventy-five thousand population....

Chemical analyses which are made public show that the city water contains from 5 to 10 bacteria per cubic centimeter, while water containing 100 to 250 is considered satisfactory for drinking purposes.

No more valuable factor in the future prosperity and development of the city exists than the quality and quantity of Springfield city water.

—*Springfield Greets You* (1919)

Ebb and Flow

Water is life. From the very flow of our bodies to the systemic operations of our blue planet, every living thing requires it. Water's very existence is a mysterious cosmology and profound luck, and its chemistry is nothing short of molecular magic. It's bigger when it freezes. It can be found as a solid, liquid, or gas *simultaneously* on Earth: Even as I write this, out my window I can see an array of water forms—thank you, Missouri weather. Sometimes called the "universal solvent," water erodes, corrodes, dissolves, carries, and, as a result, often transports minerals, nutrients, wastes, and pollutants. And while its chemistry scaffolds biology's necessary processes, it also carries pathogenic danger: Even apparently pure water can transport unseen microbes and other pollutants. Through the ages, water has, directly and indirectly, caused a great deal of human suffering and death. Though water technologies have evolved to produce remarkable improvements in human health, newer technologies now threaten the progress that has been made. So ebbs and flows the tide of help and harm.

Technology as a Saving Grace

Our Ozarks community was founded upon the simplest form of water technologies. In his book, *Jordan Creek: Story of an Urban Stream* (2008), Loring Bullard describes the journey of John Polk Campbell, credited with founding Springfield. In 1830,

Campbell decided to settle at one particular spot along Jordan Creek because [he] found a peculiar geologic feature there, a natural well "of wonderful depth." The "well" was actually a crack in the rock, a vertical cave opening downward into the low bluff. Today we would call it a "karst window," where a person could look down and literally see into a spring's underground plumbing system.

All Campbell had to do was lower that terrific technological achievement, the water bucket. But, as the population of the city increased, sources like his natural well and the city's remarkable springs became polluted and unusable. Human wastes found their way into these sources, occasionally carrying a deadly load of microbes, including cholera and typhoid. During the height of a huge cholera epidemic in St. Louis in 1849, 700 people died in one week. In Springfield, some people who got sick from the disease in the morning were dead by nightfall.

Springfield's first public drinking water source was Fulbright Spring. It was chosen because springs were considered pure and healthful, their waters purified by "filtration through the subterranean strata." Because Fulbright Spring was a few miles north of the city, in a region "sparsely settled," it was also thought to be safe from pollution. Not long afterwards, we learned that the "subterranean strata" in the karst geology of southern Missouri are not good filters at all.

By the 1880s, public water supplies began using filtration of water through sand. This simple technology vastly reduced rates of waterborne disease by removing much of the sediment and bacterial loads of source waters. A few decades later, we began using disinfection with chemicals to kill any bacteria that remained in water after filtration. Together, these two technologies make public water supplies vastly safer than they had been in the 19th century. Even in rural Ozarks areas, where most people depend on wells for drinking water, technologies have improved dramatically. Common in the old days, hand-dug wells allowed direct access to surface water or very shallow groundwater, both of which were often polluted. Rotary well drilling, first introduced into oil fields in the 1890s, allowed drillers to sink wells deep into the hard bedrock common in Missouri. Protective casings were sealed into wells to prevent the entry of surface water and shallow groundwater. Bacterial testing of wells has shown that these technological advances really do improve the safety of home water supplies.

Unfortunately, even these basic technologies, *which we take for granted in the U.S.,* are not available for many of the world's residents. According to the United Nations website, "Sustainable Development Goals,"

Water scarcity, poor water quality and inadequate sanitation negatively impact food security, livelihood choices and educational opportunities for poor families across the world. At the current time, more than 2 billion people are living with the risk of reduced access to freshwater resources and by 2050, at least one in four people is likely to live in a country affected by chronic or recurring shortages of fresh water. Drought in specific afflicts some of the world's poorest countries, worsening hunger and malnutrition. ("Clean Water and Sanitation")

Technology as an Evolving Danger

We have grown dependent on our technologies to keep us safe, and we give little thought to the safety of the water coming from our taps, because, generally, we don't have to. This is certainly not the case in much of the world, but, even here, our water technologies require constant vigilance—testing, maintenance, and repair. And even though we have robust monitoring and management programs, there are always a few cracks in the armor. Consider Flint, Michigan, where human error caused water leaving the treatment plant to be slightly acidic and able to dissolve lead from old pipes. Or Milwaukee, Wisconsin, where some lapses in water treatment, in connection with historic flooding and pollution of source waters, combined to create catastrophe with cryptosporidium, a waterborne parasite that's the leading cause of waterborne disease for humans in the United States.

These are familiar problems and pathogens, but our racing technologies in chemical manufacturing, industrial processes, health care, and intensive agriculture have created a new mix—a veritable soup—of potential problems. More troubling, the extent of these problems has only been revealed recently with the development of sophisticated monitoring and testing technologies. This hits close to home, as we grapple with the pollution of local karst aquifers with trichloroethylene (TCE), a common industrial solvent once "disposed of" in shallow ponds. Further, many streams throughout the United States have been found to contain troublesome levels of pharmaceuticals, microplastics, and personal care products, some of them hormone-mimicking compounds, which can change male fish and frogs to females and do who-knows-what to people. These chemicals are not necessarily removed by standard wastewater treatment processes, and we don't have a particularly good sense of the scope of these emerging, technology-caused issues in the Ozarks.

Practices in agriculture can contribute to the problem. Farming technology has been rapidly changing since Earl "Rusty" Butz, Secretary of Agriculture under Presidents Nixon and Ford, unleashed a policy to support technological agriculture, urging farmers to "get big or get out" and plant "from fencerow to fencerow." Today, bigger machines and more chemicals are the norm: Note the ever-increasing use of Roundup and "Roundup-ready" crops—crops genetically modified to be herbicide-resistant—coupled with GPS navigated mega-equipment and drone-surveyed fields. While technology increases the efficiency of pesticide application, the overall use of and dependence on chemicals is also increasing. Agricultural chemicals contaminate water supplies in many areas. These chemicals are difficult to remove from water, requiring sophisticated and expensive technologies such as filtration through granulated activated carbon.

Industrial agriculture is also creeping into Ozark watersheds through Confined Animal Feeding Operations, or CAFOs. There are already over a billion chickens and turkeys in the White River Watershed, and a hog CAFO threatens the quality of the Buffalo River in Arkansas. Many CAFOs use pharmaceuticals to maintain the health of genetically similar animals, an infection in which could kill an entire flock or herd. Unfortunately, the therapeutic or disease-preventative use of antibiotics for confined animals can create antibi-

otic-resistant organisms, which can cause untreatable infections *in humans*.

Will Technology Save Us?

Technology, human effort, and financial investment built our complicated water infrastructure, the sources and pipes and treatment plants that produce and deliver our drinking water. But these systems need constant vigilance—routine maintenance, repair, and replacement. In its "2017 Infrastructure Report Card," the American Society of Civil Engineers gives the following data:

> Drinking water is delivered via one million miles of pipes across the country. Many of those pipes were laid in the early to mid-20th century with a lifespan of 75 to 100 years. The quality of drinking water in the United States remains high, but legacy and emerging contaminants continue to require close attention. While water consumption is down, there are still an estimated 240,000 water main breaks per year in the United States, wasting over two trillion gallons of treated drinking water. According to the American Water Works Association, an estimated $1 trillion is necessary to maintain and expand service to meet demands over the next 25 years. ("Drinking Water")

We know the problem and most of the solutions, but will we choose to invest in our collective future? It costs money, it requires voting for taxes, and it demands thinking long-term. Aging and deteriorating infrastructure were at the root of problems in both Milwaukee and Flint—examples that provide a cautionary tale in technology failing when we fail ourselves.

Source protection—the idea that it is much more cost-effective to treat pollution at its source or prevent pollution altogether—has come of age. We have the knowledge to accomplish this, if we can only find the collective will to act upon that knowledge. Being smarter, making far-sighted choices, and not developing new technologies to address each new or evolving problem is probably a better strategy to keep people safe in the long run.

Part of the answer will certainly be new technologies, because developing technology has always been one of humankind's most important ways of solving problems. But new technologies need to be more site-specific, more tailored, more efficient, more environmental- and user-friendly, and more beautiful and satisfying. Green infrastructure is one idea gaining traction in this regard. It is much better to let plants slow and filter urban runoff than to construct treatment basins, and the plants provide us with oxygen, shade, and beauty to boot. Better than dealing with runoff is to prevent it in the first place—more plants mean more infiltration, less runoff, and restoring or mimicking the pre-development hydrology of a site. Ironically, we find ourselves circling back to the original technology of natural filtration.

The United States has a tremendous advantage: With about 4% of the world's population, the U.S. has about 7% of the world's fresh water. China, in contrast, with about 20% of the world's population, has only about 5% of the world's fresh water. Will we waste and squander this tremendous advantage? Or will the United States become a world leader in water sustainability? Will we slip to a second- or third-rate power in the environmental realm, or will we continue to be a beacon to the world? Americans have always shown that when the going gets tough, the tough get going. We will just have to get going soon if we hope to address the ebb and flow of problems and solutions in the water world.

The question is not really will technology save us, but will we save ourselves?—which leads to a further question: What does a solution of this sort look like? For starters, it means providing high-quality, place-based environmental education for every student in the region. It looks like an actualized Ozark Greenways trail network—like a green ribbon connecting all citizens in our community and protecting all of our small waterways in perpetuity. It means making decisions *as a community* based on sustainable returns on investment—considering economic, environmental, and social costs of benefits for projects both in the present and for the lifespan of the project. It looks like a community rallying around a vision of renewable energy, acknowledging that adding to global carbon emissions affects the poorest and most marginalized humans the most. And it means getting ready for a rapidly changing climate by planning for bigger droughts and floods than we have experienced in the written history of our region. It means voting for taxes to fund far-sighted, thoughtful investments in our future, and voting for leaders who will help get us there. And it means not allowing technology to distract us from our most important endeavors, or their source.

Works Cited

Bullard, Loring. *Jordan Creek: Story of an Urban Stream.* Springfield, MO: Watershed Committee of the Ozarks, 2008.

"Clean Water and Sanitation." United Nations "Sustainable Development Goals." Retrieved 24 March 2019. <https://www.un.org/sustainabledevelopment/water-and-sanitation/>.

"Drinking Water." American Society of Civil Engineers "2017 Infrastructure Report Card." Retrieved 24 March 2019. <https://www.infrastructurereportcard.org/cat-item/drinking_water/>.

Springfield Greets You: Dedicated to Our Progressive Citizens of the Ozarks. Springfield, MO: Chamber of Commerce, 1919.

Like a "Museum of Civilization":

"Legacy" Technology in MSU Special Collections and the History Museum on the Square

Selected by Joan Hampton-Porter and Shannon Mawhiney; introduction by James S. Baumlin

BY THE END OF Year Fifteen there were three hundred people in the airport, and the Museum of Civilization filled the Skymiles Lounge. In former times, when the airport had had fewer people, Clark had worked all day at the details of survival; gathering firewood, hauling water to the restrooms to keep the toilets operational, participating in salvage operations in the abandoned town of Severn City, planting crops in the narrow fields along the runways, skinning deer. But there were many more people now, and Clark was older, and no one seemed to mind if he cared for the Museum all day. There seemed to be a limitless number of objects in the world that had no practical use but that people wanted to preserve: cell phones with their delicate buttons, iPads, Tyler's Nintendo console, a selection of laptops. There were a number of impractical shoes, stilettos mostly, beautiful and strange. There were three car engines in a row, cleaned and polished, a motorcycle composed mostly of gleaming chrome. Traders brought things for Clark sometimes, objects of no real value that they knew he would like: magazines and newspapers, a stamp collection, coins. There were the passports or the driver's licenses or sometimes the credit cards of people who had lived at the airport and then died. Clark kept impeccable records.

—Emily St. John Mandel, *Station Eleven: A Novel* (2014)

In *Station Eleven*, Emily St. John Mandel describes a world ravaged by viral pandemic killing 99% of the population at a swoop. With the sudden collapse of population comes the collapse of industry, of transportation and, perhaps most dire of all, of the power grid upon which modern technology depends. For no one survives with expertise in producing and distributing the oil and gas and electricity that powers today's technoculture—in effect, throwing survivors back into a pioneer, pre-modern state.[149]

---

149. The epigraph above is taken from *Station Eleven* (New York: Knopf, 2014), p. 258. Early in her novel, Mandel gives "a partial list" of what is lost. Her list leads us to reflect on the fragility of a culture premised in cheap, abundant supplies of electricity (and of the expertise needed to provide and maintain the same):

> No more diving into pools of chlorinated water lit green from below. No more ball games played out under floodlights. No more porch lights with moths fluttering on summer nights. No more trains running under the surface of cities on the dazzling power of the electric third rail. No more cities. No more films, except rarely, except with a generator drowning out half the dialogue, and only then for the first little while until the fuel for the generators ran out, because automobile gas goes stale after two or three years. Aviation gas lasts longer, but it was difficult to come by. No more screens shining in the half-light as people raise their phones above the crowd to take photographs of concert stages.... No more pharmaceuticals. No more certainty of surviving a scratch on one's hand, a cut on a finger while chopping vegetables for dinner, a dog bite. No more flight.... No more countries, all borders unmanned. No more fire departments, no more police. No more road maintenance or garbage pickup. No more spacecraft rising up from Cape Canaveral.... No more Internet. No more social media, no more scrolling through litanies of dreams and nervous hopes and photographs of lunches, cries for help and expressions of contentment and relationship-status updates with heart icons whole or broken, plans to meet up later, pleas, complaints, desires, pictures of babies dressed as bears or peppers for Halloween. No more reading and commenting on the lives of others, and in so doing, feeling slightly less alone in the room. No more avatars. (p. 31)

In the epigraph above, Clark Thompson—one of the few survivors of Georgia flu—has taken up residence in what had been an airport. It started as a whim: Early on, while survivors were still reeling from shock, Clark "placed his useless iPhone on the top shelf" of the Skymiles Lounge glass cabinet case (p. 254). The narrative continues:

> What else? Max had left on the last flight to Los Angeles, but his Amex card was still gathering dust on the counter of the Concourse B Mexican restaurant. Beside it, Lily Patterson's driver's license. Clark took these artifacts back to the Skymiles Lounge and laid them side by side under the glass. They looked insubstantial there, so he added his laptop, and this was the beginning of the Museum of Civilization. He mentioned it to no one, but when he came back a few hours later, someone had added another iPhone, a pair of five-inch red stiletto heels, and a snow globe.
>
> Clark had always been fond of beautiful objects, and in his present state of mind, all objects were beautiful. He stood by the case and found himself moved by every object he saw there, by the human enterprise each object had required. (pp. 254-55)

Clark's Museum grew to have two uses. One was to teach children born after the pandemic what the "old" modern world had been like. Another, closer to heart, was to preserve the physical beauty of objects that had lost their mechanical function. Having surrendered their purpose, they transformed into art.

Such is our own relationship with the artifacts and technologies of bygone ages: We marvel at their makers' ingenuity—at their combinations of metal and glass and Bakelite plastic. We admire their design, at times a heavy Art Deco, at times an elegant Art Nouveau. Studying them, we peer into worlds that preceded us.

Shelves of working recording instruments (2019).
MSU Special Collections and Archives.

Though more formal than Clark's, we have our own museums and archives for much the same purpose. This present photo album focuses on instruments of recording and projection. The instruments gathered in collections of the History Museum on the Square are largely "dead," perhaps repairable but otherwise functionless. We admire them, nonetheless, for the beauty adhering in the artifacts themselves. The "legacy" instruments gathered in MSU Special Collections and Archives are different, in that they remain functional—and vitally so, for the future of our increasingly data-driven culture.

The innovations informing Gutenberg's printing press (*ca.* 1440) remain among the greatest technological achievements of Western European culture. By their means, the written word was disseminated and preserved. As a technology, the moveable-type printing press has worked so well that books printed five hundred years ago remain legible, preserved in their contents. Assuming that they stave off fire, water, mold, worms, and neglect—the banes of old books—Gutenberg's legacies should last another five hundred years.

The *printed* word has been foundational to Western knowledge and culture. Now, as ink is being replaced by electrons and other modes of audio-visual-digital recording, the preservation of data has become less certain, paradoxically. In audio recording, consider the progression from wax cylinders to discs (of different material—from shellac to vinyl—and of different playing speeds, from 78 to 45 to $33\frac{1}{3}$ rpm) to magnetic wire reel to plastic magnetic tape to cassette tape to 8-track to CD to DVD, etc. Or consider the evolution of computer memory from punch cards to reel tape to floppy disks (of different sizes: $5\frac{1}{4}$" and $3\frac{1}{2}$") to flash drives to external hard drives to the "cloud." Of these, the "cloud" seems to offer a

means of storage/preservation superior to the printed text: Over time, we shall see. But compare the printed text—which, well cared for, lasts a millennium and more—to data stored on CD, whose life expectancy stands at a decade or so. In most media storage formats, data erosion is inevitable. Worse, the loss of functioning instruments has trapped much valuable information on cylinders, reel tapes, and floppies. To release and reuse this information, we need to keep all such "legacy" instruments alive. Otherwise, much of the work of previous decades will fall useless, like the battery-dead laptop in Clark Thompson's "Museum of Civilization."

A selection of recording and projection instruments used locally follows.

Keystone View Company Magic Lantern glass slide projector (after 1900). *Courtesy of MSU Special Collections and Archives.* Used for presentations at Missouri State Normal School; the Lantern remained in use through the late 1930s, even after the school purchased a motion picture projector (*ca.* 1919).

▲ Edison Ediphone dictation machine (after 1917). *Courtesy of the History Museum on the Square, Springfield, Missouri.* The Ediphone recorded sound onto hard wax cylinders.

▶ Right column, top to bottom: Webster-Chicago wire reel recorder (ca. 1945). Webster-Chicago wire reel recorder (1950s). Edison Voicewriter dictation machine (1950s). *Courtesy of the History Museum on the Square, Springfield, Missouri.*

II. From the Later Twentieth Century to the Present

"Fishing on Lake Springfield," power station in background (1979). *The Business Collection, History Museum on the Square, Springfield, Missouri.*

Radio Signals to Rocket Engines:
The Making of Camp Crowder
Mike O'Brien

From linking military units with electronic communications to bridging the space between Earth and the moon with rockets, one small Ozarks community played vital roles in two of the biggest endeavors of 20th century America.

Tucked in the southwest corner of the state, Neosho was a quiet farming center with about 5,000 residents in the early 1940s when the U.S. Army hurriedly built a training facility that suddenly swelled the local population more than tenfold. Then in the 1950s and '60s, Neosho earned the nickname "Spacetown USA," as the powerful engines that propelled U.S. rockets were developed, tested, and manufactured there.

"I don't know if there is any other little town in America that underwent the transformations that Neosho did," says local historian and author Kay Hively. "This was an isolated community that was descended upon by all sorts of people from all over the country—from all over the world, really."

The U.S. Army Signal Corps Arrives

Although Neoshoans weren't aware of it at first, the Army took aim at their neighborhood early in 1941 when impending war worries prompted a stealthy nationwide search for a suitable site to expand military capabilities in radio, teletype, and telephone networks. The Army's Signal Corps was headquartered at Fort Monmouth, New Jersey, but little room existed there for the major training facilities that war planners envisioned. Newton County had open land, plus good access to highways (U.S. 60 and 71) and railroads (Frisco and Kansas City Southern). Discreet surveying confirmed the choice.

In May 1941, it was publicly announced that the Army had targeted Neosho for a major cantonment (military camp). In mid-August, the Army was granted eminent domain over about 9,000 acres of farmland just south of town. In September the name was declared Camp Crowder, after the late Maj. Gen.

a.

Enoch Herbert Crowder, a native Missourian who as the Army's judge advocate general had authored the Selective Service Act of 1917 and overseen the first years of the draft it created.

The pace of development at Camp Crowder in the autumn of 1941 was dizzying, despite being hampered by October rainfall totaling more than nine inches, triple the usual amount. As the U.S. actually entered World War II following the December 7 attack on Pearl Harbor, some 20,000 construction and support workers were busy around the clock erecting the more than 700 buildings called for in initial plans. They paved miles of roadways, laid miles of water, gas, and sewer lines, and strung miles of electric and telephone lines. Streets were named after famous Missourians—Daniel Boone, Mark Twain, Neosho native Thomas Hart Benton, et al.—and the main entrance to the post was christened Lyon Gate after Gen. Nathaniel Lyon, the Union commander killed at the Civil War battle at Wilson's Creek near Springfield.

The first trainees began to arrive in February 1942. That same month, word was received from Washington that even more construction was authorized to accommodate an increase in Camp Crowder's population from the originally anticipated 35,000 to as many as 50,000 soldiers. The land appropriated by the Army also expanded to a sprawling 66,000 acres, one-sixth of Newton County. Eventually more than 1,500 buildings were constructed, including barracks, kitchens and mess halls, 16 chapels, a 700-bed hospital, 15 infirmaries, three dental clinics, scores of classrooms, six movie theaters, a 5,000-seat fieldhouse, and several gymnasiums, warehouses, vehicle service centers, and post exchange (PX) stores.

In addition to tens of thousands of servicemen, the post became home to hundreds of members of the Women's Army Auxiliary Corps (WAAC)—later simplified to Women's Army Corps (WAC)—who took on jobs ranging from clerks to truck drivers to surgical assistants. And in addition to civilian employees, the cantonment's workforce was augmented by captured German soldiers. About 2,000 POWs were confined at Camp Crowder over the course of the war, paid 80 cents per day to perform construction and housekeeping chores.

Camp Crowder's mission broadened during the war to include training for medical corpsmen, dental technicians, mechanics, cooks, and other needed skills. However, the chief objective of Camp Crowder remained training and turning out communication specialists. Recruits sent to Neosho were chosen for their aptitude to master the complexities of electronic communication—but they also were expected to be physically fit and effective soldiers, and therefore had to undergo the same basic training in fighting skills as other infantrymen.

C.

highest scores in our company of 200 men. Many of those people hadn't ever seen a gun before.... They had to repeat the course.

William T. Bluhm, in his published memoir *Signaling the French*, recounted a letter he wrote while training at Camp Crowder in 1943-44:

I spent last night at the infiltration course ... a stretch of land, ours being 100 yards long, that is supposed to simulate a battlefield. We crawl on our bellies while they fire live machine gun bullets over our heads. They use tracer bullets, and I could see the damn things flying around in the air. Every once in a while an explosion of a land mine would go off, and you feel the ground shake. I'll probably never need to use this experience, but we've got to be prepared for anything and everything.

"You may have erroneously gathered the impression that because you are in the radio school and are learning to become radio operators that you now are classified as non-combatants," sternly cautioned a Signal Corps handbook handed to trainees upon arrival at Crowder. "If you assume this attitude you are very wrong. Men, do not live in a fool's paradise. The specialized training we give you does not bring with it any immunity whatsoever from any of the hardships and peril that is the lot of all soldiers."

"They get you to obey orders without question," wrote Pvt. Russ Georgeson in a letter to family back home in rural Wisconsin in 1942:

A lot of close-order drill, and obstacle courses we run through. We started out with five-mile marches, then 10-, 15- and finally a 20-mile march with full pack and rifle. At the rifle range we had to qualify by shooting a certain score. They talked about the marksmanship of American farm boys. A fellow from northern Michigan and I shot the

After the first couple of weeks of intense basic indoctrination, recruits' days were divided—half devoted to continued infantry training, half to the curriculum of the Signal Corps Replacement Training Center (later renamed the Central Signal Corps School).

One of their first challenges was to learn the dit-and-dah language of Morse Code. Trainees spent hours with headphones clamped over their ears, listening to machine-generated Morse transmissions, translating the tones into letters and numbers in their heads, then transcribing them onto paper using typewriters.

Instructors monitored the students' progress. Some trainees quickly achieved the primary goal of being able to "copy" at least 15 words (75 characters) per minute, with 25 words (125 characters) per minute considered ideal proficiency. They were steered toward becoming radio operators, learning how to use a key to tap out the dits and dahs as well as to receive

Morse sent by others. The students also practiced receiving code visually via blinking signal lights.

Trainees who showed themselves to be more adept with a typewriter keyboard were trained as teletype operators, practicing until they could hammer out error-free text at a 35- or 40-word-per-minute clip. Those with mechanical aptitude were taught to maintain and repair the complicated teletype machines. Still others learned the intricacies of telephone switchboards, with the most athletic men assigned to be "pole monkeys" who clambered up and down utility poles to string and maintain phone lines.

In contrast to the modern communication technology, a special Signal Corps unit at Camp Crowder bred and trained carrier pigeons. Homing birds had been used for centuries to carry messages, and they continued to be considered a reliable tool through World War II. As many as 13,000 pigeons were housed at Camp Crowder. The birds were trained to swiftly find their way back to their home roosts. The best were said to be able to travel an astounding 600 miles in a single day under favorable wind conditions. Army pigeoneers boasted that 90% of the messages sent by birds during World War II reached the intended destinations.

d.

The most celebrated pigeon at Camp Crowder was Kaiser, a male captured from the Germans during World War I. He fathered more than 75 offspring for the Signal Corps before dying in 1949 at the record age of 32. Mounted by Smithsonian Institution taxidermists, Kaiser is enshrined on display there today. The Army closed down its homing pigeon units in 1957.

The main focus of high-tech communication training at Camp Crowder was radio, a technology then less than a half-century old. While commercial broadcast radio gained immense popularity with the general public during the 1930s, the Signal Corps had spent the latter half of that decade pushing development of gear for point-to-point communication.

Transmitters available for deployment in the early 1940s were crude and heavy by today's standards, employing vacuum tubes and requiring clumsy batteries or noisy generators for power in the field. Some were refrigerator-size behemoths weighing hundreds of pounds, capable of sending voice and Morse signals thousands of miles on shortwave bands. Others were shorter-range and more portable, designed to be carried in backpacks into battle so that commanders in the rear could direct troops on the front lines.

Trainees learned how to operate each type of radio on the airwaves and how to perform routine maintenance and simple repairs. They were taught procedures to relay precise target coordinates to bombers flying high overhead and artillery gunners miles away. They were schooled in techniques of communicating information effectively and efficiently to comrades while revealing minimal useful information to enemies who might be eavesdropping. They even learned how to destroy their radios with hammer, axe, grenade, or fire to keep usable gear from falling into enemy hands if overrun.

Recruits with foreign language skills or aptitude were keenly sought by the Signal Corps to facilitate communication with allied military personnel. Language classes and exercises were conducted at Camp Crowder in French, Russian, Norwegian, Greek, and Chinese.

A reported 181,155 Signal Corps trainees were sent from Camp Crowder to World War II battle zones, along with 31,255 Crowder-trained combat medics. While most of the trainees were Army infantrymen or airmen, the Signal Corps also admitted members of the Marines, Navy, and Coast Guard to the communication school.

Following the surrenders of Germany and Japan, in September 1945 Camp Crowder became one of the Army's "separation points" to process veterans out of their wartime duty into civilian life or to assist them in re-enlistment in the peacetime Army. The camp's role in that operation ceased at the end of that year after some 12,000 soldiers were processed.

In 1946, Camp Crowder's population dwindled. By summer the post was manned by only a skeleton crew. In January 1947, the Army announced it was vacating the cantonment, with more than 1,000 buildings to be auctioned for transfer or scrapping. Thousands of acres were offered for sale, with prior owners given first opportunity to repurchase their former land.

The Local Citizenry Responds

The news that Camp Crowder was being shuttered drew mixed reactions from residents of Neosho and surrounding communities. In the introduction to his book *From Camp Crowder to Crowder College*, Neosho historian and author Larry James put it this way:

> The construction of Camp Crowder had a greater effect on life in Neosho and Newton County than did any other single event.... Many individuals had an opportunity to work at good wages for the first time in many years.... When construction was completed, many opportunities presented themselves to citizens of the area. With the need

e.

for housing during the war years, families were able to rent spare bedrooms to supplement their incomes. Local businesses profited, and many jobs, for both men and women, opened at the camp.

Another student of Neosho history, Wes Franklin, agrees: "The impact Camp Crowder had on the Neosho community was both immediate and long-lasting. Neosho's population was multiplied many times almost overnight. Neosho's current water treatment plant is a Camp Crowder structure built for the camp. Many of the water lines around the edges of town were Camp Crowder pipes."

Adds Kay Hively: "The camp brought Neosho into the 20th century. The town had to grow up overnight. It was an amazing transformation. Lots of people who just knew farming and were struggling to get by suddenly were able to get jobs that paid well for the time."

However, Hively, whose book *Red Hot and Dusty: Tales of Camp Crowder* continues to sell thirty-six years after first publication, notes some negative aspects

as well: "Those citizens who were vacated from their homes and farms went both willingly and unwillingly. Feelings ran in every direction. For some, the removal was a scar that has never fully healed."

University of Missouri sociologist Lucille T. Kohler spent months living in and studying Neosho during the first two years of Camp Crowder's existence. In *Neosho, Missouri, Under the Impact of Army Camp Construction*, a 1944 report published by MU, Kohler observed that Neosho before 1941 "had escaped the full impact of technological and social change, and thus managed to prolong a sense of solidarity." However, when the Army moved in and began the condemnation of land for the building of the camp, "a still greater degree of solidarity resulted."

"Social, economic and historic bonds were intensified and became stronger for the people of the locality than any separative forces they may have experienced in the past," Kohler wrote. "Not only was the taking of

the land a threat to 'what belongs to us' but in a more profound sense to 'what we belong to.' The common interests of place and folk were disturbed."

Most of the acreage had been small farms, many held in the same family for generations. While some landowners were willing to give up their property after struggling through the Great Depression, many were bitter over being forced from their homes and disappointed at the appraised values. "Hostility became so active that when government appraisers appeared, some of them were met with hoes and shotguns," reported Kohler.

Compared to some other Ozarks communities, Neosho was a "low-wage town, anti-union, with male workers before the coming of the camp earning $17 to $20 and women $12 to $18 a week," according to the MU report. The construction boom and Camp Crowder's civilian jobs that followed doubled or tripled the wages for many residents.

However, the unprecedented inflow of money tempted exploitive members of the community to gouge. Prices for food, clothing, and other necessities were inflated. Kohler talked to a Neosho garage mechanic and his wife who were forced to move to a town 20 miles away when rent on their small house went from $8 to $35 per month. Two teachers returning for the reopening of school in September 1942 found their apartment rent more than doubled, from $17.50 to $37.50.

Accommodations grew tight. "Many of the strangers were forced into adaptive living arrangements in bunk houses, tents, and trailers," Kohler found. "A local couple in a four-room house rented a room to a family, and then a second to another family. When a third appeared, they were offered the garage, and it was accepted. Then two trailers asked for space in the yard, making a total of six families to one bathroom…. A housewife who arranged cots for 12 men in a 12-foot room was visited by other housewives who wished to acquire the technique or to marvel."

Reports criticizing fancy prices for primitive lodging—chicken coops, in some cases—began to appear in newspapers in larger cities in Missouri and surrounding states. Neosho civic leaders were embarrassed by the bad publicity. The city council passed an ordinance limiting annual rental rates to no more than 12% of the assessed valuation of the property. However, by then complaints had reached Washington, and the wartime Office of Price Administration sent a federal rent administrator to Neosho to keep a watchful lookout, according to Kohler.

Following the war, the torrent of outside money dried up. While the loss was bemoaned by those who had profited to one degree or another from the presence of Camp Crowder, other locals expressed relief. Traffic congestion eased. Instances of drunken driving, public drunkenness, and peace disturbance were greatly reduced. Gambling and prostitution waned.

The latter two vices had been particularly vexing to law enforcement agencies, and also to the Army. Maj. O. E. McKensey told a Neosho group that the rate of incidence of venereal diseases among Camp Crowder soldiers was 107 out of every 1,000 men who were tested in 1942. As the officer noted, VD cases spike in the summer months when "great numbers of 'tourist' streetwalkers move into the area in trailers," according to a report in the *Neosho Daily News*. McKensey said the Army no longer took a "hush-hush" approach to public discussion of the problem and was combating it proactively.

The newspaper published an editorial in the spring of 1945 applauding the cooperation that had developed between Camp Crowder authorities and Neosho police to control prostitution. However, the editorial warned, prostitution "continues to be the most serious problem of health and police officials," with cases of gonorrhea and syphilis regularly reported.

Meanwhile, Neosho School Superintendent R. W. Anderson complained in a speech at a meeting of the neighboring Monett Kiwanis Club that schoolgirls who attended dances and other activities at Camp Crowder USO clubs were being lured into "immoral activities," either by visions of romance or in return for money or other material rewards. Anderson said junior high school girls were the most susceptible, "because they are at an age when they are emotionally unstable and don't know what it is all about."

Joplin, 20 miles northeast of Neosho, was a popular weekend destination of soldiers on leave from Camp Crowder. The Newton County prosecuting attorney was quoted by MU's Kohler as lamenting the closure of Joplin's "red-light district" in 1942 because it had "relieved our problem here (in Neosho)." The prosecutor said he had been "working closely with my colleague in Joplin because when he cracks down there, some of the girls drift over here; and when I crack down here, some of ours show up there."

Despite enforcement efforts, prostitution and gambling persisted throughout the war years. In July 1945, the month before the surrender of Japan ended the war, the *Joplin Globe* reported that 100 women and girls were arrested by Joplin police in what the newspaper characterized as "the biggest drive against prostitution since construction of Camp Crowder." A third of those arrested were found to be infected with venereal diseases and were transferred to a hospital in St. Louis for treatment. The Newton County prosecutor showed Kohler a closet in his Neosho office filled with casino-quality gambling devices—dice tables, cards, poker chips, etc.—confiscated in raids on illegal clubs.

When the customer base of soldiers evaporated in the later 1940s, the professional gamblers and the imported prostitutes moved on, much to the relief of law-abiding local residents. However, traditional—and legal—romantic relationships also blossomed because of Camp Crowder.

"Several local girls married soldiers and then moved away to where the soldiers had come from," says Larry James. "Crowder took a lot of the women away from here." Other soldiers were visited in Neosho by sweethearts from their faraway hometowns, and some of those couples decided to marry before the men were shipped off to war.

Russ Georgeson, the farm boy from Wisconsin, was one. In an interview preserved at the Wisconsin State Military Museum, he recalled:

> Ruth came down and we just made the decision that we wanted to be married before I went away, at a chapel right on the Camp Crowder grounds. The chaplain had an assistant who was one of the witnesses, but he had to send for another witness because he needed two. He got a fellow who was working KP doing dishes because he could also play the organ. It was a small ceremony, just Ruth

and me, the chaplain and the two witnesses. I hadn't been in the service long enough to make any really good friends to have in the ceremony, so that was it. There was a lieutenant who was the platoon commander who was quite a nice fellow and generous. He let us use his car for the weekend. So we went into Joplin for our honeymoon.

A fictional couple who supposedly fell in love at Camp Crowder were Rob and Laura Petrie, the characters played by Dick Van Dyke and Mary Tyler Moore in the popular 1960s TV comedy series, *The Dick Van Dyke Show*. Both Van Dyke and the show's head writer, Carl Reiner, had been stationed at Camp Crowder during World War II. They worked the post into a couple of the show's scripts. According to the program's story line, Laura and Rob met when she visited Camp Crowder as a dancer in a USO troupe.

Camp Crowder also was the inspiration for "Camp Swampy," where perennial private Beetle Bailey is stationed in his long-running newspaper cartoon strip. Creator Mort Walker was at Camp Crowder in 1943. In 1950 he began producing the "Beetle Bailey" strip, which grew to be one of the most endearing and enduring cartoon features in newspaper history. Although Walker died in 2018, "Beetle Bailey" continues to appear in more than 1,000 papers daily, penned by Walker's two sons, Brian and Greg, who worked alongside him during his last years.

About the time Walker was launching his cartoon feature, the Army was considering reactivating Camp Crowder to meet military training needs of the Korean War. Early in 1951, it was announced that Camp Crowder was to be reopened as a reception and evaluation center for new recruits. Beginning in July of that year, fresh recruits spent about a week getting medical checkups, undergoing aptitude testing, and being issued uniforms and other essential gear before being sent on to other bases, where they would begin actual basic and advanced training. Over the next two years, about 100,000 men were processed through Camp Crowder.

In early 1953, the Army announced that Camp Crowder would become the site of a disciplinary barracks for soldiers found guilty in courts martial of having committed mostly minor crimes. Longtime Ozarks Congressman Dewey Short, whose 7th District territory included Newton County, used his influence as chair of the House Armed Services Committee to push for making Crowder a permanent training facility for military police. Short, a Republican, succeeded in getting the name changed from Camp to Fort Crowder—but when he was defeated by Democrat challenger Charlie Brown in 1956, the Army backed away from the plan to train MPs at the fort. The post was reduced to "standby" status and seemed on the verge of being abandoned altogether.

However, before he was ousted from Congress and from his seat on the Armed Services Committee, Rep. Short had steered the Air Force toward Neosho as it sought a site for rocket development.

W. Baisley   Professor von Braun   E. Wright   May 17, 1961

"Spacetown USA"

In 1956, ground was broken at Fort Crowder for construction of a $13 million factory to build and test rocket engines. The initial plan was for the Air Force to team with California-based manufacturer Aerojet, but by 1957 Aerojet had been pushed aside and the Air Force's new partner was Rockedyne, a division of North American Aviation.

Over the next dozen years, some 3,000 rocket engines were produced at what officially was designated Air Force Plant No. 65. Early production was focused on powering military missiles, such as the Thor, for delivery of weapons. However, when the space race got seriously underway in the 1960s, with the U.S. and U.S.S.R. vying to become the first nation to send men to the moon and back, design shifted to larger engines, including those for the pioneering Mercury and Gemini spaceflights. The crowning achievement was producing the powerful engines for the massive Saturn rocket that launched the Apollo missions to the moon.

Although much of the Rocketdyne work was cloaked in secrecy, Neosho became widely known as "Spacetown USA" during the 1960s. The town welcomed highly trained scientists, engineers, machinists, and other skilled workers needed to turn out the complex rocket hardware. High-level military brass, political leaders, and space industry luminaries such as Wernher von Braun visited Plant No. 65 to check on progress.

"When they tested those engines, you could hear the roar all over town," recalls Larry James. "Actually, you did more than just hear them—you felt them, too. Anything on your wall that wasn't fastened securely would vibrate off and fall to the floor."

The rocket engine manufacturing brought another infusion of money into Neosho. At its peak, Rocketdyne employed some 1,300 workers and was said to have pumped more than $30 million into the local economy annually.

h.

Rocketdyne operations wound down in the early 1970s, but the facilities were taken over by Teledyne, a high-tech conglomerate that manufactured turbine engines, among other things. In addition to building its own powerplants, Teledyne also won contracts to overhaul and upgrade military jet engines.

Then in the 1990s, Teledyne gave way to Sabreliner, another aviation engine manufacturer. Finally, from 2003 to 2015, Premier Turbines overhauled helicopter engines at Crowder, although on a much smaller scale than in the heyday of Rocketdyne.

The Local Citizenry Responds (Again)

With its "Spacetown" era now in its past, Neosho has settled into a less glamorous but still vibrant lifestyle. Counting about 12,000 residents today, the town's largest employer is the La-Z-Boy furniture manufacturer with 750 on the payroll, followed by Nutra-Blend animal feed producer (440 workers) and several smaller companies.

"Much of our growth continues to be the result of the legacy of Camp Crowder and the U.S. government's investment in infrastructure in Neosho," acknowledges Michael Franks, economic development leader for Neosho and Newton County, who points out that the modern 1,000-acre Neosho Industrial Park is located on what once was a portion of Camp Crowder, with roads, water, and wastewater treatment facilities retasked to industrial development.

Crowder College occupies another significant portion of the former military base. The two-year community college, founded in 1963, enrolls more than 5,000 students and has gained widespread notoriety for student projects demonstrating the efficiency and performance of solar-powered vehicles. And more than 4,000 acres of the former fort is devoted to a training facility for the Missouri National Guard.

"Neosho would not be the same community it is today if it had not been for Camp Crowder, and

Rocketdyne brought a lot of brilliant minds and skilled labor to town," says Wes Franklin. He agrees with Franks that those remarkable developments of the 20th century established a firm foundation for a prosperous 21st century. "Neosho is waving a big flag to businesses," he says. "We're set up. We have the location. We have the infrastructure. We're ready for you."

Sources

A Camera Trip through Camp Crowder, Missouri. St. Louis, MO: Everett Schneider Publishing Co., 1943.

Basic Notes. Camp Crowder, MO: U.S. Army Signal Corps Central Signal Corps School, 1943.

Bluhm, William T. *Signaling the French: Adventures of a World War II American Army Team.* Bloomington, IN: iUniverse, Inc., 2008. [Available for $15.95 from Amazon.com, or by calling iUniverse, 1-800-288-4677.]

Camp Crowder. St. Louis, MO: Southwestern Bell Telephone Company, 1943.

PROOF LOAD 3500 LBS.

PROOF LOAD 3500 LBS.

k.

Field Manual FM11-80: Signal Pigeon Company Handbook. Washington, DC: United States War Department, 1944.

Hively, Kay. *Red Hot and Dusty: Tales of Camp Crowder.* Neosho, MO: K. Hively, 1983. [Available postpaid for $10 from the author, 600 W. Hickory St., Neosho, MO 64850.]

James, Larry. *From Camp Crowder to Crowder College.* Neosho, MO: Newton County Historical Society, 2006. [Available postpaid for $19 from the Newton County Historical Society, P.O. Box 675, Neosho, MO 64850.]

Kohler, Lucille T. *Neosho, Missouri, Under the Impact of Army Camp Construction: A Dynamic Situation.* Columbia, MO: University of Missouri Press, 1944.

Low Speed Radio Operators Handbook. Camp Crowder, MO: U.S. Army Signal Corps Central Signal Corps Replacement Center, 1942.

"Neosho Man Describes Deplorable Conditions Here to Monett Club." *The Neosho (Mo.) Daily News,* Nov. 28, 1942.

Radio Operator's Ready Reference. Camp Crowder, MO: U.S. Army Signal Corps, 1942.

"Ramblings." *The Neosho (Mo.) Daily News,* Jan. 5, 1945.

In addition to the sources cited, in-person interviews were conducted with Neosho historians Kay Hively and Larry James, and with Lee Ann Murphy of the *Neosho Daily News* staff. Additionally, the author exchanged correspondence with Neosho historian Wes Franklin and with Michael Franks, chief executive officer of GRO Neosho.

Images

a. A few of the scores of two-story barracks rising out of the mud in the autumn of 1941 as farmland near Neosho is converted to the sprawling new U.S. Army training base named Camp Crowder. *Photo courtesy of Larry James.*

b. In an Army Signal Corps class, Camp Crowder trainees listen to Morse Code transmissions on headphones, translating the dits and dahs into letters and numerals and then transcribing with typewriters. *Photo courtesy of Larry James.*

c. Recruits learn how to climb poles and install telephone lines between them—a part of the curriculum of the U.S. Army Central Signal Corps School in the early 1940s. *Photo courtesy of Larry James.*

d. Camp Crowder trainees operate a field radio setup powered by a portable electric generator (at left) under the watchful eyes of two Army Signal Corps instructors. *Photo courtesy of the Missouri State Archives.*

e. A member of the Signal Corps' pigeon unit releases a homing pigeon, one of thousands of birds bred and trained at Camp Crowder to carry messages in war zones. *U.S. Army Signal Corps photo.*

f. An aerial view of Camp Crowder during its early days. Eventually, the post grew to cover 66,000 acres of Newton County outside Neosho, with some 1,500 buildings constructed in the 1940s. *Photo courtesy of the Missouri State Archives.*

g. Along with the tens of thousands of male soldiers and recruits, several hundred members of the Women's Army Corps were stationed at Camp Crowder during World War II. *Photo courtesy of Larry James.*

h. During a visit to Neosho in May of 1961, German rocket pioneer Wernher von Braun (center) discusses rocket engine manufacturing with Rocketdyne plant manager Earnest Wright (right) and W. J. Baisley. *Photo courtesy of Larry James.*

i. Engines destined to power space-bound rockets are built in Air Force Plant No. 65 on the Fort Crowder grounds in the early 1960s. The factory was a cooperative venture between the U.S. Air Force and Rocketdyne. *Photo courtesy of Larry James.*

j. A rocket engine is readied for firing on a test stand at Fort Crowder in the early 1960s. When such engines were tested, the roar was heard and tremors were felt throughout nearby Neosho. *Photo courtesy of Larry James.*

k. Five rocket engines built in Neosho jut from the first stage of a Saturn V rocket being assembled near the launch pad in Florida. Neosho's Rocketdyne engines powered the first and second stages of the Saturn V on all Apollo flights, including the missions to the moon. *Photo courtesy of Larry James.*

Broadcasting the Ozarks:

Media Images from MSU Special Collections and the History Museum on the Square

Selected by Tracie Gieselman-Holthaus

a.

b.

c.

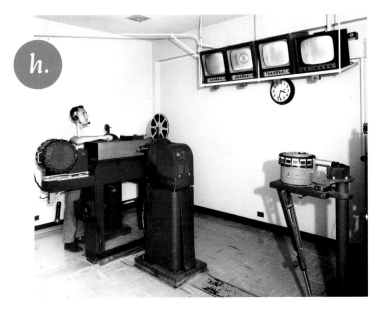

Images

a. KGBX Radio Station (*ca*. 1945). *Entertainment Collection, courtesy of the History Museum on the Square, Springfield, Missouri.*

b. KTTS Television Station (after 1953). *G. Pearson Ward Collection, courtesy of the History Museum on the Square, Springfield, Missouri.* In March 1953, KTTS-TV came on air as Springfield's first network station.

c. KYTV Television Station (after 1953). *Business Collection, courtesy of the History Museum on the Square, Springfield, Missouri.* In October 1953, NBC affiliate KYTV (now KY3) became Springfield's second network station.

d. Man standing by Zenith console (*ca*. 1960). *G. Pearson Ward Collection, courtesy of the History Museum on the Square, Springfield, Missouri.* The framed TV ads to his right show the CBS Saturday evening lineup: *Rawhide, Have Gun, Will Travel,* and *Gunsmoke.*

e. "Introduction to Literature," KTTS-TV program (1960s). *G. Pearson Ward Collection, courtesy of the History Museum on the Square, Springfield, Missouri.*

f. "Teen Dance Time" with Jess Wade (*ca*. 1960). *G. Pearson Ward Collection, courtesy of the History Museum on the Square, Springfield, Missouri.* KTTS-TV host Jess Wade presents a record player to the winning dance couple.

g. Max Baer by KTTS-TV camera (1969). *G. Pearson Ward Collection, courtesy of the History Museum on the Square, Springfield, Missouri.* During the Branson/Silver Dollar City filming of the *Beverly Hillbillies,* Max Baer (a.k.a. "Jethro Bodine") touts the show's sponsors.

h. Technologies on display in the KTTS-TV film room (1960s). *G. Pearson Ward Collection, courtesy of the History Museum on the Square, Springfield, Missouri.*

i. KOLR-10 Newsbeat broadcast with local news anchor, Mike Peters (1980s). *Ozark Public Television Collection, MSU Special Collections and Archives.* In 1971, Springfield's KTTS-TV changed its call letters to KOLR-10.

j. KOZK-TV studio recording equipment (*ca*. 1975). *Ozark Public Television Collection, MSU Special Collections and Archives.* First broadcast in 1975, KOZK is a PBS station (a.k.a. Ozarks Public Television) currently housed on the Missouri State campus.

k. Ozark Public Television crew member (1980s). *Ozark Public Television Collection, MSU Special Collections and Archives.*

Ozark Jubilee:

A Hillbilly Variety Show Breaks Into the Big Time on Live Nationally Broadcast Television
Thomas A. Peters

The hillbilly variety show form of professional entertainment—a mixture of country songs (including singing and playing music on instruments like the guitar, fiddle, and banjo), comedy routines, square dancing, and such matters—has been around since the days of vaudeville, and it continues to this day. The Ozarks' own Weaver Brothers and Elviry developed, expanded, and fine-tuned a hillbilly variety act that became very successful on the vaudeville circuit, including the East Coast of the U.S. and even Europe. From the 1920s and '30s onward, live hillbilly variety shows on stage and as radio broadcasts have been popular. The *National Barn Dance* in Chicago, the *Grand Ole Opry* in Nashville, *Korn's-a-Krackin'* in Springfield at the Shrine Mosque, the *Big D Jamboree* at the Sportatorium in Dallas, and the *Louisiana Hayride* in Shreveport are notable examples of hundreds of such shows that sprouted up from New York to California. The hillbilly variety show tradition has continued

through the *Hee Haw* television program in the late 1960s and early 1970s to the present-day live shows in Branson, Missouri, Pigeon Forge, Tennessee, and elsewhere.

While country music had been played, sung, and enjoyed for centuries at various places of the British Isles and the American upland South in a mode that could be characterized as amateur and diffused, the adoption and diffusion of phonograph records, and particularly of radio in the early 1920s created the technological conditions that enabled country music to emerge as a professional form of entertainment. Yes, country music as an "industry" is still less than one hundred years old.

In the early 1920s, Ralph D. Foster was a go-getting young man in Joplin, Missouri, south of Kansas City. His long career with radio began when he had the idea to start a small, low-powered radio station in an 8' x 12' corner of the tire store he and his partner oper-

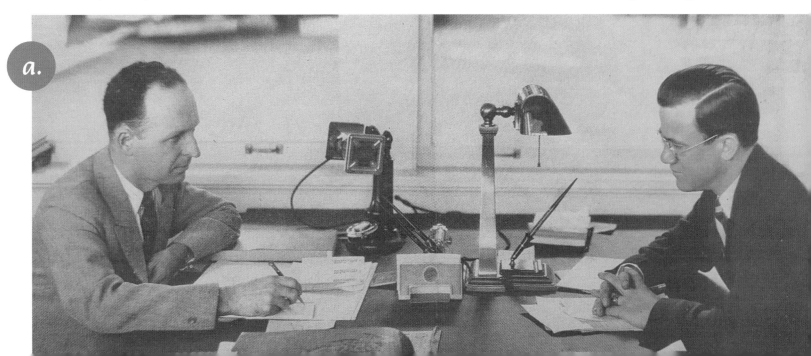

a.

ated; though a lark, the idea was to drum up more tire business. These were the early years of radio, when a listener's receiver set required arduous assembly, considerable tweaking, and was as likely to be run off of a bulky battery as an electrical outlet. Ralph's radio programs tended to be spontaneous and sporadic, often interrupted when a customer entered the tire store. To help fill the programming void, Ralph and Jerry eventually began to sing and perform other entertainments over the radio. Thus "the Rubber Twins" were born.

b.

In the early 1930s, Ralph decided that radio showed more business promise than selling tires, so he relocated to Springfield, Missouri, and doubled down on radio, operating both KGBX and KWTO in Springfield. This move—far from any competing metropolitan areas—was a stroke of entrepreneurial genius. In Missouri and surrounding states, most other cities of similar size had already established permanent commercial stations: Springfield was his for the taking. But *two* stations under one ownership

in the same media market was one station too many for the FCC. (In 1934, the Federal Communications Commission replaced the Federal Radio Commission in regulating market competition and radio frequencies.) Compelled in 1944 to break up his duopoly or "twinstick," Ralph decided to keep KWTO. He slowly built up a stable of local performers who would come into the studios at various times throughout the day to play live over the radio. Most of these programs lasted fifteen minutes, and often the performers played "for exposure"—that is, for low or no wages.

The impact of radio on the Ozarks has been tremendous. Receiving wireless radio broadcasts in homes in the hills and hollers of the Ozarks was a disruptive, transformative technology. Suddenly the world entered one's living room. With its down-home mix of news, weather, country music, farm reports, and sports, KWTO-AM 560 ("Keep Watching The Ozarks!") became an extremely popular and influential radio station in the Ozarks, but the emergence of national radio networks exposed Ozarkers to many other nationally broadcast shows, including the network portion of the weekly *Grand Ole Opry* show. KWTO had to compete with those other nationally broadcast hillbilly variety shows.

As radio entertainment developed into the 1940s, Ralph and his Springfield entrepreneur-business partner, Si Siman, decided to try producing a weekly hillbilly variety show that would be broadcast via KWTO one night a week. Then the stable of performers would tour the region on other nights of the week, if local organizations would sponsor an event in nearby towns, such as Marshfield, Mansfield, Monett, and Ava. They decided to call the program *Korn's-a-Krackin'*, a riff on a recently popular Broadway show called *Hell's-a-Poppin'*, which Ralph and Si had seen in Kansas City. The Weaver Brothers and Elviry, Bill Ring, Lennie Aleshire and Goo-Goo Rutledge, and many others performed on *Korn's-a-Krackin'* during its six-year run in the 1940s. For a time, the show was broadcast nationally over the Mutual Broadcasting System. Interest in the show eventually declined, in part because it lacked a nationally known master of ceremonies and headliner.

Television followed close on the heels of radio. By the late 1920s, television transmissions had been successfully demonstrated, but the onset of the Great Depression, followed by World War II, resulted in a long latency period between the development of television as a technology and its broad adoption and diffusion. By the late 1940s and early '50s, during the postwar boom years, radio networks had become strong, radio programs were quite sophisticated, and advertising revenues were substantial; but television was beginning to make cracks and dents in the world of radio. Forward-thinking radio executives like Foster and Siman in Springfield knew that television presented tremendous opportunities. The challenge was to develop and deliver a winning formula of compelling television programs. The hillbilly variety show offered promise.

Beginning in the late 1940s, various attempts were made to bring the hillbilly variety show to television. Most of the shows were short-lived and didn't draw large audiences. Michael Saffle (2011, p. 85) summarizes some of these early attempts:

> 1948 (summer): *Hayloft Hoedown*, broadcast by ABC in Philadelphia during the summer of 1948, was the very first in a long line of variety shows that foregrounded rural musical acts.
>
> 1948-1949 (December through July): *Saturday Night Jamboree* was broadcast live from New York City, which was a hotbed of interest in hillbilly music at that time, soon to be followed by folk music.

> 1948-1950: *Village Barn*, a square-dancing and audience-participation show emanating from a night spot in Greenwich Village in New York City, broadcast for two years, from May 1948 to May 1950.
>
> 1949: The *ABC Barn Dance*, modeled after the long-running network radio broadcast, National Barn Dance out of Chicago, lasted less than a year.
>
> 1950: *Windy City Jamboree* was broadcast on the Dumont Network, which existed as a network only from 1946 to 1955.
>
> 1951-1959: NBC broadcast *Midwestern Hayride*, also known as *Midwest Hayride*.
>
> 1953: *Old American Barn Dance* was broadcast on the Dumont Network.
>
> 1955: *Pee Wee King Show*. Pee Wee was a popular country crooner at that time. He may have been the first country and western performer to wear a nudie suit on stage. He was from a Polish-American family from Wisconsin.
>
> 1955-1956: The *Grand Ole Opry* was broadcast nationally by ABC-TV.

Once the technologies of any broadcast communication channel are made workable and economically scalable, the big challenge becomes finding, developing, and delivering lots of compelling content. This is true of the printing press, radio, television, websites and blogs, all the way to the present darlings of broadcast media, Facebook, Twitter, and Instagram. In the late 1940s, KWTO radio was trying to hone a winning formula of compelling entertainment content. They tried to hire big-name country music performers, but Ralph and Si had a difficult time retaining them for long. Many of them would be lured to the *Grand Ole Opry* or would move to other radio stations and other geographic markets when the demand for local appearances in the Ozarks began to wane.

One example of KWTO hiring a big-name act was the Carter Sisters, Mother Maybelle, and Chet Atkins, who were performing as a team in the late 1940s and early '50s. KWTO offered them a lucrative contract, so they all came to Springfield in 1949. Their 7 a.m. show was sponsored by Biederman's Furniture and their noon show by MFA Mutual Insurance Company. Their 3:45 p.m. show was known as Red Star Flour's "Cornfield Foll-ees Time!" According to an article in the November 1949 issue of the *KWTO Dial*, "Mother Maybelle, her three lovely daughters and guitarist Chet Adkins [sic] have more personality than Dale Carnegie, more fun than a cross-eyed pigeon on a merry-go-'round, and they've got music coming out of their ears!"[150] After less than a year at KWTO, this group left for Nashville and the *Grand Ole Opry*.

Foster and Siman soon had to confront the persistent, confounding problem that had challenged them for years: They needed an established country and western star to host and carry the show. They needed that star to thrive and stay in Springfield. They decided to approach Red Foley, who already was a big, established country music star. For years Red had hosted the Prince Albert Tobacco segment of the *Grand Ole Opry*, which was a very popular weekly national radio broadcast.

Red had started his career with John Lair's Cumberland Ridge Runners, playing the bass violin on the WLS *National Barn Dance* out of Chicago, the most successful of the early barn dance radio shows that cropped up around the nation, including such unlikely places as New York City and California. The *Big D Jamboree* in Dallas was, well, big, as was the *Louisiana Hayride* out of Shreveport. The *Grand Ole Opry* in Nashville eventually became the premier hillbilly vari-

150. Published monthly from 1941 through 1951, the *KWTO Dial* has been digitized by MSU Special Collections and Archives, using a print collection provided by local musicologist, Wayne Glenn.

RADIO HOME of KWTO.

This magnificent Radio Home of KWTO is one of the most complete and outstanding of its kind anywhere in the country. It is located right in the heart of Springfield, on U.S. Highway 66, just across the street from the Abou Ben Adhem Shrine Mosque, which has the largest auditorium in Southern Missouri.

c.

Central Studio

News Room of KWTO

Announcer's Studio

ety show, with instrumentalists, vocalists, comedians, and square dancers.

As Red's career began to take off in Chicago, he was paired with an entertainer named Lulu Belle to form a comedy team. For this particular act, Red's stage name was changed to Burrhead. It was not uncommon at that time for individual entertainers and groups to work under various stage names, often depending on the sponsor of the program. Lulu Belle and Burrhead were a big hit. Lulu Belle patterned her on-stage dress and demeanor after Elviry from the Weaver Brothers troupe, an amazingly successful hillbilly vaudeville show that emerged from Christian County, Missouri in prior decades.

Despite his on-stage success, Red had a few off-stage problems in the early 1950s. His second wife, Eva Overstake, who earlier had broken up the Lulu Belle and Burrhead duo because she was jealous of their onstage chemistry, committed suicide in Nashville. Also, the *Grand Ole Opry* management team did not like having big-time stars on their show. They preferred an ensemble cast. Red was becoming too popular. Thirdly, Red had a drinking problem. So, during a weekend visit from Si Siman, over a bottle of whisky at the Andrew Jackson Hotel in downtown Nashville, Red agreed to relocate with his third wife, Sally Sweet, to Springfield and host the *Ozark Jubilee*.

Thus the *Ozark Jubilee*, the first successful nationally televised weekly hillbilly variety show, was born and quickly rose to prominence.[151] It began as a nationally broadcast radio show over the ABC radio network, then added a local television broadcast by local TV station KYTV. Late in 1954, Si finally convinced the fledgling ABC television network to broadcast the show nationwide, beginning in January 1955. When

151. For recent discussions, see Rita Spears-Stewart (1993), Wayne Glenn (2005), Kaitlyn McConnell (2017), and Thomas A. Peters (2017). This present essay is part of the author's larger book project on the *Ozark Jubilee*.

d.

Red arrived in Springfield in April 1954, he stayed temporarily with the Siman family, until he could find a home for Sally and himself.

Everything seemed to be coming together, but a technical problem arose. Although KYTV could receive national broadcasts, they did not have the capacity to initiate and upload a live national network broadcast. As Si trenchantly observed, "We can suck, but we can't blow." As the network, the telephone company, and local technicians worked to solve this problem, during the first few months of 1955 the national broadcast originated from the studios of KOMU in Columbia, Missouri. For several months, every Saturday morning the entire *Ozark Jubilee* production team and performers would load onto a bus for the long ride to Columbia. They often would stop at Buford Foster's restaurant, the Night Hawk,

in Camdenton for breakfast. Buford, a local entrepreneur with his eyes open, quickly discerned that local square-dancing troupes, both the Lake of the Ozarks Bullfrogs (adults) and Tadpoles (kids), might be able to perform and excel on the television show.

The *Ozark Jubilee* was an amazing amalgam of local and national talent. It was one of the first nationally broadcast television programs of any type, if not the first, to solve a problem that vexed the new medium: While some shows appealed either to men, or to women, or to children, no show had proven it could appeal to all three groups. Early viewer studies of the *Ozark Jubilee* revealed that its audience was almost an equal mix of all three groups, and the average number of viewers per TV set was higher than many other shows.

Regional talent buoyed the *Ozark Jubilee*. All of the various square-dancing teams that performed on the *Jubilee* over the years—the Bullfrogs, the Tadpoles, the Promenaders, the Wagon Wheelers, and others—came from the Springfield and Lake of the Ozarks areas. Much of the musical and comedic talent consisted of people born in the greater Ozarks area, or who were residents of the Ozarks during the late 1950s during the show's run. They included Slim Wilson, Speedy Haworth, Smiley Burnette, Zed Tennis, Bill Ring, the Philharmonics, Luke Warmwater, Buster Fellows, Lennie Aleshire, Goo-Goo Rutledge, and many more. Some natives and current residents rose to national prominence through their performances on the *Ozark Jubilee*. These included Porter Wagoner and Brenda Lee.

Many nationally known country and western performers made guest appearances on the *Ozark Jubilee*, including the Carter Sisters, Ernest Tubb, Johnny Cash, Buck Owens, Minnie Pearl, Merle Travis, Patsy Cline, George Jones, and Cowboy Copas.

Over the years, the *Ozark Jubilee* changed its name (first in 1957 to *Country Music Jubilee*, then in 1958 to *Jubilee USA*) and its duration (ranging from 30 to 90 minutes), but it almost always had a Saturday nighttime slot. People would drive to Springfield from many surrounding states to be part of the live audience. When the demand for seats was particularly high, usually in the summer months, the *Ozark Jubilee* would run two shows on a given evening, one for national broadcast and one to entertain the overflow crowd. The production team and cast continued to put out live nationally broadcast television and radio programs week after week, with no summer hiatus, for nearly six years.

On the final, 297th *Ozark Jubilee* broadcast, on September 24, 1960, several interesting things occurred. First, Red Foley strongly encouraged viewers and listeners to write to their local stations, expressing their opinions that the *Ozark Jubilee* should not be canceled. Letter-writing was a tried and true method for gauging the size, demographics, and reactions of listening and viewing audiences, all the way back to the early days of radio in the 1920s and

'30s. The beauty of broadcast media, such as radio and television, was that they easily and quickly reached far-flung audiences that numbered in the millions. The problem for station management, program sponsors, and advertisers was that it was difficult to know the size, demographics, and reactions of these "invisible" audiences. The volume of mail received served as a barometer. Even at the final hour of the *Ozark Jubilee* in September 1960, Red was calling for a letter-writing campaign to try and save the show.

Another amazing aspect of that final show was that Ralph Foster sang a song, apparently the only time he had done so during all the years of KWTO, *Korn's-a-Krackin'*, and the *Ozark Jubilee*. By doing so, he harkened back to the early days of radio in St. Joseph in the '20s, when he was a member of the Rubber Twins, strummin' and singin' to entertain folks and to ply his trade in Firestone Tires.

References

Glenn, Wayne. 2005. "Jubilees, Clear Lakes, and Good Rockin' Tonight." In *Ozarks' Greatest Hits: A Photo History of Music in the Ozarks*, pp. 332-97. Springfield, MO: W. Glenn.

KWTO Dial. 1941-1951. Missouri State University Special Collections and Archives. Retrieved 30 March 2019. <https://digitalcollections.missouristate.edu/digital/collection/KWTO>.

McConnell, Kaitlyn. 2017. "Looking Back at the *Ozark Jubilee*." *Ozarks Alive!* March 19, 2017. Retrieved 30 March 2019. <http://www.ozarksalive.com/looking-back-ozark-jubilee/>.

Peters, Thomas A. 2017. "*Ozark Jubilee*: Country Music Springfield Style." *OzarksWatch: The Magazine of the Ozarks* Series 2, vol. 6. no. 1, pp. 37-44.

Saffle, Michael. 2011. "Rural Music on American Television, 1948-2010." *Music in Television: Channels of Listening*, ed. James Deaville, pp. 81-101. New York: Routledge.

Spears-Stewart, Rita. 1993. *Remembering the Ozark Jubilee*. Springfield, MO: Stewart, Dillbeck & White.

Images

a. "The Rubber Twins," Jerry Hall and Ralph Foster. *Photo from the KWTO Dial magazine (publ. 1941-1951), courtesy of Meyer Communications.*

b. *Korn's-a-Krackin'* performers. *Photo from the KWTO Dial magazine (publ. 1941-1951), courtesy of Meyer Communications.*

c. "Radio Home of KWTO" (*ca.* 1947). House on St. Louis converted by radio technologies. *Courtesy of the Bob Piland Collection.*

d. The Promenaders performing on stage at the Jewell (undated). *Courtesy of the Bob Piland Collection.*

e. KWTO program director Lou Black, Red Foley, Si Siman, and script writer Don Richardson on the set of *Ozark Jubilee* (*ca.* 1955). *Photo courtesy of Shirley Jean Haworth.*

f. Springfield's downtown Jewell Theatre, home of the *Ozark Jubilee* (*ca.* 1950). *Photo used with the permission of the* Springfield News-Leader.

Building the James River Dam and Lake Springfield (1956-1957):
Photos from the M. E. Gillioz Collection, History Museum on the Square
Selected by Tracie Gieselman-Holthaus

a. M. E. Gillioz (center), bridge and road builder, photographed at Fellows Lake and Dam construction site (1955). *Gillioz Family Collection, MSU Special Collections and Archives.*

b.

Dam construction on the James River (1956).
Courtesy of the History Museum on the Square, Springfield, Missouri.

c.

Dam and bridge construction on the James River (1956).
Courtesy of the History Museum on the Square, Springfield, Missouri.

d. Dam and bridge construction on the James River (1956).
Courtesy of the History Museum on the Square, Springfield, Missouri.

James River Dam and Bridge (*ca.* 1957).
Courtesy of the History Museum on the Square, Springfield, Missouri.

e.

Springfieldians at Work, 1950-1980:
An Album of Building, Buying, and Selling
Selected by Tracie Gieselman-Holthaus

133

f.

h.

g.

i.

j.

k.

135

Images

a. Colonial Baking Company (1950). *Business Collection, courtesy of the History Museum on the Square, Springfield, Missouri.*

b. Lily Tulip Cup Corporation (1951). *Business Collection, courtesy of the History Museum on the Square, Springfield, Missouri.*

c. Aerial view, Kraft Foods Springfield Plant (1953). *Business Collection, courtesy of the History Museum on the Square, Springfield, Missouri.* Home of Kraft Macaroni and Cheese and Velveeta.

d. Ash Grove Lime and Portland Cement Company in Galloway (undated). *Business Collection, courtesy of the History Museum on the Square, Springfield, Missouri.*

e. Ash Grove Lime Kiln in Galloway (1954). *Business Collection, courtesy of the History Museum on the Square, Springfield, Missouri.*

f. "Ramey's, Sunshine and Glenstone" (undated). *Business Collection, courtesy of the History Museum on the Square, Springfield, Missouri.*

g. Crank's Drug Store, Commercial Street (undated). *G. Pearson Ward Collection, courtesy of the History Museum on the Square, Springfield, Missouri.*

h. First floor, Heer's Department Store (undated). *Heer Collection, courtesy of the History Museum on the Square, Springfield, Missouri.*

i. "Everybody likes the Frigidaire. You will find them at Heer's" (undated). *Heer Collection, courtesy of the History Museum on the Square, Springfield, Missouri.*

j. Hiland Dairy food processing plant (undated). *Business Collection, courtesy of the History Museum on the Square, Springfield, Missouri.*

k. Processing milk at Hiland Dairy (undated). *Business Collection, courtesy of the History Museum on the Square, Springfield, Missouri.*

Stargazing in the Ozarks:

Images from Baker Observatory

William B. Edgar and Robert S. Patterson, with James S. Baumlin

From our home on Earth, we look out into the distances and strive to imagine the sort of world into which we are born. Today we have reached far out into space. Our immediate neighborhood we know rather intimately. But with increasing distance our knowledge fades, and fades rapidly, until at the last dim horizon we search among ghostly errors of observations for landmarks that are scarcely more substantial. The search will continue. The urge is older than history. It is not satisfied and it will not be suppressed.

—Edwin P. Hubble, "Not to Be Served But to Serve" (1953)

Edwin Hubble, by his inspired use of the largest telescope of his time—the 100-inch reflector of the Mount Wilson Observatory—revolutionized our knowledge of the size, structure, and properties of the Universe. He thus became the outstanding leader in the observational approach to cosmology, as contrasted with the previous work that involved much philosophical speculation. Although he regarded himself primarily as an observational astronomer assembling and analyzing empirical data, Hubble, from the very beginning of his astronomical career, took the widest possible view of the relationship of his investigations to the general field of cosmology. Indeed, Edwin Hubble advanced the astronomical horizon on the universe by steps relatively as large in his time as those taken by Galileo in his studies of the solar system.

—N. U. Mayall, "Edwin P. Hubble: A Biographical Memoir" (1970)

It's in astronomy, arguably, and in astrophysics that the Ozarks has made its greatest contributions to science. We're talking about the life and work of Edwin P. Hubble (1889-1953), for whom NASA named its first orbiting observatory—the Hubble Telescope ("About the Hubble Telescope"). In a fitting tribute, a replica of the telescope sits in front of the courthouse in his hometown of Marshfield, Missouri ("Take a Trip").

One of Hubble's contributions was his determination of the distance to Messier 31 (M31), the Great Spiral Nebula in Andromeda. Before Hubble's time, this cloudy-looking object (hence *nebula*, Latin word for "cloud") was thought to be a component of our own local system of stars, gas, and dust—the Milky Way. In fact, everything in the sky was thought to be part of the Milky Way. This nebula can be seen with the unaided eye during moonless fall evenings. But a small telescope reveals a curious spiral structure.

In the 1920s, after his appointment to Mount Wilson Observatory in southern California, Hubble used the observatory's 100-inch (2.5-meter) reflecting telescope to measure the distance to M31. He would achieve this by finding the distances to objects believed to be contained within the Nebula itself, among which were the "standard candles" known as Cepheid variable stars. The premise is simple: If you know how bright an object truly is, then you can calculate how far away it must be to look as dim as it does.

Henrietta Leavitt had recently shown that there was a way to learn the true brightness of a Cepheid. It turned out that the longer the period of brightness change, the greater the intrinsic, or true, brightness. Once the so-called "Period-Luminosity Relation" was calibrated for the Cepheids, it was just a matter of Hubble finding the period of light variation of several Cepheids seen within M31 to lead to its distance.

In order to find the periods of the Cepheids, Hubble took many photographs, over time, of the outer reaches of M31's spiral form. From these he was able to graph the brightness change of a Cepheid and mark off the interval of time between successive peaks in brightness. Hubble had to make long exposures on the primitive photographic plates used in the early 20th century. Coupled with the requisite darkroom work, this made for a long, laborious project.

Although the Cepheid Period-Luminosity Relation was not well calibrated at the time, Hubble was still able to show that the distance to M31 was immense … millions of light-years! With this result, Hubble was able to deduce that M31 is a galaxy in its own right and not some local component of the Milky Way. By determining the distance to M31, Hubble was able to establish that *galaxies* permeate the universe, separated by vast gaps of empty space. Thus, he elevated M31 to the status of the Great Spiral *Galaxy* in Andromeda.[152]

M31 (Andromeda Galaxy), using the Baker Observatory 8" Schmidt Camera. *Courtesy of Robert S. Patterson and MSU Department of Physics, Astronomy, and Materials Science.*

✳ ✳ ✳

At Missouri State University's Baker Observatory, we continue Hubble's work, studying Cepheid variable stars using some of the latest technology. In astronomy these days, CCD detectors are used to record data and images. These CCDs are similar to the detectors used in smartphone cameras but manufactured to much more stringent requirements. When used to record the brightness of stars, they are much faster, more stable, and more accurate than traditional astronomical photographic techniques. This is especially true for the photographic methods Hubble used a century ago.

At Mount Wilson, Hubble used the 2.5-meter telescope to make photographic exposures of one-half to one hour or so. Today at Baker Observatory, we can use a telescope five times smaller in diameter to record stellar images as accurately as Hubble did, but in just one half-minute. In addition, the data are already digitized and ready for computer analysis, without hours of laborious darkroom processing.

152. Following this success, Hubble turned his attention to a related problem. Astronomers had long noticed that the light from these nebulae was redder than it "should" be. This was probably caused by the stars' motion away from the Earth, much like the sound of a siren dropping sharply once the siren's source goes past a listener (Lemonick, p. 126). As Michael Lemonick writes, "Hubble and his assistant, Milton Humason, began measuring the distances to these receding stars and found what came to be known as Hubble's constant: The farther a galaxy is from Earth, the faster it is racing away. This means that the universe as a whole is rapidly expanding" (p. 126).

This discovery provided verification of Einstein's General Theory of Relativity. Published in 1915, Einstein's theory predicted that such expansion or a contraction was occurring, much to the dislike of Einstein himself, who had "uglied up" his theory with a "cosmological constant" so as to predict a stable universe (Lemonick, p. 126). Now this was no longer necessary: General Relativity had been right after all. In gratitude, Einstein visited Hubble at the Wilson Observatory, making Hubble a popular science superstar (Lemonick, p. 131).

One type of star that has been studied at Baker Observatory is known as a "yellow supergiant" (YSG). These have similar properties to the Cepheids that Hubble used to find the distance to M31. They have about the same surface temperatures as do the Cepheids and are just as large in diameter, but they aren't pulsating and, therefore, aren't changing in brightness. These are called non-variable YSGs.

To investigate why the non-variable YSG didn't seem to be pulsating, nearly one hundred stars were selected for high-precision CCD photometry at Baker Observatory. Some of these "program stars" were observed frequently over a few weeks, while others were observed now and then over several years. Although about 70% of the target stars didn't show significant brightness variation, about 30% did show low-level brightness change of less than 0.03 magnitude.

Comparison with the theory of stellar evolution for stars of this type indicates the most likely reason for the occurrence of the non-variable YSG. During the course of their development, these stars go through episodes of stability, when their pulsations die down temporarily, leaving them non-variable. Later as they age, they begin to pulsate, changing brightness as they once again become Cepheid variable stars. Through this and other projects, we add to the knowledge given us by our great fellow Ozarker.

Here in the Ozarks, it's important that we keep Hubble's legacy alive. And we have the means, fortunately, in Baker Observatory. Established in 1977, the William G. and Retha Stone Baker Observatory has had its telescopic eyes on the cosmos ever since. The observatory has grown into a functional lab and teaching facility, featuring three permanently mounted telescopes: a brand-new 0.5-meter (20") diameter Corrected Dall-Kirkham Astrograph telescope from

b. Baker Observatory by night. *Courtesy of Robert S. Patterson and MSU Department of Physics, Astronomy, and Materials Science.*

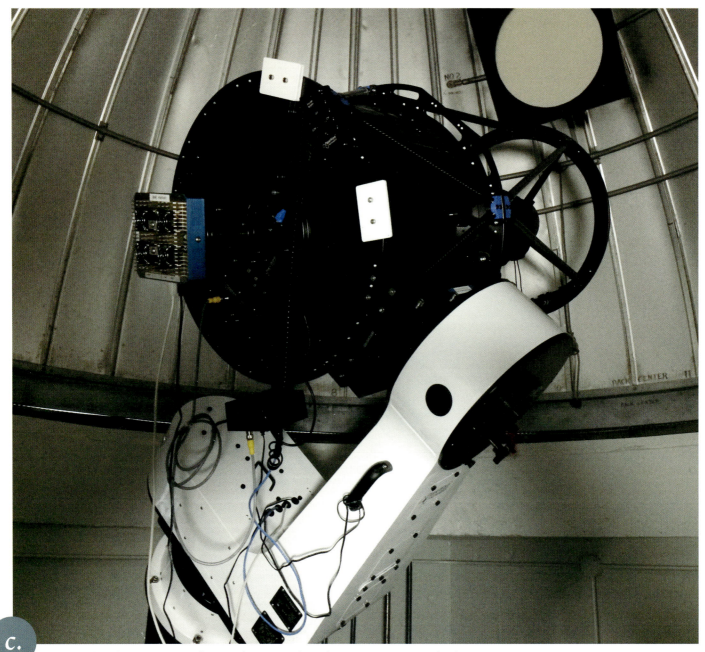

c. Robotic Schmidt-Cassegrain reflector telescope, Baker Observatory. *Courtesy of Robert S. Patterson and MSU Department of Physics, Astronomy, and Materials Science.*

PlaneWave Instruments; a 0.4-meter (16") diameter robotic Schmidt-Cassegrain reflector; and a 0.36-meter (14") diameter Schmidt-Cassegrain hybrid. The first two telescopes are under complete computer control. Although the 0.5-meter is operated by humans, the 0.4-meter is autonomous and runs itself. All three telescopes are used for faculty and advanced undergraduate student research.

A unique pleasure of the Ozarks "deep countryside" is its clear view of the evening sky: Get far enough away from urban light pollution and, behold!—the Milky Way swims into view. An urban-dwelling third of the world's population can no longer see the Milky Way with the naked eye. This is true in urban Greene County, certainly: Having grown up in an age of light pollution, most city-dwelling youngsters (and even some adults) will have heard of the Milky Way while never seeing it for themselves.

So, if you're interested in doing some high-powered stargazing, take a night trip to Webster County, site of Baker Observatory. Public observing nights are scheduled throughout the year. For directions and special instructions (where to park, what to bring, what to wear), visit the observatory website ("Public Observing Night").

M42 (Orion Nebula), using the Baker Observatory 8" Schmidt Camera. *Courtesy of Robert S. Patterson and MSU Department of Physics, Astronomy, and Materials Science.*

References

"Baker Observatory." MSU Physics, Astronomy and Materials Science Department. Retrieved 28 April 2019. <https://physics.missouristate.edu/BakerObservatory.htm>.

Lemonick, Michael. "The Century's Greatest Minds: Scientists and Thinkers." *Time Magazine*, March 29, 1999, pp. 124-131. Located in the Edwin Hubble Information File at the Library Center's Local History Collection, Springfield Greene County Library District.

Mayall, N. U. "Edwin Powell Hubble, 1889 to 1953: A Biographical Memoir." *Biographical Memoirs*, Vol. XLI. New York: Columbia University Press, 1970, pp. 175-194. Located in the Edwin Hubble Information File at the Library Center's Local History Collection, Springfield Greene County Library District.

NASA. "About the Hubble Space Telescope." NASA Scientific Visualization Studio. Retrieved 28 April 2019. <https://svs.gsfc.nasa.gov/Gallery/index.html>.

"Not to Be Served but to Serve: Hall of Famous Missourians." Located in the Edwin Hubble Collection at the Library Center's Local History Collection, Springfield-Greene County Library District.

"Public Observing Night at Baker Observatory." MSU Physics, Astronomy and Materials Science Department. Retrieved 28 April 2019. <https://physics.missouristate.edu/public-observing-night.htm>.

"Take a Trip." Retrieved 2 April 2019. <http://takemytrip.com/2016/08/marshfield-missouri-hometown-of-hubble-telescope-namesake/>.

Innovating in the Ozarks:
Springfield's Downtown IDEA Commons and the Model of
Collaborative Innovation and Entrepreneurship
Allen D. Kunkel

From the 1920s to the turn of the century, the Missouri Farmers Association milling facility presented the rural Midwest with supplies that helped feed America. Seven years after its doors were closed, the MFA building was brought back to life in the form of Jordan Valley Innovation Center. Livestock feed was once shipped from the middle of downtown Springfield. Today, cutting-edge technology and innovative collaboration are a major export.

—Jordan Valley Innovation Center, "About" (2019)

Individuals and organizations, through a combination of vision, commitment, relationships, and risk, have transformed the Ozarks. As demonstrated throughout this book, this combination will continue to spur innovation and technologies for decades to come. One bright example of this is Missouri State University's IDEA Commons, an innovation park developed in downtown Springfield. A plan that began in 2007, IDEA Commons is a district where ideas thrive through the intersection of Innovation, Design, Entrepreneurship, and the Arts. The vision for this district is to stimulate innovation and technologies that will create jobs, provide educational opportunities and, ultimately, change the world. Among other services, the district provides office and lab space to private businesses that, collaborating with university faculty, staff, and students, aim to create, accelerate, and advance new technologies and startups.

The IDEA Commons district began with the devel-

opment of the Roy Blunt Jordan Valley Innovation Center (JVIC), which opened in 2007. The vision for JVIC began through a relationship between Missouri State University and Brewer Science, a local micro-electronics manufacturer with a global outreach; this collaboration continues today, some twelve years later. Showing commitment and a tolerance of risk—useful qualities in 21st century entrepreneurship—the university turned a vacant and dilapidated building, the MFA Feed Mill on historic Boonville Street, into a state-of-the-art research and technology commercialization facility. Private companies, known as Corporate Affiliates, occupy about half of JVIC. The university's two research centers take up the other half: These are the MSU Center for Applied Science and Engineering (CASE), which is "committed to the development and support of advanced materials research," and the Center for Biomedical and Life Sciences (CBLS), "committed to the development and support of advanced biotechnology industries" ("About").

The JVIC model was created to facilitate local partnerships in business and academics: Brewer Science and a number of other companies had long sought to collaborate with a university in research and product development, but always found it difficult to navigate typical higher education policies, particularly regarding patents and "intellectual property." Missouri State University, through its forward-thinking leadership, developed a model wherein the companies would own and control the intellectual property developed in collaboration with JVIC. This one inno-

a.

Springfield MFA Feed Mill, Boonville Street (1940). *Courtesy of the History Museum on the Square, Springfield, Missouri.*

b.

Renovated JVIC building (2016). *MSU Photographic Services, courtesy of Allen D. Kunkel.*

vative aspect of the JVIC model opened the door for business partners to develop new technologies and products in a collaborative "technology accelerator" model, yielding medical devices, pharmaceuticals, electronic sensors, nanotech devices, composite aircraft components, and numerous other products.[153] University staff and faculty contribute to this collaborative research. And students contribute, as well: By their participation, students are provided unique opportunities for interdisciplinary experiential learning by working on real-world projects in a corporate lab environment.

153. Partnerships are always changing. As of this writing, the current JVIC team includes Brewer Science and Mercy Research as Senior Corporate Affiliates; Springfield Innovation, Inc., Stratum Nutrition, and the MSU College of Natural and Applied Sciences as Associated Organizations; and Clinvest Research, Dynamic DNA Laboratories, ElectroCore, National Institutes of Health, IDF, sHEALy Wellness, Physical Sciences Inc., Tuning Element, National Institutes of Health, and Engineer Research and Development Center as Industry and Government Partners ("Partners").

The IDEA Commons innovation district is currently anchored by three university properties: JVIC, Brick City, and the Robert W. Plaster Free Enterprise Center. Shortly after JVIC opened in 2007, the university began to occupy space in a private renovation project on West Mill Street, known as Brick City. This development, which is now owned by the university, is home to the university's Art and Design Department, as well as to the UMKC PharmD program and private advertising firm The Marlin Company. Innovation and new technologies are also influencing the work at Brick City. For example, new technologies in digital software and electronic advertising are driving the work that The Marlin Company provides to Fortune 500 companies. In addition, the students in Brick City embrace innovation and technology in their digital art and design projects.

c.

Buildings on West Mill Street, prior to renovation (undated). *Historic Springfield Collection, MSU Special Collections and Archives.*

d.

Renovated Brick City Art and Design Buildings (2016). *MSU Photographic Services, courtesy of Allen D. Kunkel.*

In 2013, the university opened the Robert W. Plaster Free Enterprise Center, with the eFactory as a primary anchor of the facility. The eFactory is a technology-focused business incubator aimed at assisting startup companies and serving as the front door to the Springfield region's entrepreneurial ecosystem. This building on North Jefferson Avenue—formerly a turkey processing facility—is also home to the Cooperative Engineering program, supporting civil, electrical, and mechanical engineering students.

The eFactory has continued to develop numerous programs to serve the entrepreneurial technology community. These include an accelerator program that makes equity investments in new technology startups. In addition, the eFactory supports a number of technology-focused community organizations to leverage the growing network of local entrepreneurs. Finally, the eFactory hosts the Greater Ozarks Centers for Professional Studies (GO CAPS), a multi-track program for high school juniors and seniors who are interested in either Business and Entrepreneurship or Information Technology and Software Solutions. We

expect that this future generation of technology leaders will make life-changing impacts on the Ozarks.

The university's existing assets in IDEA Commons—JVIC, the eFactory, and Brick City—have created excitement for the future of innovation and technology development in the Ozarks. The success of the IDEA Commons vision has evolved into the next phase of development, which will also have lasting impacts. This next phase of development will begin with an expansion of the eFactory and the Cooperative Engineering Program at the Robert W. Plaster Free Enterprise Center. This expansion includes additional startup space for the eFactory incubator, enabling it to meet current demands. The eFactory business incubator has been fully occupied for over a year and has a waiting list of clients. This expansion will provide the additional capacity to meet the future technology development needs of startup companies focused on medical, advanced manufacturing, logistics, and related technologies. In addition, the project will include an expansion of the mechanical engineering program.

Robert W. Plaster Free Enterprise Center (2016). *MSU Photographic Services, courtesy of Allen D. Kunkel.*
Home of the IDEA Commons eFactory.

Beyond the eFactory expansion, the larger overall IDEA Commons expansion project will be a transformational project for downtown Springfield, with long-lasting impacts on technology development. This project includes a major expansion of JVIC, whose existing facility is near capacity; continued demand will support new research and technology development through additional corporate affiliates. In addition to JVIC, the IDEA Commons expansion also involves a private, class A office building, which will cater to technology companies that want to be located in downtown Springfield and near young professionals and college students. A major part of technology development is being able to attract and retain talent needed to support technology companies. The IDEA Commons expansion project will also include supporting uses, such as open green space and a parking garage. With the completion of these projects, downtown Springfield will be a greener, more accessible, more attractive, and more exciting place to work, visit, and live.

The IDEA Commons expansion project has also renewed discussion to "daylight" Jordan Creek, which currently is located underground in a box culvert traversing six blocks through downtown Springfield. In preparation for this major project, which is being led by the City of Springfield, the university has secured additional properties in the area that will allow a continued long-term expansion of the IDEA Commons innovation park. Although there are no current redevelopment plans for these properties, it is assured that they will support the development of technology and innovation.

In reflecting on the past twelve years since the vision for IDEA Commons was developed, one can identify the major cause of its success. Yes, the project

f.

Analyzing materials in the cleanroom, JVIC (2016).
Starboard & Port Photography, courtesy of Allen D. Kunkel.

g.

Biomedical research, JVIC (2016).
Starboard & Port Photography, courtesy of Allen D. Kunkel.

grew through a combination of vision, commitment, relationships, and risk. But it's the JVIC model—the innovative partnering of business and academics in a modernized, shared urban space—that made the project work. IDEA Commons rests in a unique community collaboration among Missouri State University, the Springfield Area Chamber of Commerce, the Springfield Business Development Corporation, the City of Springfield, City Utilities, and the local business community. IDEA Commons is a place for ideas—for collisions between scientists and artists and entrepreneurs and other creative talent. As IDEA

Commons grows, no one knows what the results will be; but readers can rest assured that there will be continued significant impacts on innovation and technology in the Ozarks.

References

"About." Jordan Valley Innovation Center, Missouri State University. Retrieved 6 April 2019. <https://jvic.missouristate.edu/about/>.

"Partners." Jordan Valley Innovation Center, Missouri State University. Retrieved 6 April 2019. <https://jvic.missouristate.edu/partners>.

Building Businesses and Building Community, Ozarks-Style:

The O'Reilly, Herschend, Morris, Darr, and Askinosie Families

James S. Baumlin and Thomas A. Peters

Invention is what happens when you translate a thought into a thing.... Innovation is what happens afterward.

> —Pagan Kennedy, *Inventology* (2016)

Each generation of Ozarkians has had its great men and women: In the years of early settlement, there's John Polk Campbell and his wife, Louisa Cheairs Campbell; in the years after Civil War, there's Sempronius H. Boyd, who helped bring the railroad into town; for *fin de siècle* culture, there's Rose O'Neill (of Kewpie fame) and R. Ritchie Robertson (Springfield's "Star-Spangled Scotsman," who developed the local public music programs); during that spectacular decade of the 1920s, there's John T. Woodruff, who brought America's "Mother Road," Route 66, through downtown Springfield; in the decades before, during, and after World War II, Ralph D. Foster entertained us with radio; in the 1950s, Foster joined Si Siman in entertaining us with television. Each generation enabled the next, building infrastructure, attracting investments, innovating in technologies and business models.

Writing in 2019—already two-decades deep into the 21st century—we find the present moment rooted in the past. For today's Springfield was fashioned largely in the postwar years of the 20th century. It was in the '50s, '60s, and '70s that it grew from a town into a small city. In our commerce, communication, transportation, structures of economy, entertainments, and urban-social environment, we look to the future, even as the past surrounds us with reminders. In many respects, the family businesses that helped build late 20th-century Springfield-Ozarks continue their efforts into the 21st century.

In acknowledging recent contributions of local entrepreneurs, we know others worthy of mention. There's David Glass, one-time CEO of Walmart and owner of the Kansas City Royals; there's Jack Stack of SRC Holdings (formerly Springfield Remanufacturing Co.); there's Jack Gentry of Positronic; there's the late Fred McQueary, president of McQueary Bros. Drug Co.; and there's the late real estate magnate, John Q. Hammons. Others we could name but lack space, for which we apologize.

The family businesses that we are highlighting have more in common than their Springfield-Ozarks roots. In each case, their success rests in an innovation that revolutionized a technology, product, or business model; they started, grew, *and stayed* in the region while expanding their markets nationally and globally; and their corporate philosophies have included community service, environmentally sustainable practices, and educational outreach. *They give back*, in other words, to the community that has blessed their endeavors.

Bringing Auto Parts to the People: The O'Reilly Family

An S&P 500 company with 2018 revenues of $9.54 billion ("O'Reilly Auto Parts"), O'Reilly Auto Parts has grown into a behemoth of the American "automotive aftermarket industry" ("The O'Reilly Culture"), with 5,200-plus stores in forty-seven states, including Alaska and Hawaii.

If post-World War II Springfield is built on the automobile, then Springfield's O'Reilly family can

a. Charles and "Chub" O'Reilly with employees (1957). *Courtesy of the O'Reilly family and O'Reilly Auto Parts.*

take credit for a good piece of that growth. In the 1930s, automobiles were competing against streetcars and railroads—and winning. And Charles Francis O'Reilly took advantage of that shift in transportation technologies: Moving from St. Louis to Springfield, he worked for Link Motor Supply, becoming manager in 1932 ("The O'Reilly Story"). His son Charles H. "Chub" O'Reilly joined him and, together, they "made Link the predominant auto parts store in the region."

A threatened corporate reassignment led father and son to strike out on their own, opening O'Reilly Automotive, Inc. "on December 2, 1957 with one store and 13 employees at 403 Sherman in Springfield" ("The O'Reilly Story"). Their business grew, swiftly and steadily:

> [S]ales totaled $700,000 in 1958.... By 1961, the company's volume had reached $1.3 million—the combined volume of O'Reilly Automotive Distributors, a division formed to serve independent automotive jobbers in the area. In March of 1975, annual sales volume rose to $7 million, and a 52,000-square-foot facility at 233 S. Patterson was built for the O'Reilly/Ozark warehouse operation. By that time, the company had nine stores, all located in southwest Missouri. ("The O'Reilly Story")

In identifying its success, the corporate website is too modest by half, for the family business had tapped into an unexploited market: Those "independent automotive jobbers" who served smaller garages.

It was in 1978, however, that the company "made its billions" by developing a "dual-market strategy, maximiz[ing] sales both in the professional 'do-it-for-me' business and in the retail 'do-it-yourself' market" ("The O'Reilly Story"). It was a simple idea, focusing once again on an unexploited market: *People wanted to work on their own cars*, and the O'Reilly stores obliged, supplying them with parts.

While the automotive industry is rarely named among the nation's "green technologies," the O'Reilly company does its part:

> Our business is one that fits naturally with environmental sustainability. We Live Green at O'Reilly Auto Parts by helping you keep your cars running efficiently, extending miles per gallon and extending the life of your car. Each day, our Team Members conserve resources and reduce operating costs through recycling and intentional energy conservation. We offer convenient drop-off locations for used oil, oil filters, batteries, radiators and other parts as well as other programs that remove hazardous waste from our environment. Our environmental awareness includes actions taken across the spectrum of our operations from Corporate Offices, Distribution Centers and Stores to our Delivery Fleets and Solar Project initiatives. ("Living Green at O'Reilly Auto Parts")

A Book, a Cave, a Lake, and a Highway: The Herschend Family Builds Branson's Silver Dollar City

In its corporate website, the Herschend Family Entertainments Corporation (HFE) declares its reputation as "the nation's largest family-owned themed attractions corporation." It's a reputation well deserved. To the region's residents and visitors, HFE is best known for its Branson theme park, Silver Dollar City.

In 1989, Peter Herschend—son of Hugo and Mary Herschend, lease-holders of Marvel Cave and builders of Silver Dollar City (SDC)—contributed an essay to *OzarksWatch*. In it, he lists five events that shaped the

b.

The Herschends—Sherry, Jack, Peter, and JoDee—celebrating the 50th Anniversary of Silver Dollar City (2010). *Photo by Joshua Clark, courtesy of the* Branson Tri-Lakes News.

"life, culture, and economy of Ozark Mountain Country during this century," distinguishing it from other "less blessed parts of the Ozarks." First was the book, Harold Bell Wright's *Shepherd of the Hills* (1907): "It was to *our* hills that Harold Bell Wright came," writes Herschend, "and it was about *our* hills that he wrote his story, *which would become our story*" (emphasis added). More than the mythos of its hills and hollers, Wright's best-selling novel gave the region its ethos—its hill folk character-types, who seem born out of the land itself. It was a mythos, and an ethos, ready-made for commercializing: a people and a place frozen in time, forever expressing the pioneer Ozarks.

Before the theme park, there was the cave. In 1951, having purchased a 99-year lease on the property, Herschend's parents began Marvel Cave's transformation into a "modern" tourist attraction, installing electric lighting, covered concrete walkways, and a cable car—innovative for its time—carrying visitors "up and out" of its depths.[154] With these changes, visits turned from a six-hour wilderness-adventure in spelunking into a clean, safe, 90-minute guided tour. Herschend's second region-shaping event—the filling of Table Rock Lake in 1960—increased regional advertising for tourism, which fed into the third of Herschend's "chain of important events." This third, "the advent of Silver Dollar City," "started as a fledgling attraction (it operated in 1960 only as an adjacency to Marvel Cave) but has grown to be an industry with a national reputation."

It's curious to think of the current theme park "only as an adjacency," a means of entertaining the lines of visitors as they awaited their cave tours. One might call that "a happy accident." But, by building an 1880s mining town on top of the cave itself, the Herschend family had transformed "mere nature" into commodified culture. Soon, the cave became the "adjacency," replaced in popularity by the town-replica and its pioneer citizenry: Park employees dressed and spoke "in character" (that is, *as hillbillies*) while pursuing traditional tasks—blacksmithing, woodcarving, glassblowing, quilting, cooking. It was an "authentic" experience of a rural past largely lost to the rest of modern "citified" America, though preserved somehow here in Ozark Mountain Country.[155] (And it doesn't hurt that the theme park has combined cultural image-making with the latest technologies in rides and roller coasters. The "postmodern amusement park," which SDC has grown into, allows visitors to step in and out *of simulated worlds*.)

"The fourth factor," as Herschend describes it, was a product of "big technology"—of dynamiting and heavy machinery: It was "the mostly unheralded and unsung effect of 'new' Highway 65 from Springfield to the Arkansas state line. When that highway opened in the mid '60s, all-of-a-sudden Ozark Mountain Country became much closer to everywhere." In Herschend's list, the taming of nature (both the river-turned-lake and the cave-turned-amusement "ride") came through earth-moving machinery that reshaped the terrain, making it accessible (and amenable) to tourism. The technologies of media-advertising "spread the word" nationally. And the technologies of speed—specifically, of automobile transportation—capped SDC's success. As Herschend writes,

> The travel time that the new Highway 65 cut on a trip from Springfield to Branson was not of great significance—perhaps 15 minutes. But the perception of proximity (and ease of access) that the new, straighter road gave, made all the difference in the world to people who lived in St. Louis, Kansas City, Tulsa, and beyond.

---

154. In planning their retirement to the Ozarks (where they had been vacationing since the 1940s), Herschend's Chicagoan parents realized the cave's potential for tourist income. Their plans, clearly, required an infusion of technology—of lighting and mechanization—that changed the cave's *use*, as well as its clientele: It would belong to the tourist now, not the spelunker.

155. A further transformation of Silver Dollar City "from a small local business to an enterprise of national standing," writes Herschend, "occurred in 1967, when one of the then top-of-the-television-rating shows, *The Beverly Hillbillies*, came to Silver Dollar City for five days of filming. The impact of that publicity was felt almost instantly and continues even today, because of the hundreds of re-runs of episodes of *Beverly Hillbillies* throughout the country." In this case, the transformative technology was electronic media—nationally broadcast television.

He's right: To this day, *the automobile drives the economy* of Ozark Mountain County.

The fifth and last of Herschend's "major factors" will be obvious to Branson visitors today: it's "the collective impact of what is now called the music show industry." Contemporary Branson is, indeed, successful as a "collective"—as a synergy of music shows and theme parks and cave tours and lakeside resorts, each taking its turn as tourists change lanes and lines, moving from one attraction to the next.[156]

That the family's theme park would grow in time more famous than the cave beneath it bears witness to the power, not of "nature tamed" (the cave, the lake, the hill-cut highway), but rather of the mythos and ethos of the Ozark Mountain Country, itself—which brings us back full-circle to Harold Bell Wright. Elsewhere, Herschend spoke of that fateful year, 1960, when SDC was built atop Marvel Cave and opened for business. In her book on Branson's famous families, Arline Chandler recounts one such conversation:

> Pete Herschend … stated that three things simultaneously came together in 1960 to propel tourism in Branson, Missouri…. "Table Rock Lake filled…. The Shepherd of the Hills Homestead started its outdoor drama based on Harold Wright's book. And Silver Dollar City opened. Those three things were not planned to coincide. They simply happened." (p. 10)

These may well have been coincidental, but the Herschend family surely knew ahead of time of the Homestead and its plan for dramatizing Wright's *Shepherd of the Hills*. The Herschends didn't steal the Homestead's fire, didn't recreate Old Matt, Dad Howitt, Mad Howard, and the rest of Wright's characters; their mining

town was not Mutton Hollow. But the *world* of Silver Dollar City would be interfaced with *The Shepherd of the Hills*, such that visitors could step *outside* the theater and *into* the place—the mythos and ethos—of Ozark Mountain Country, experiencing its old ways "close up." And this, surely, was deliberate. We've already quoted Herschend in this regard: "It was to our hills that Harold Bell Wright came, and it was about our hills that he wrote his story, which would become our story."

Bass Boats and Wilderness Resorts: Fishing with Johnny Morris and Family

Ranked twelfth in Forbes' 2019 list of "America's Most Reputable Companies," Bass Pro Shops has turned its founder and owner, John L. "Johnny" Morris, into a billionaire, currently one of Forbes 500 Richest in the World.

Call it a Springfieldian's love affair with the country, lakes, and fish—fish especially. In his twenties, Morris fished the Ozarks lakes professionally. Being a bass fisherman—and a professional one at that—gave him insight into the tools and technologies, the tricks and techniques, the likes and dislikes, the wants and, above all, the needs of the aspiring "bass master." Bass fishing had its devotees back in the 1960s and '70s, but it had yet to explode into today's nationwide popularity. As much as any human booster, the largemouth bass has beckoned tourists. Indeed, it fits into a chain of events that helped invent the regional tourist industry: The damming of Ozarks rivers built the lakes that held the fish that attracted the tourists who needed the gear (poles and tackle and boats—boats especially) that someone with enterprise and real expertise would offer for sale.

In 1972, Morris started small, selling lures and bait out of the back of his father's Brown Derby liquor store. He knew the gear, being himself a "pro,"

---

156. Besides Silver Dollar City, HFE owns Showboat Branson Belle, White Water, and Dolly Parton's Stampede. Clearly these share, rather than compete for, the region's tourist dollars.

c.

Johnny Morris with fellow anglers at the Bass Pro Shop grand opening (1972). Courtesy of Bass Pro Shops Group.

and offered his own improvements on the lures and tackle then available. But his breakthrough came in kit-building: He knew how "the pros" liked to outfit their boats and provided an affordable package to hobbyists. Outdoors writer Steve Price explains:

Because not all bass boats, outboard engines, trolling motors, trailers, and other accessories were available from a single location, fishermen had to purchase them separately and then pay once again to have someone, most often the boat dealer, install them to create a workable fishing machine

out of different parts. Everything changed in 1978, when Bass Pro Shops expanded its already successful fishing tackle business, based in Springfield Missouri, to include boats.

All Bass Pro Shops founder Johnny Morris did was begin offering his aluminum Bass Tracker boats as a complete, ready-to-fish package, including boat, motor, custom-designed trailer, trolling motor, and sonar electronics. He single-handedly created one-stop shopping for bass boats. Not only that, prices were published nationally and the boats were available at that price at dealerships throughout the United States. *It totally revolutionized boat buying,* and quickly became the industry-wide standard. Bass Tracker boats are still being sold the same way today, with total sales now well above 400,000. (Price, p. 218; emphasis added)

But Morris's entrepreneurship was just beginning. Perhaps his greatest successes lay in "destination retail" and "immersive shopping," having turned Bass Pro Shops Outdoor World—the Springfield "Granddaddy" store and Bass Pro national headquarters—into "the number one tourist destination in Missouri, attracting four million families, sportsmen and women, and outdoor enthusiasts every year" ("About Us").[157] It's the "Outdoor World" *brought indoors,* starting with the technologically state-of-the-art Wonders of Wildlife National Museum and Aquarium: Declared "the largest, most immersive wildlife attraction in the world," Wonders of Wildlife boasts "the greatest collection of record-setting game animals ever assembled" ("About Us")—a taxidermist's heaven-on-earth. Add the NRA Sporting Arms Museum and the

___

157. And the brand has gone national: "Today," declares the corporate website, "Bass Pro Shops has nearly 200 retail stores and marine centers across North America and 40,000 associates[, who] welcome more than 200 million visitors annually" ("About Us").

National Archery Hall of Fame, and you get far more than a store: You get—as advertised—an immersive *experience*.

As a world-class "wilderness resort," Big Cedar Lodge (among other Morris family properties) puts an exclamation point on claims made here: Bass Pro grew out of a series of innovations in technology (Bass Tracker boats), business strategy (immersive shopping), and nature/culture-commodification (Ozarks-themed Wonders of Wildlife). Like other families who've made their fortunes locally, the Morrises are staunch conservationists, having established the Dogwood Canyon Nature Park, "a 10,000-acre nonprofit wildlife preserve" on Table Rock Lake just southwest of Branson. And Springfield's "Grand-daddy" store hosts the John A. and Genny Morris Conservation Education Center, "home to The Wonders of the Ozarks Learning Facility (WOLF) School, a comprehensive outdoor learning school that serves as a national model for outdoor education." On this subject, we'll give Morris himself the final words:

> The people of our company believe very strongly that the future of our industry, the sports we serve and the sports we personally enjoy are absolutely more dependent upon our conservation efforts ... than anything else. It is far more important than any catalog we mail, any new store we open, or any new products our vendors create. ("About Us")

Turning Throw-Aways into Wealth: The Darr Family

William H. Darr was born on Monday, March 9, 1931 on a farm south of Ellington, Missouri—a small town then, with a population of 655, and a small town still, though its current population has risen to 987. Ellington is located in Reynolds County in the eastern section of the Missouri Ozarks. When Bill and his siblings were growing up, times were difficult and money was scarce, but the Darrs had faith, family, the farm and, through persistence and hard teamwork, food on the table. Life was simple, with few of today's amenities.

After working several decades in the egg industry, Bill decided to pursue his dream. The fundamental, sparkling idea that has propelled his life and work—an idea shared by Darr's family and his associates in American Dehydrated Foods, Inc.—was to "re-purpos[e] unused egg protein into a highly digestible, balanced source of essential amino acids for the pet-food industry" (ADF: "About Us"). As the corporate website continues, "Mr. Darr's leadership set ADF's standard for delivery of uncompromising quality in its products and services," including "premium-quality Chicken Protein, Chicken Broth, and Chicken Fat from a USDA-inspected facility" (ADF: "About Us"). A long-time leader in Hazard Analysis Critical Control Point System (HACCP), ADF has used "HACCP principles ... to ensure product quality and safety. Most products are EU-certified so that they can be imported into European Union countries and Canada" (ADF: "About Us").

Restated in lay terms, Darr's industry-changing innovation lay in taking byproducts from food processing, such as in egg and poultry production—byproducts that historically have been perceived as having little or no value—and then, through specialized processing that is responsive to the needs of customers, transform these one-time throw-aways into quality ingredients with significant nutritional and market value. As declared on the ADF website, American Dehydrated Foods and its associates "are committed to transforming and marketing agricultural materials into superior value-added products and services which lead to developing long-term relationships with customers and suppliers that provide profits for continued business growth" (ADF: "About Us").

d. Bill and Virginia Darr in front of the Darr Agricultural Center (2015). *Courtesy of MSU Photographic Services.*

e. A young Bill Darr learning the industry at Henningsen Foods (*ca.* 1960). *Photo courtesy of Thomas A. Peters.*

The basic idea that Bill Darr articulated, fleshed out, pursued, and implemented on a large scale is becoming one of the remarkable, sustainable economic drivers of 21st century food processing. In all areas of business, entrepreneurs are finding new ways "to repurpose" and extract value from previously shunned byproducts. Bill Darr pioneered the way for agribusiness. In early 2019, he sold the businesses he had built up over the decades to a German corporation for $900 million.

Now he, his wife, their children, and their grandchildren can concentrate on their other passion: giving back to communities and region they love, its students especially. Declaring a mission "to empower at-risk youth to overcome barriers to opportunity," the Darr Family Foundation (DFF) is known Ozarks-wide for its grants and scholarships. As Tom Carlson notes,

Bill and Virginia especially enjoy helping young people. They provide scholarships themselves to kids from their hometown in Ellington and to MSU students majoring in agriculture. Carol Silvey with the Community Foundation of West Plains says that Darr and Virginia embody the best principles of philanthropy. They stay in touch with the kids they give scholarships to. And whenever they are honored for a gift they always seem humbled and surprised, Silvey says. For Darr, he says his greatest pleasure is seeing people do well. ("Hard Work")

As a lasting legacy, there's the Darr Agricultural Center, a major facility and gift to Missouri State University. With good reason, Missouri State's William H. Darr College of Agriculture is proud to bear the Darr family name.

"It's Not About the Chocolate, It's About the Chocolate": The Askinosies Practice Fair-Trade Pricing, Profit Sharing, and Open-Book Accounting

Situated on Springfield's "Historic C-Street," Askinosie Chocolate finds itself in ethereal company, named one of twenty-five businesses in Forbes' 2016 List of Small Giants. Celebrating "companies that favor greatness over growth" ("Small Giants"), it was the list's first year, placing Askinosie in Forbes' inaugural class. In a Forbes web article, Bo Burlingham describes this latest entrepreneurial category:

> All of the companies on this list have had opportunities to get as big as possible, as fast as possible. Growth is good, but the leaders of these companies have had other, nonfinancial priorities as well, such as being great at what they do, creating great places to work, providing great service to customers, making great contributions to their communities and finding great ways to lead their lives. The wealth they create, though substantial, is a by-product of success in these other areas. ("The Best")

In the preface to his book, *Meaningful Work*, founder Shawn Askinosie makes an obvious first point: "I started," he writes, "a small-batch chocolate factory at the forefront of the American craft chocolate boom" (p. xv). Superficially, one might say that he got lucky, having followed a "foodie trend" hunch that proved

true. But his success, surely, is "not about the chocolate"—not, that is, about the product *per se*. Suited to business in the 21st century, the Forbes selection criteria are given below. And Askinosie Chocolate *has them all*. In each case, writes Burlingham,

> —The company has been acknowledged as outstanding by those who know the industry best.
> —It has had the opportunity to grow much faster, but its leaders decided to focus on being great rather than just big.
> —It has been recognized for its contributions to its community and to society.
> —It has maintained its financial health for at least ten years by having a sound business model, a strong balance sheet and steady profit margins.
> —It is privately owned and closely held.
> —It is human-scale, meaning frontline employees have real interaction with top leaders. ("The Best")

And there's "one other factor," according to Burlingham: "It's what I refer to as mojo, the business equivalent of charisma. When a leader has charisma, you want to follow him or her. When a company has mojo, you want to be connected with it. You want to buy from it, sell to it, work for it" ("The Best"). That's something that Shawn Askinosie has aplenty: *mojo*. He had it as a lawyer when successfully defending murder cases. And *he needed it* when a change of heart pushed him to a change in vocation, from defense lawyer to artisanal *chocolatier*.[158]

---

158. There is much spirituality in his book, which is intended "for entrepreneurs at heart or in practice—or both. It is for those who are searching for their own personal meaning in their work, and seeking to transform that meaning into a vocation" (p. xv). The basic formula for entrepreneurship remains, though Askinosie takes a wide-angled vision: "The path to vocation is smoother," he writes, "if we can survey the landscape in our neighborhood, workplace, city, or world at large and take stock of existing

Within our list of local companies, his is the only one to have been founded in the 21st century (in 2007, to be exact, when he sold his first chocolate bar). But, joined by his daughter and business partner, Lawren, his is already a multigenerational family business with plans to stay in Springfield while making sales nationally and gathering its resources globally.

Askinosie acknowledges Jack Stack—fellow Springfieldian and founder of SRC Holdings—for having taught him "open-book management," where employees "have a stake" in the business outcome: Stated in a nutshell, "if employees share in the profits and the outcome of a company's success, they will feel greater ownership in the process leading up to the end result" (*Meaningful Work*, p. 94).[159] Unabashedly, he pursues the "shared stake" model in his Commercial Street factory (whose dozen employees generate

$2 million in revenue annually). As important, this same model applies to global partnerships. "As far as I know," writes Askinosie, "we are the only chocolate makers who are also the direct importers of cocoa beans in the United States. And we're the only chocolate makers in the United States directly trading on several continents and importing our own cocoa beans" (p. 96). The power of such a model is best told in story, not in abstractions:

> When we began purchasing cocoa from the Mababu farmer group in Tanzania, most of them had never tasted chocolate. It's sad, but true. They were harvesting a commodity, until we came along. Seven years later, they're cocoa bean experts....
>
> They want to give us better cocoa because they know that we will come back again, that we pay a high price for quality, that we profit share, and that we care about them. A couple of years ago we won a silver medal at the International Chocolate Awards for our 72 percent Mababu, Tanzania, Dark Chocolate Bar. We had a beach party with our farmers to celebrate this distinction and presented them with a plaque to commemorate the honor. The news was a surprise to the group and they celebrated with clapping, cheering, and yelling. After a few moments the noise died down; I will never forget when the chairwoman, Mamma Mpoki, stood up with a twinkle in her eye and asked me, "What do we need to do to win gold?" (pp. 92-93)

One of the more inspiring aspects of the Askinosie model is its commitment to social justice "at home" in Springfield *and* Tanzania, among other global partners.[160]

---

needs" (p. 14). And he addresses the reader directly: "Your vocation may well lie at the intersection of your talents, what the world needs, and your passion" (p. 20).

159. "I was aware," Jack Stack writes in his book, *A Stake in the Outcome*, that "you can make more money if you own the company," and also that "you can create wealth for the people you work with at the same time as you're creating it for yourself. *I didn't see why those two goals had to be in conflict*" (p. 27; emphasis added). He continues:

> It was going to take everyone to make the business successful, and I believed everyone should share in the rewards. Nor did I think I was giving anything up by sharing equity. It seemed to me that we'd all wind up richer in the long run if everybody was an owner. Why? Because I thought our people would perform at a higher level if they had an ownership stake. I thought they'd give us an edge that our competitors didn't have.
>
> ...What better goal can you have than for everybody to be a winner? I was looking for pride of ownership, authorship, a sense of accomplishment, satisfaction, meaning in life—not just for me, but for all the people I was working with. (*A Stake*, pp. 27-28)

Together, Stack and Askinosie show how Springfield's entrepreneurs "lead the way" nationally in enlightened business models and trade practices.

---

160. Note that we *did not* write, "and 'abroad' in Tanzania." It's within "the global village" that Askinosie does business, and therein lies the 21st century future of entrepreneurship grounded in social justice. In *Meaningful Work*, he explains:

Father and daughter at a coffee plantation, Tanzania (2014). *Courtesy of the Askinosie family.*

In so many ways, this "small giant" of a company turns the traditional business model on its head. What's business about, if not about growth: about "managing" employees, cutting costs, increasing one's market share and profit? *It's about "the bottom line," right?* In this context, the company motto—"It's not about the chocolate, it's about the chocolate"—seems a *non sequitur.* Implicit, however, is the recognition that it's not the product or the profit only, but also the process and the principle that matter. As the company website declares,

> For us, this confounding phrase means that we have a Zen-like approach to our business: we hold the craft and quality of our chocolate in almost equal balance with doing as much good as we can in the world. Ultimately though, our first priority is to craft the highest quality and best-tasting chocolate possible, because we recognize without that, we would not be able to do all of the other

There's a global movement under way in which all manner of workers are seeking meaning and dignity in their work. It's a global shift in mind-set. Acknowledging, embracing, and implementing this perspective on work will mean your organization will not only survive, but thrive. Second, businesses should engage employees in work that matters. Third, a corporate calling will result in a better product or service. (pp. 26-27)

stuff. "What stuff?" you ask! Well, any of it—we could not share profits, involve students in our business through Chocolate University, undertake community development projects in our hometown and our farmers' communities, and much more, if we did not create great chocolate that people love. ("Learn")

In an interview with Susan Adams, Askinosie makes this point explicitly: "There are many ways to grow besides scale. We can grow in net profitability, by increasing salaries and wages for the people we work with, and in the ways we impact the lives of our farmers" ("Second Acts").

As much as any family business cited in this survey, the Askinosies have a strong commitment to communitarian service. What distinguishes them is their local-global perspective. Their "Chocolate University," for example, "is an experiential learning program with a worldwide reach for local students":

> The goal is to inspire students through the lens of artisan chocolate making to be global citizens and embrace the idea that business(es) can solve world problems. We involve neighborhood students from Boyd Elementary, Pipkin Middle School and local high schools in our business through visits to their classrooms, field trips to our factory, updates from origin, visits to origin and much more. The goal is not for these students to become bystanders during a lecture on chocolate making. It's to provide them with a hands-on experience that takes them from the inner workings of the factory to the cocoa bean farms across the world. ("Chocolate University")

By the way, the family's business model may be 21st-century, but its manufacturing technologies are definitively old-school.[161] To make artisanal chocolate, one must *relearn* the old techniques on century-old machinery. Playing an Askinosie-esque word game, we can say, "Not everything new is new."

Works Cited

"About Us." Herschend Family Entertainment. Retrieved 6 March 2019. <http://www.hfecorp.com/about-us/>.

ADF: "About Us." American Dehydrated Foods. Retrieved 6 March 2019. <https://www.adf.com/about-us>.

Adams, Susan. "Second Acts: Why Shawn Askinosie Dumped His Law Practice to Make Chocolate." *Forbes*, 9 May 2017. Retrieved 9 March 2019. <https://www.forbes.com/sites/forbestreptalks/2017/05/09/second-acts-why-shawn-askinosie-dumped-his-law-practice-to-make-chocolate/#7e4c76b10156>.

Askinosie, Shawn, with Lawren Askinosie. *Meaningful Work: A Quest to Do Great Business, Find Your Calling, and Feed Your Soul*. New York: Tarcher Perigee, 2017.

Burlingham, Bo. "The Best Small Companies in America, 2016." *Forbes*, 27 January 2016. Retrieved 9 March 2019. <https://www.forbes.com/sites/boburlingham/2016/01/27/the-best-small-companies-in-america-2016/#3270bde6610c >.

Chandler, Arline. *The Heart of Branson: The Entertaining Families of America's Live Music Show Capital*. Charleston, SC: History Press, 2010.

"Chocolate University," "Learn." Askinosie Chocolate. Retrieved 8 March 2019. <https://www.askinosie.com/learn/chocolate-university.html>.

Carlson, Tom. "Hard work, focus drive Darr's

---

161. As Shawn tells Adams, "the first piece of equipment I bought was a 100-year-old, 6,000-pound chocolate mill from Germany called a *melangeur*. It took me more than a year to cobble together all the equipment I needed" ("Second Acts").

empire." *Springfield News-Leader*, August 30, 2014. Retrieved 8 March 2019. < https://www.news-leader.com/story/news/business/2014/08/31/hard-work-focus-drive-darrs-empire/14834299/>.

Darr Family Foundation. Retrieved 8 March 2019. <https://darrff.org/>.

Herschend, Peter. "An Advocate Looks at Ozark Mountain Country: An Ozarks View." *Ozarks-Watch* vol. 3 no. 4 (1990): 22, 32.

"O'Reilly Auto Parts." Wikipedia accessed 9 March 2019. <https://en.wikipedia.org/wiki/O%27Reilly_Auto_Parts>.

"The O'Reilly Culture," "The O'Reilly Story," "Living Green at O'Reilly Auto Parts." O'Reilly Auto Parts. Retrieved 8 March 2019. <https://corporate.oreillyauto.com/>.

Price, Steve. *The Fish That Changed America: True Stories about the People Who Made Largemouth Bass Fishing an All-American Sport.* New York: Simon and Schuster, 2014.

"Small Giants: The Best Small Companies of 2018." *Forbes.* Retrieved 9 March 2019. <https://www.forbes.com/feature/small-giants/#7723870d4612>.

Stack, Jack, and Bo Burlingham. *A Stake in the Outcome: Building a Culture of Ownership for the Long-Term Success of Your Business.* New York: Doubleday, 2002.

"Your Adventure Starts Here." Bass Pro Shops. Retrieved 6 March 2019. <https://www.bass-pro.com/shop/en/about-us?cm_sp=FtrBPSS-Mar2019_FT#sec-conservation>.

I Live in Springfield and I Work in Tokyo:

Practicing the Five Principles of Entrepreneurship

Daniel D. Goering

In 2016, I was homeless, living in a small van inside a soon-to-be-demolished parking garage in Tokyo, Japan. I was finishing part of my doctoral program in organizational psychology during the day. At night, I was creating a predictive analytics software to prevent burnout, depression, and suicide in the workplace. I was Western Unioning home what little money I could to my wife and four children—the oldest of whom was five and the youngest only a few months old—who were living in Kirksville, Missouri. To this day, I am not sure if my colleagues at the University of Tokyo, Japan's most prestigious university, are aware that their American colleague was commuting to the lab each day from his van parked in a garage off some shadowy side street.

I had not been homeless the entire time I was in Tokyo; in fact, it was only for six months, halfway through my one-year "visiting fellowship" as a researcher at the university's Graduate School of Medicine. For the first half of that assignment, our family of six lived in a roughly 500-square-foot apartment. My wife had been eight months pregnant with our fourth child when we moved there. The move required an overseas flight, a long and sleepless flight. She gave birth to our fourth child there in a Japanese hospital. She spoke no Japanese at the time. All of these are tokens of her resilience and fortitude. About four months after the birth, due to health complications with the baby, we decided that it was best for the family to head back home to the U.S. while I finished publishing my research and beta-testing my burnout-prevention app. United as always, we headed home to Missouri before our temporary separation.

Back stateside, with great help from family and friends, I spent two weeks getting my wife and kids settled into their new home in Kirksville. From Kirksville's tiny airport, I began the flight back to Tokyo with several layovers. Although I have spent almost as much time in Tokyo as in any other city in the world (except for my hometown), the flight back felt foreign to me: I was alone; none of the crying children on the plane were mine; there were no emergency potty breaks. I knew this meant only that my wife had an increased burden back home. To relieve some of the burden, I saved money and sent it back home to her. To do so, I moved out of our Japanese apartment and sold our furniture and other belongings on Craigslist and "SayonaraSale.com." I delivered the furniture from our apartment in what would soon become my home: a silver 2003 Nissan Serena minivan.

Being homeless was an odd experience, a mix of liberation, transformation, frustration, and humiliation. It was liberating, because I had no bills to pay (in Japan, you typically prepay auto insurance and inspections for a two-year period). It was liberating, in that I sensed a weakening tie to society and self and an increasing tie to something transcendent. It was transformative, in that I began to see things in a way I had not seen before: the frail line between success and failure, the impermanence of one's mortal life and legacy, the thin thread upon which the threat of chaos looms over us all. It was frustrating and humiliating as well, though I shall elaborate on these later. In sum, the experience provided intellectual and spiritual insights which I am still processing to this day.

A Japanese minivan and its American inhabitant (2016). *Courtesy of Daniel D. Goering.*

Despite the challenges living in the van posed, my wife and I felt it was important to stick to the vision of finishing my research and turning that research into real solutions to help people working in Japan. My wife and I both were convinced that the research I was engaged in would help scholars' understanding of how to improve the workplace. We also shared a passion to free these research findings from the Ivory Tower and *apply* them in the workplaces in Japan. To apply them, I used my training as a meta-analyst to design a predictive analytics software that could predict and prevent work-related psychological problems such as burnout. I incorporated that software into a self-help app to help the everyday people who had been struggling in their work lives.

Back when we were living in the apartment, we had seen such people every day at what had been our nearest train stop, Shimosa-Nakayama Station. Train stations such as ours—those located primarily in suburban locales on the outskirts of Tokyo—were relatively common places of suicide for people who had been overworked and burned out. Such suicides typically occurred on Monday mornings, at the height of the commuting hour. I have been told that, on the day they jump, people often wear their work clothes like any of the other hundreds of Mondays on their way to the office. Our hope was that my app would prevent people from burning out through predictive analytics: that it would sense the precursors to burnout, ask four or five questions to diagnose the root causes, and provide customized output to treat it prophylactically, leading them through a set of easy and precise steps to stay healthy. I continued to work on developing and implementing this app in Tokyo while my family remained in Kirksville. I would do my research during the day in the lab and then work on developing the algorithms and app during the evening.

After getting a prototype of the app completed, I made appointments with managers I knew in a large HR services provider. During college, I had worked there as an intern. I presented the idea to them as an additional service that they could sell to their clients. The Japanese Ministry of Health, Labour, and Welfare had passed legislation only months before requiring companies of fifty or more employees to do an annual "Stress Check" on employees. These checks are meant to find and help stressed-out employees. The Stress Check is a self-reported diagnostic that essentially asks an employee to indicate whether or not they are depressed. If an employee reports being depressed, then the company pays for the employee to see a clinician to "fix" the employee's mental health.

The Stress Check had been criticized by prominent researchers in Japan as being a notable yet insufficient step in reducing Corporate Japan's stressful work environment. Employees are required to report on their own mental health condition by answering point-blank questions about their own mental health status, which most people would be reluctant to answer candidly. As such, this government-mandated check put the onus of diagnosing and reporting one's own work-related stress on the back of the employee, which, ironically, was itself added stress on employees. Finally, the check assumed that being burned out or depressed was due to employees being mentally weak—in essence, if an employee reported being depressed, it was presented as the *worker's* fault and responsibility to get the help the worker needed; this effectively ignored the role the workplace conditions

and management have on employees. Japanese workplaces, in my opinion, are some of the most dreadful places to work (at least as far as modern societies go), requiring long hours of feigned productivity to make the boss and more senior employees look good, followed by hours of *de facto* mandatory drinking together to build team comradery. Many Japanese workplaces are designed with 20th century "production-line thinking" trying to solve 21st century, information-age problems.

In short, the government's mandated Stress Check was well-intentioned garbage. Everyone knew it, but everyone played along, pretending that it wasn't. This kind of "just go along with it and don't ask questions" thinking can be typical of organizations with high power distance (i.e., those with rigid and strong hierarchical structures), such as is common in Japanese culture, and even in some American institutions such as the U.S. military. Thus the misery of Japan's corporate work life persisted, fundamentally unchanged, but at least the bureaucrats and politicians could say that they were trying.

I saw my software as a way for companies to know the scientifically precise changes they needed to make to their HR that would increase their profitability while simultaneously improving employee well-being. In fact, employee well-being and company profitability are often intricately linked. My software finds a company's greatest HR strengths and weaknesses, and resolves "people problems" before they become too costly. It then guides a company to make simple improvements with intelligent, meta-analytically precise solutions to prevent future problems and more strongly link its HR to increased profitability. Originally, however, I was focused only on preventing burnout. I pitched my app as a way to prevent burnout before it happens and save a company the potentially tens of thousands of dollars that employee burnout can cost in company medical fees, mandatory paid time off, and lost productivity. In addition, it would help improve the employees' lives. I had convinced the managers at the HR-services provider that the software had great merit, and they agreed to show the app to a few clients.

After showing the app to the HR managers of client companies, it was clear there was no interest in the app by HR managers. This was discouraging, and I asked to go back and interview those HR managers to learn more specifically *why* they did not have interest. During my interviews with them, they mentioned how their job as HR managers was fundamentally concerned with *compliance*; and while on a personal level they wanted to use the app to improve the well-being of employees, it was difficult to make a case to upper management. Naturally, I asked what would be helpful to them to make a case to upper management. They offered suggestions about other issues they struggled with, such as "engagement" (that is, how to get employees to be more motivated in their work). They mentioned instances of "deviance," how employees would steal things from the office or misuse company resources for personal benefit. They also mentioned how they are often surprised when high-performing employees quit the company.

"If there were a way to predict if or when our top performers were going to quit and prevent it from happening, that would be super valuable to me and to upper management," one HR manager, Nakajima-san, said to me. "Of course preventing burnout is important, but honestly it is not necessarily the first thing that comes to my mind. It isn't the most pressing problem for me in my job."

I took this feedback back to the company that had introduced me to these HR managers. I said that I could use my database to adjust the app to make these predictions as well. I could do it by myself and it would take several months, or, if the company was willing to partner with me, I could get it done sooner.

Unfortunately the company said that, while they supported and liked the idea, they weren't ready to put any money behind it. However, one of the people in the meeting, Tateno-san, loved the idea and introduced me to a programmer and entrepreneur friend of his, Sota-san.

After several meetings together, Sota and I decided to partner together, fifty-fifty. He would run the day-to-day business side of things and synchronize the predictive algorithms I had calculated using meta-analytics with my HR database onto a web-based user interface. We continued to reach out to other HR managers at larger companies—using our combined professional networks—to get feedback on the system prototype as we continued to develop it. Around this time, I was wrapping up my research program at the university and had more time to focus on the development of this web-based workplace improvement software.

Sota and I continued to get feedback from managers at large companies, several of them Fortune 500 companies based in Tokyo. We used the feedback to keep improving the prototype; however, despite several repeat meetings where the managers seemed pleased and excited to use the product, none of them were actually signing on to use it. They liked the concept; they liked the idea; they liked the interface and saw the value in it. Yet there were often budget constraints: The HR managers had already allocated their budgets for that year toward separate initiatives, including spending on the mandated Stress Check. We were just spinning our wheels.

I still believed in the idea of using predictive analytics to proactively prevent "people problems," and so we continued to talk to HR managers. At one medium-sized company, we were introduced to the CEO, Ms. Suzuki. I told her the types of predictions we could make with the use of meta-analytics coupled with my database. She mentioned that she had prob-

b. A typical morning subway ride in Tokyo (2016).
Courtesy of Daniel D. Goering.

lems with dysfunctional turnover (the loss of talented employees) and with promotions within the company. She also proposed something new to us: "Could you predict whether or not a team will work well together? Could you predict whether a leader will be the most effective for a team out of a pool of candidates? We have a lot of trouble determining who to put on what teams and who should lead those teams. If you could do that, then I'm ready to sign on right now." I told her I would go back and think through whether or not we could do something like that. After a few days thinking through the analytics models, I developed a very basic prototype for her. She was thrilled, and we had our first customer!

Listening to Ms. Suzuki's insights was pivotal for the development of my HR software, because it revealed to me two primary things. First, it revealed to me that we had been talking to the wrong people all along. We had been approaching HR managers, particularly those at large companies, and HR managers were not the decision-makers. They were constrained by their budgets and specific initiatives that had been planned one or two years prior; and the larger the company, the longer things would take. Sota and I were just two people, and we did not have the resources to do this kind of "elephant hunting." We needed to talk to CEOs, mostly at smaller-sized companies. The second thing Ms. Suzuki's insights revealed to me was that I had been offering a solution to a problem that I *thought* companies had (i.e., rampant burnout), but in fact they did not. Or at least they did not know they had it, or it was not their most painful problem. I figured that if I could get into the workplace to help solve issues related to turnover and talent management, then I could also advise management on how to improve the workplace to reduce employee burnout.

With this knowledge in mind, Sota and I decided to experiment with my hypothesis. He continued to approach the HR departments at larger companies; meanwhile, I was going to start focusing on talking to CEOs of smaller companies. Our second success came soon thereafter, from a friend and mentor of mine, Nate. Nate was an inspiration to me. He had started a solar energy company in California. After the 2011 Great East Japan Earthquake and subsequent Fukushima nuclear power plant disaster, Nate realized that Japan was making great strides toward renewable energy sources. Nate moved quickly to be one of the first large-scale solar energy providers in Japan, taking advantage of the Japanese government's twenty-year fixed rates to purchase photovoltaic (solar powered) energies into the electric grid. One success led to another, and now Nate's company is one of the leading renewable energy producers in Japan.

I approached Nate about my software. I asked him the types of problems and pain points he had in his company. He mentioned that his company had grown steadily to nearly fifty employees in four years, and it was getting large enough that "people problems" were beginning to crop up despite that, relative to his competitors, his company paid handsomely, particularly for high performers. He had recently lost some talented employees. He had also made some bad hires, people who interviewed well but were not a good fit for the high-performance demands of his company. To continue growing at his current pace, he would need a way to predict the fit of employees. Similar to Ms. Suzuki, Nate had also had some issues with promoting people to leadership positions: Some people seemed ready but then floundered when they actually got into the leadership role.

Nate and I continued conversations. He gave feedback on how to improve the interface and user experience of the software. We made adjustments to fit his and others' needs and would make a new iteration of a prototype. We would test that out and then again make adjustments as needed. Nate continued to mentor me, introducing me to other executives whose companies could benefit from my software. He was also a champion of it and of me, even after I returned home.

Nate signed on to use the software, and we had an interesting experience the first time we used it. I went over the results with him, indicating where his greatest HR risks and strengths were now and where they would be in the near future. I ran simulations to determine with scientific certainty what his best courses of action were to take. One of those actions involved an employee whom my software predicted was stealing from the company. When I mentioned this to Nate, he balked and said, "No, this guy? He's like a little old Japanese grandpa-type guy. He's one of our on-site engineers. No way." I advised him to at least look into it, because the software had predicted it as high-risk, based off of this individual's and others' responses to perceptions of the work, perceptions of leaders, and individual personality traits. I wasn't sure *what* or *how* he was stealing, but the system predicted that some kind of shenanigans were going on.

The next day, Nate notified me that he had not done anything about it, but out of the blue the supervisor of the high-risk employee had called him up only two hours after Nate and I had talked. The supervisor mentioned that they needed to let go of this employee, because the employee was skipping his job in order to run his own side business. "Your system really works!" he said. While the timing of the supervisor's phone call to Nate was serendipitous, this nonetheless demonstrated how powerful and precise my software is.

This was a relief to me, because, although I knew the system worked in theory, I had not seen it actually "take flight" until that moment. "It flies! It actually flies!" I remember thinking to myself. It was exhilarating! The analytics scan not only saved Nate's company money from an unproductive employee who was collecting a paycheck while neglecting his job, but it

prevented potential, costly legal issues. For instance, if the employee used Nate's company equipment to run the side business, Nate's company could potentially be liable for any damages from that side business. This and other successes we had soon thereafter showed we had a proven, fully functioning product that could help improve the workplace with powerful predictive precision. I have had several more instances where an executive was initially dubious about an HR issue flagged in our scan that turned out to be spot-on. "Your system really works," was a great pick-me-up.

Around Christmastime, I was getting ready to return back home and reunite with my family. Sleeping in the van had actually become routine, and I was thankful to have somewhere to go that was out of the wind. Sota and I continued to refine the software and interface, and we had partnered with Ms. Suzuki's company to help us sell the product. She invited me to demonstrate the software at a seminar in front of 100 or so of their clients. The seminar was to be held a few months after I was to return back home. My research was under review at a journal whose editor-in-chief was the original developer of the concept of burnout, Christina Maslach, who was also famous for stopping the notorious Stanford Prison Experiment. My software was a proven concept, albeit still rough around the edges in its user interface. I remember packing up my belongings in a duffel bag and thinking to myself, "Not too shabby for a homeless guy."

I returned back to my wife, twenty-five pounds lighter and gray hairs now in my beard. My youngest boy did not recognize me for some time. He would cry and recoil when I got him out of his crib in the mornings. I continued to work with Sota every week—and on some weeks everyday—via Skype, iMessage, and email to coordinate our activities related to the predictive analytics software. I finished up my doctoral program in the summer, then headed back to Tokyo for a few weeks to keep business moving forward. I started my position as an Assistant Professor of Management at Missouri State University at the end of that summer, in August.

Now I continue to teach at the intersection of management and entrepreneurship. My research continues to focus on using meta-analytics to answer some of our toughest management questions. My classes and my research have a strong emphasis on application and implementation. I continue to work in Tokyo—via Skype and other media—on a regular basis, almost entirely in the evenings after the kids' bedtime when I am at home and it is late morning in Japan. I teach full-time at the university during the day in the U.S., so Sota runs the day-to-day operations in Japan while I continue to listen to customer feedback and make improvements to the analytics and analytic output. I use my experiences in Japan to breathe life into the theories that live inside the books used to teach students. I also travel to Tokyo on occasion in the summer when school is not in session. In this way, I truly live in Springfield and I work in Tokyo.

The Five Proven Principles of Successful Entrepreneurs

My experience starting up a business in Japan actually follows a pattern that research has shown is common to many successful business startups. Below, I provide a brief overview of what makes businesses and entrepreneurs successful. Granted, there are no silver bullets; but after decades of research, there are a few key findings. I share these with my students, and I would like to open them up to readers here, as well. The first portion focuses on "who" successful entrepreneurs *tend* to be. The perennial "nature vs. nurture" debate has the same answer as it always does, even when it comes to entrepreneurship: *It's complicated, and it's a little bit of both.* Ultimately, it appears that the "who" of successful entrepreneurs—sorry to spoil the surprise—is less relevant than the *"what"* of it. That is, even if a person has a different personality from the prototypical "successful entrepreneur," that is fine; as long as people can follow some of the proven principles common to most successful entrepreneurs, that is relatively more important. Therefore, the second portion focuses on *what* successful entrepreneurs typically do. Although I may not be "a success" yet relative to some of the Ozarks' best known entrepreneurs, I see my startup as a success by my own measure, and I still think it's "Not too shabby for a homeless guy." Therefore, to breathe life into these proven principles of successful entrepreneurs, I draw from my experiences working in Tokyo.

The "Who" and the "What" of Entrepreneurship

First, let us consider the "who." Who are entrepreneurs? What kinds of people are likely to become entrepreneurs and succeed at doing it? Oftentimes, when people think of entrepreneurs, images of creative geniuses and quirky inventors come to mind. Or perhaps one may think of successful entrepreneurs as high-powered executive types who have equal parts of prescience of the future, creative genius on how to build that future, and boldness to make their vision of the future a reality. These images are romantic and, at best, only partially accurate *some of the time.* In fact, what we know now about entrepreneurship is that it has less to do with prescience and more with perseverance; far less is attributable to genius than to pragmatism. *Successful entrepreneurs are simply those who found something useful (and not necessarily novel) to other people and then brought it to them at a price they were willing to pay.*

Although such romantic images of entrepreneurs as bold, fearless, outgoing, and passionate go-getters still persist in the collective consciousness, the objective reality is slightly less glamorous. Research in this area[162] indicates that, while people who are more comfortable with high-risk situations (i.e., high-risk propensity) are more likely to *become* entrepreneurs, people who like to avoid risky situations are slightly more likely *to be successful* as entrepreneurs. Furthermore, people who are more level-headed and even-tempered (i.e., high "emotional stability") are more likely both *to become* entrepreneurs *and be successful* at it.

What about extraversion? Do you think the more extraverted one is, the more likely that person is to be an entrepreneur? If so, you're right, but not to any practically meaningful level: How extraverted a person is matters only about 2.5% in terms of being likely to become an entrepreneur, and it matters a little less than 1% in terms of how likely someone is to be successful as an entrepreneur. Basically, the science tells us that the relationship between extraversion and entrepreneurship is statistically "*not 0*," but, for all practical purposes, the relationship between the two has no meaningful effect. That is good news for people who are low in extraversion (i.e., so-called "introverts" or "ambiverts") who want to be successful entrepreneurs; they are *de facto* at no disadvantage (although they are also not at any advantage, either).

I am focusing here primarily on personality traits known as "The Big Five," which are the most commonly studied personality traits in organizational psychology today. Broadly speaking, personality traits describe a person's general tendencies to act, think, or feel a certain way. Under The Big Five model, personality traits are conceptualized on a continuum, so a person is not "an extravert" per se, but a person has "high," "medium," or "low" extraversion relative to other people. A person with high extraversion then is someone who is more likely to be outgoing, friendly, excitement-seeking, or cheerful when compared to others. Therefore, under this model, it is technically not accurate to describe someone as "an extravert" or "an introvert," but rather as being low, medium, or high in extraversion. As noted, the trait of extraversion has little real bearing on how successful an entrepreneur will be. However, the two Big Five personality traits that *do* seem to matter in the most meaningful ways are "conscientiousness" (i.e., how reliable, hardworking, and orderly a person tends to be) and "openness to experience" (i.e., whether a person tends to hold broad interests and uses unconventional thinking). People with these two personality traits are more likely both to become entrepreneurs *and* be successful at it.

I personally have high levels of extraversion. This is perhaps partly a function of being the ninth child in a family of thirteen where there were *always* people around. It is also simply a matter of who I am innately, at least according to my Mum, referring to my early childhood. My high levels of extraversion made living in the van very difficult for me psychologically. I had this inborn tendency to want to talk to people mixed with a nagging self-consciousness and shame of being homeless. I am thankful for smartphones that allowed me to communicate with my wife back home, even in the darkness of my van "bed" to help me channel some of those conflicting emotions. I have medium levels of "conscientiousness," meaning my levels of conscientiousness put me at no statistical advantage over other entrepreneurs. Strengths that I do have, as far as personality traits related to successful entrepreneurship, are my high levels of "openness to experience." This means I enjoy variety and different ways of viewing the world. These help me see different ways to solve problems. People with this trait tend to be more successful as entrepreneurs in general. I note these to show how knowing yourself can help you be a more

162. See H. Zhao and S. E. Seibert (2006), The big five personality dimensions and entrepreneurial status: A meta-analytical review, *Journal of Applied Psychology* 91(2), 259; and H. Zhao, S. E. Seibert, and G. T. Lumpkin, (2010), The relationship of personality to entrepreneurial intentions and performance: A meta-analytic review, *Journal of Management* 36(2), 381-404.

successful entrepreneur. Knowing who you are can help you know some of the "ingredients in your mystery basket," so to speak; knowing your own personality can help you know where your personal strengths and weaknesses lie and how to manage them.[163]

All of this research is important in understanding *who* is likely to break out of "the 9-to-5 mold" and become an entrepreneur and, of those people, who is likely to be successful at doing so. That said, I want to emphasize here the relative importance of the "who" of entrepreneurs. These findings are by no means finalistic nor fatalistic: People who have narrow interests can and still do become successful entrepreneurs. People who are bombastic and prone to take high risks can still become wildly successful. Just the same, some people win the lottery while others get struck by lightning. It is important to keep in mind that we are speaking in terms of probabilities and likelihoods. All this is to say that researchers have determined the *percentage* that the "who" relates to being a successful entrepreneur, and people who may not fit the ideal mold of an entrepreneur can take comfort in this statistic: The "who" of being an entrepreneur only explains about one third of whether someone is likely to become an entrepreneur and be successful at it. To be precise, researchers estimate that a person's personality explains between 30% to 40% of variation in becoming a successful entrepreneur. So what accounts for the rest of the equation?

In a nutshell, the remaining 60% to 70% is the "what" of being a successful entrepreneur. As mentioned, if *who* a person is accounts for about a third of the equation predicting if that person will be a successful entrepreneur, then the remaining two-thirds

is about *what* they do to accomplish that. This is the part of what people can actually do today to increase their chances of being a more successful entrepreneur. While there are several theories about what entrepreneurs can do to increase their chances, I have condensed these down into five proven principles that most anyone can implement. For my academic colleagues: This is, admittedly, a simplification; it is, however, a useful and applicable one. The five principles are: *Experiment, Identify Early Adopters, Connect to an Ecosystem, Effectuate*, and *Iterate!* I go into detail about each below.

Principle 1. *Experiment*.

The first principle is *experiment*. Ultimately, despite all of our advances in science, the truth is that we simply do not know if a business idea is going to be successful or not. There are myriad reasons why this may be, one primary reason being that people's tastes, needs, and problems are changing all the time. This constant change makes it hard for the slow, grinding wheels of science to catch up. "Experiment" means you cannot assume that you have the right solution or the right business idea. In fact, you cannot even assume you have the right *problem* or the right group of people. So what do successful entrepreneurs do?

They experiment. And so must you. You must go out and test your ideas in the market through a repetitive process of trial and error. For example, I originally set out to use my expertise in meta-analytics to create a burnout-prevention application for individual employees. I thought that HR managers in Japan would love to buy an app that would help prevent their employees from burning out. However, through several interviews with several HR managers, it became increasingly clear that I had the right solution but for the wrong problem and the wrong group of people. The right solution was using big data and predictive meta-analytics to foresee HR problems before they became problems. The HR managers

---

163. Note that I'm drawing my vocabulary ("extraversion," "conscientiousness," "openness to experience," etc.) from the Five Factor Model of Personality (a.k.a. "The Big Five"). For those interested in a free assessment of their personality traits using this instrument, please visit the following site: <http://www.personal.psu.edu/j5j/IPIP/ipipneo300.htm>.

thought this would be amazing—they could prevent people problems from occurring. However, *contrary to my assumptions*, burnout was not the most painful problem companies were having. Bad hires, misfit employees, talented employees quitting, sinking motivation among employees, people failing in leadership positions: While burnout was also a problem, *these* were the most painful problems they faced. My first hypothesis—that HR managers' most painful problem was employee burnout—was proved to be incorrect. Accordingly, I used their input to pivot and adjust the algorithms to predict what they really wanted to know.

Eventually, it also became clear that HR managers at large Fortune 500 companies were not the right place to start. There was too much bureaucracy for a two-man startup; besides, the budgets for HR initiatives seemed to have been decided at least a year in advance. Through another trial-and-error experiment, I began talking to executives, like Nate. Executives of small- to medium-sized companies wanted a way to predict if a person would be a good fit for a job, a team, the company; they wanted to know if high-performing persons were likely to quit in six months and what the company could do to keep them onboard and happy beyond simply throwing money at them. (The reasons people quit jobs is complicated, with pay being only a part of the equation for most personality types). Through these interviews, and through Nate's mentorship, I realized that small- or medium-sized companies had grown enough that they were beginning to encounter "people problems," but they did not have the tools yet to handle their growing workforce. Hence, *they were actively looking* for solutions to the growing number of people problems they were encountering in their growing business. Furthermore, executives at firms of this size could make decisions more quickly. Thus, my hypothesis was supported: *I now had the right solution to the right problem for the*

right group of people.

This trial-and-error process is important, because each of us sees the world differently. You may have a solution to a problem that *you* think you have, and that solution may do a great job of solving your problem ... *for you*. However, you cannot assume that other people have the same problems or needs as you, or that they want your personal solution. Therefore, the first thing that successful entrepreneurs do is *experiment*.

Principle 2. Identify *Early Adopters*.

Another behavior common to many successful entrepreneurs is that they begin by serving a niche of people, gain their footing, and then expand from there. This is what is commonly referred to as a company's *early adopters*. Early adopters are those who adopt your product, service, or other business idea early on and tend to champion it to others. *But how do you know who those people are?*

Early adopters might be grouped together in different ways. They can be those who share similar demographics (e.g., age, sex), socioeconomic status (e.g., middle class), common interests (e.g., the outdoors, racing), occupations (e.g., HR managers), life stage (e.g., parents, college students), or one of many other attributes. It is important to experiment to find who these people are. A quick way to check if someone is an early adopter or not is to ask yourself these three questions:

Does this person have the problem or
    need my business is solving or providing?
Does this person *know* they have this problem
    or need?
Is this person *actively looking* for a way to solve
    this problem or fulfill this need?

If "yes" to all three, then this person is likely to be one of your early adopters. The next step is to discern what common attributes tie together all of your early adopters. Then experiment to determine if you are correct. If correct, then focus on these people as your initial target market.

Finding your early adopters is important, because most small businesses and startups have finite, if not scarce, resources. A little work up front will help you focus your resources on the people most likely to buy from you with the least amount of persuading on your part. Oftentimes entrepreneurs try to be everything to everybody all the time. I see this in many of my students' initial business ideas. I think this may be due to the ways that movies and books represent successful entrepreneurs. We hear about the wild success of Facebook, for example, and how everyone from grandparents to young kids use Facebook for so many various uses. However, this is misleading, because we are looking at Facebook *now* as one of the biggest companies in the world, and not how it began and grew.

Facebook is so ubiquitous now that it is hard to remember that Facebook actually, in its beginnings, was quite exclusive in its niche. From one of the earliest archived pages of Facebook's original website (https://web.archive.org/web/20050806011211/http://facebook.com/), you can see how it was limited to connecting people exclusively at colleges. One of the reasons for Facebook's success, I believe, is that it started out by focusing intensely on its early adopters: *college students*. Only those who had an ".edu" email address were allowed to register. College students wanted ways to search for people in their same school or in their same classes. They were hungry for ways to connect with friends of friends and look up people who were at a party the week before but whose name they had forgotten. Facebook, or "TheFacebook" as it was called originally, solved these problems for college students. College students in 2005—I was one of them, in fact—had the problems, knew they had the problems, and were looking for ways to solve the problems that Facebook solved. Joining Facebook as a college student in 2005 was a no-brainer. After gaining wild success with college students, Facebook had a proven concept to then expand further.

Nate helped me understand that I too should focus on a niche—on the people who really feel the pain the most. Following his advice, I discovered that my early adopters are executives at small- or medium-sized companies. He helped me realize that it was people like him, CEOs of small- to medium-sized companies, who have people problems, who know they have these problems, and are looking for ways to fix them. I now help executives in Japan and in the U.S. know what their people problems are now and in the near future, how badly these problems are hurting their bottom line, and then help them solve and prevent these problems to improve their profitability.

Principle 3. Connect to an *Ecosystem*.
The next thing that many successful entrepreneurs do is get connected to an entrepreneurial *ecosystem*. An ecosystem is important, because it is difficult to succeed without the aid and insight of other non-client individuals. Entrepreneurial ecosystems vary by region and industry, so the health and rigor of an ecosystem will depend on where you do business (both geographically and by industry). The ecosystem that exists for my startup in Tokyo is quite different from the ecosystem that exists in the Ozarks—and, frankly, the entrepreneurial ecosystem in the Ozarks is amazing!

There are several resources available to entrepreneurs in the Ozarks. The eFactory (https://efactory.missouristate.edu/), for example, offers training, mentorship, and other modes of support to local entrepreneurs. They also have an annual cohort of mostly tech ventures that go through their accelerator program. The Jordan Valley Innovation Center (JVIC) helps entrepreneurs with prototyping their products. The Small Business Development Center, in cooperation with the U.S. Small Business Administration, provides one-on-one consultation to Ozarks entrepreneurs, offering various analytic tools (including feasibility studies) in cooperation with faculty at Missouri State University's College of Business, where I teach. There are several local SCORE chapters throughout the Ozarks (www.score.org), where entrepreneurs can seek advice from volunteer business experts on topics that run the gamut from marketing to financing to law. Furthermore, there are faculty and staff from several universities in the area, including Drury University and Missouri State, who are all eager to make students and community members alike successful in their new ventures. There is access to funding through private investors, banks, and investment groups. Furthermore, there are several successful entrepreneurs in the area who stimulate the local economy and act as role models for other ambitious entrepreneurs. There are also resources specific to women entrepreneurs, such as Rosie (https://www.rosiesgf.com/), and minorities, such as Minorities in Business (https://sgfmib.com/). *This* is what a healthy ecosystem looks like. Hence, many successful entrepreneurs embed themselves within an ecosystem.

Principle 4. *Effectuate.*

The next thing many successful entrepreneurs do, or something they understand, is the idea of *Effectuation.* Though abstract in theory, *effectuation* can be illustrated by the reality cooking show, *Chopped*, whose chef contestants are given a basket of mystery ingredients and are judged by celebrity chefs on the quality of the meal they make with those ingredients in a short period of time. Now, conventional thinking on entrepreneurship begins with the end and then works backward: We want to create the next Facebook or the next big tech venture to "go IPO," so we gauge where we are now relative to this final outcome and then make a plan for how to get there. This is similar to the con-

ventional cooking shows (such as Julia Child's) where a famous chef tells the audience the dish they will be making. The chef then proceeds to gather the ingredients and cook the meal. The meal has been planned far in advance and everything prepared, ready for the chef before the cameras start to roll. This is similar to traditional business planning.

However, the world in which most entrepreneurs find themselves is much more chaotic and the ingredients are sparse and the time is scarce. On *Chopped*, similarly, the chefs do not know the ingredients beforehand, the kitchen is not one where they ordinarily cook, and they are being scrutinized throughout the entire process by judges, all within a short and frenzied timeline. The chefs in this scenario are doing what they can with what they've got, and it usually isn't much. In similar situations, the entrepreneur learns to effectuate: Resources are limited, so you have to make the best with what you have available to you. This is important, because entrepreneurs need to compete with the behemoth companies that *do* have access to greater resources and *can* reliably plan based on established financials, for example. In a manner of speaking, many entrepreneurs are competing with the "Julia Child" of their industry, so we need to be resourceful and smart with what we have available.

I think the best way to do the best with what you've got is through an idea, seemingly unrelated, called "Servant Leadership." Scriptural in foundation, the concept of Servant Leadership emphasizes service to others and lifting up others instead of one's self. We all have our "baskets of mystery ingredients," and I think one of the best ways to engage in effectuation is to ask myself, "How can I best serve others with the talents and resources I have?" *And then do that.* You'll have to pivot as you experiment and learn from people, but it's a wonderful place to start. For example, I began my journey into entrepreneurship by asking myself this very question. In a moment of reflection, I felt that creating a burnout-prevention app based on predictive meta-analytics would be how I could best serve people. Through experimenting, I learned from HR managers and executives that the best way to serve those people was by solving a related, albeit slightly different, problem that they were—and are—having. At that time, my own "mystery basket" contained ingredients of being homeless, but it also contained ingredients of being a researcher at Japan's top university. I knew what I had. I accepted it. And then I used it the best I could to serve others.

My "ingredients" now include being a professor, an entrepreneur, a husband, and a dad, with two business partners who handle the day-to-day operations. This allows me to focus primarily on teaching and serving the Springfield community while continuing to develop technologies to benefit workplaces in Japan and globally. This is how I am able to live in Springfield and work in Tokyo. Of course, everyone's "ingredients" are different, and that is precisely the point of effectuation: Take inventory of what you have available to you and serve the best way you can with what you have.

Principle 5. *Iterate!*

The final thing that many successful entrepreneurs do is *iterate!* I place the exclamation point there, because you are going to need energy, and a lot of it, to get through this entrepreneurial process successfully. My HR predictive meta-analytics software was not my first attempt at serving others through a new business. I have half a dozen other ideas that I attempted, though these are now buried somewhere in my soul. Very few people get it right the first time. The process of *creating value for others*—and that's how I'm defining entrepreneurship—requires perseverance in the face of rejection and sometimes even humiliation. However, keep experimenting, repeating, and iterating. Eventually, the hard work and perseverance will pay off.

As I've noted, the two personality traits that do seem predictive of successful entrepreneurship are "conscientiousness"—being hardworking and achievement-oriented—and "openness to experience." Certainly, principles 4 and 5 above are grounded in these traits. A person needs to keep working hard and needs to be open to others' viewpoints. As I'm describing it, entrepreneurship is more than a money-making activity: It's a way of being-in-the-world, tied intimately to one's personality. In its deepest soul-sense, entrepreneurship is not just what one *makes for others*, but what one *makes of one's self*.

Above, I told *what I did* to complete my year's study in Tokyo. Let me briefly revisit that story, to tell you *what I learned*.

Back to the Minivan

Being homeless in Tokyo was liberating and transformative. It was also frustrating. It was frustrating to find private places for daily hygiene. I would alternate between nearby Family Mart and 7-11 convenience stores to brush my teeth and use their restrooms, which in Japan are usually immaculate and well-maintained. I would shower at the university's gym. Sometimes I would do a day's worth of laundry in the gym shower and dry it in one of the lockers; I would retrieve it after work, before the gym closed. Other times I would use the laundromat. Hauling laundry to and from the laundromat was not so bad; however, I would make several trips to the van, timing each to make sure no one saw me put the laundry into baskets I had hanging in the van, covered with a sheet. Being a large American male, getting dressed in a small Japanese minivan was also frustrating. After getting dressed, I always peeked out the corner of the window to make sure there were no onlookers nearby; I wasn't sure what they might think of a big bearded white guy exiting a van that had, just moments before, been rocking back and forth and pumping up and down. I wanted to avoid that kind of embarrassment.

And living in the van was humiliating. Despite my best efforts, there were people, I think, who had caught on to me. There was a security guard from a building across the alleyway. He would train his eyes on me as I walked past and keep me in his gaze as I entered my van. I moved my van to a different part of the garage and changed up my routine to minimize the chances of running into him and others. And hygiene was humiliating at times, despite my efforts. I was kicked out of the university gym twice for being a "nuisance," due to stinky gym clothes. The Japanese are very particular about hygiene and highly sensitive to personal cleanliness in consideration of others. This cultural emphasis on physical and ritual cleanliness is tied to Shintoism, Japan's ancient spiritual tradition. It is also due to so many people—about 127 million people—packed into a mostly mountainous area about twice the size of Missouri. So I understood why I was kicked out. It was humiliating.

Asking for help was difficult. Intellectually, I know it should not be, because everyone falls on hard times to some degree at some point in their lives. Emotionally, however, it was difficult to accept help; I did not want to inconvenience anyone or be a burden. At what point is it acceptable to ask other people—other people who are trying hard to meet their own goals—for help? The situation revealed to me some of my innermost beliefs about myself and others. For instance, I could not, at some deep level, remove the idea that I had gotten into this mess, so I was obliged to get out of it as best as I could. Yet I tried to reconcile this with my beliefs about grace and second chances. Pragmatism beat out intellectualism in the end, though, and on one sweat-soaked and sleepless night in the van, at my wits' end, the pragmatic part of me screamed that it was fine to accept help. So I went to my church and asked for help.

I had a brief stint in a hostel paid for by my church. Hostels—not *hotels*—are essentially co-ed dorm rooms, typically with a dozen or so 20-year-olds traveling from all around the world. I enjoyed swapping stories with so many other wanderers, but my stories were always in past tense it seemed, from when I was a young, single, 20-year-old, carefree traveler. I did not share with my fellow roommates that I was currently a 33-year-old homeless father of four who had hit some hard times. Plus, I found myself waking up when many of the travelers were just heading to bed—partially due to their jet lag, and partially because they were free-spirited college students on summer vacation. After ten days or so in the hostel, I decided it was not going to work out. While the van felt like a sauna (I had the windows mostly blocked for privacy reasons, in case police or security guards tried to look in), I at least could shut my lights off when I wanted to.

I returned to the van. Tokyo has a latitude similar to Fayetteville, Arkansas, and so its summers are hot and humid. I would get dry ice from the grocery store—they gave it out for free if you bought prepackaged sushi—and use a battery-operated fan from the "100 Yen Store" (like a dollar store) placed behind the dry ice to blow cooled air on me at night. It worked for half the night until the batteries died or the cheap fan busted. The other half of the night I would sweat beads in the intense humidity of late summer.

A long-ago friend of mine had asked repeatedly to come visit, and through his inquiries he had deduced that I was homeless. He kindly opened up his house to me for about a month until the weather cooled down a little. I lied and told him I had found a place to live, because I was certain I was being a burden to his pregnant wife. I am very grateful for their help, which got me through the hottest weeks of summer.

I suppose the only way I know how to sum up living in the van is that I valued the freedom of it all, despite the frustrations and hidden humiliation. The paradoxes that living in the van allowed me to see helped me understand the world better. On my nightly walks from the university back to the garage, I would pass by many other homeless people and drunk "salarymen" (white-collared workers of Corporate Japan), and then meet up with a Rolls Royce or Ferrari at the very next crosswalk. The parking garage where I had stayed was located in one of the most affluent parts of Tokyo. In fact, there was a nearby Ferrari showroom and a Lamborghini dealership within a five-minute walk of where I parked my van. Yet only a little farther out, I would also pass the train bridges and overpasses where many homeless people had laid makeshift mattresses of corrugated cardboard. Their locations protected a person from the elements, but they had to compete with the pigeons and the mess pigeons made. Many of these people, I noticed, would be watching shows on iPhones or iPads. Meanwhile, I would walk home from my job at Japan's most prestigious university—second only to Harvard in terms of the number of alumni in CEO positions at Fortune 500 companies—back to my minivan, a relative luxury compared to cardboard. When meeting people for the first time I would tell them the district of Tokyo where I lived, and they were often impressed and thought it fitting for a University of Tokyo researcher. Sometimes I would add, "Well, I live in Azabu, but *in a van* in Azabu." The reply would usually be a laugh followed by, "I imagine rent is pretty expensive there, huh?"

My tentative conclusion of it all is that life is like the show *Chopped*. We all have our own unique "mystery baskets" and a brief period of time to make the best life we can from those ingredients. I am no more qualified to judge the quality of the meals the chef contestants make on *Chopped* than I am the lives the Ferrari drivers, salarymen, and homeless alike make on those streets. We are all working with our own mystery ingredients, and the first step is to know what's in the basket. Then each of us can know how to best serve each other.

Being the Medium:
Social Media and Identity in the Ozarks
Lanette Cadle

Love it or hate it, social media is now an integral part of daily life, as shown through research from the Pew Foundation's American Family Life and Internet Project; the Ozarks is no exception to this shift in how identity—and, yes, community—is formed. It's true that it can be difficult to know who one is interacting with online. It is also true that, just like IRL (in real life), there are knaves and scamsters, trolls and villains. However, also like life face-to-face, there are well-wishers, artists and writers, and just plain helpful folk ready to help, well, you. Whether through a forum, blog, comments section for a news article, Twitter, Facebook, YouTube, Flickr, Tumblr, Pinterest, or something yet to appear, and whether one uses a smartphone, desktop, laptop, tablet, or a device yet unknown, sharing through digital spaces is now as central to one's identity as being able to communicate effectively in person or in print; and the Ozarks is no exception to this. Of course, there are always Luddites reacting to any new technology, but the Pew internet usage numbers do not lie: Growing from the base percentage of 46% in 2000, in 2016 90% of all adults used the internet, and that included 67% of adults 65 and older, which is a considerable rise from 2000's 12%. The social media numbers are also significant and show a faster upward curve from 2008 to the 2016 data: 69% of all adults used social media and 34% of adults 65 and older (Anderson and Perrin 2017). With these numbers in mind, the idea of an identity—and community—online makes sense, and Ozarks culture embraces it.

The Ozarks has long been a home for writers, artists, and musicians—a haven for creativity. What social media brings to this picture that's new is how it gives literally everyone, not just professionals, the chance to communicate and create for the widest possible audience. Today, literally anyone can be a YouTube star and being witty on Twitter or in your blog can lead to a book contract. If it doesn't, that's fine, too. Content creation is no longer just for traditional media outlets like newspapers or television or only for the rich or those in cultural centers like New York or Los Angeles—all that is needed is internet access and the desire to communicate or create.

Social media by its very nature matches the universal desire to communicate with an unprecedented ease of access. What this means for the person next door or the laptop tapper at the coffee shop using free Wi-Fi is an outlet where each can build an identity, much like artists or writers do with their art. The building of identity through a mass of small interactions over time is one way to describe *ethos*, a term that encompasses a sense of self with a reputation built over time. It is also an accurate barometer for how social media functions. Social media builds ethos and can be seen as a way to accelerate the natural pace that ethos is limited to IRL (in real life), a.k.a. through face-to-face interactions. One example of a celebrity social media maven is Wil Wheaton, a.k.a. Wesley Crusher from *Star Trek: The Next Generation* and more recently, as the meta-version of himself on *The Big Bang Theory*. He does not have 3.22 million Twitter followers (as of January 31, 2017) simply because of leftover fandom from a long-ago television series. He built a new and more compelling ethos through social media posts that showcase a tech-savvy individual who has a lot of things to say about the intersections of geek culture and technology. In addition to that, he tweets with a great deal of wit and style—elements greatly prized in the social media world. Beginning with his blog, he was one of the first to move on to Twitter and use it well. The success of his book based on his blog, *Just a Geek* (2009), would not have happened without the careful groundwork and continuing growth of the Wil Wheaton self through social media. In social media—and this is a broad generalization across several types of social media past, present, and future—reputation is gained by sharing first, being consistently first and accurate, and being a consistent and reliable source for shared interests. Wil Wheaton is a good example of these principles in action.

Another way to create positive ethos through social media is more prosaic: simply and consistently sharing details about one's life over time. An excellent example of someone who has done this well is the author Neil Gaiman. An author in genres ranging from comics to television to novels, Neil Gaiman is a major presence on social media. As I note in "Shadow or the Self: Neil Gaiman and Social Media" (2019), he "creates a shadow self that gives fans the access they crave to his inner processes and daily life. This means making selections, much like a curator faced with a massive archive must select pieces that form a cohesive exhibit" or the incremental process of "a shadow self, an embodiment that both is and is not the real Neil Gaiman" (p. 150).

Social media done well, as by Wheaton, Gaiman, and millions of others less well known, creates ethos, something that can be seen as a unique and personal authorship—selecting some details, obscuring others. Viviane Serfaty writes of this function in her book on blogging and early bloggers, *The Mirror and the Veil* (2004), when she asserts that blogging "establishes a dialectical relationship between disclosure and secrecy, between transparency and opacity" (p. 13). She adds,

The paradox lies in the invisibility seemingly enjoyed on the Internet by both writers and readers. Thanks to the screen, diarists feel that they can write about their innermost feelings without feeling humiliation and identification, readers feel that they can inconspicuously observe others and derive increased understanding and power from that knowledge. (p. 13)

Serfaty also points out that "the very action of bringing something to light renders other areas even more opaque, so that the screen is transformed into a mirror onto which diary-writers project the signifiers of their identity in an ongoing process of deconstruction and reconstruction" (p. 14). This should not be surprising. A good blogger is a good writer, and good writers select what they show and what they withhold.

In addition to its mirror-like qualities, social media also acts as a virtual *polis*—a place where politicians, opinionmakers, and the infamous trolls roam. Citizens who write a letter to the editor for their local newspaper know they are sharing their ethos, their character in public. Users of popular forums such as *Something Awful*, *Slashdot*, *DeviantArt*, or even the commenters for the online version of the *Springfield News-Leader* know that they are sharing with a wide audience; in fact, that is the point. Whether from the viewpoint of the journalist or the private citizen, the fear of the de-professionalization of journalism is now old news. Blogs, Twitter, YouTube, Vimeo, and whatever shows up next down the digital highway exist and are fully formed; users don't have to ask permission from anyone to spread news. In fact, they never had to—digital social media simply gives them efficient venues. In this digital present, social media has successfully flipped what used to be a top-down broadcast into a collaborative stream of constant micro-blasts that give news, analysis, and advocacy from the ground up. What is lost in overall professionalism by some is gained by sheer numbers and the natural fallout of sustained ethos over time. In other words, reliable users gain readership/followers by their past performance. What is gained is a major shift from news controlled by corporate interests to news disseminated at the point of interest to others interested—no top-down, no expert vs. masses, and no editorial filters.

It is wrong, though, to see social media as merely a less efficient or, worse, less ethical way to disseminate news. The primary purpose of social media is not to serve as a news source replacement or even a re-mediated newspaper. No, the primary purpose of sharing news on social media is to build ethos for the news-sharer, plain and simple. Social media is another way people build their online persona, and how they show the identity that they choose, whether it be good, bad, or ugly—even if the choice is not a conscious one. Granted, some individuals or groups may construct an identity that includes the trope of "intrepid reporter," but for most, the sharing of news is an exchange of what has been called "cultural capital" in a "gift culture," where webs of like-minded people share pieces of themselves through their interests.

José Van Dijck (2004) was one of the first theorists to write about this concept; years later, social media scholars still see social media and, to some extent, the internet itself as a gift culture, a way the internet works when all is well. Of course, the downfall of a gift culture is what happens when the unscrupulous take what was freely given and re-identify it as theirs with no reference to the prior giver, better known as plagiarism. On one end, there are gift culture participants who get it right: They acknowledge the original source by naming them in-text and giving a permalink to the original post (they may even give a permalink to the home page as well). That is how to treat a gift—make it possible for those who wish to read the full post to get there quickly and easily through description and hyperlinks. On the other, dicier, end is the idea that

"everything is free on the internet," an idea that, when innocently expressed, is based on a misunderstanding of how copyright works—thinking that if there isn't a little "c," there is no copyright and the content is thus public domain. In reality, the opposite is true. Creation equals copyright—no symbol needed. Public domain cannot be assumed; it must be detailed and marked. The epitome of the idea that "everything is free on the internet" is the "scraper," a version of a spider app that crawls the web in search of keywords. When it finds the chosen subject, it then "scrapes" the article and publishes the whole piece on its own site. This is ethically challenged on several levels. Even when the scraper cites the source, publishing the entire post takes away reader incentive to go to the original site, thus breaking down the positive give-and-take of a gift transaction. Of course, some scrapers do not cite at all, which is the most obvious form of plagiarism. This theft not only hurts the writer/content creator involved; it leads to a breakdown of innovation and creativity, due to the fear that one's work or, in digital terms, one's identity, will be stolen.

The ease of use and open communication for all that makes social media, especially blogs and forums, so very appealing also makes them attractive to trolls and content thieves. However, just like in the print publication world, that does not take away the drive to communicate and create. Social media fame—or, in the common phrase, "being a star on YouTube"—is within reach for anyone with a little moxie and the willingness to tweet or, in the case of YouTube, to post video daily.

The Ozarks has its tastemakers and talented, those who have something and share it. One example is Jeff Jenkins, a local comedian and founder of the Skinny Improv, now Springfield Improv Theater. Under the username @theskinnyguru on Twitter, he is gaining presence with 2,020 tweets and 879 followers as of March 2019. He tweets local appearances and gives links to blog posts. Another comedian making a name on social media is Sarah Jenkins, @jenksie, best known locally for the television show, "The Mystery Hour." As of March 2019, she has 6,635 tweets and 835 followers. Visitors to her Twitter stream will learn about events like the "Rated SGF" film festival and read random witticisms like the March 21, 2019 tweet, "Oddly endearing: A grown man walking out of the office building with a lunchbox. I hope he had a baggie full of carrot sticks"—or this one, from March 13, 2019: "What if Facebook is found guilty of things and they have to give us all, like, $0.12 in compensation? Justice!"

From the creative writing world, Danielle Evans, former Visiting Professor of English at Missouri State University, is a nationally known fiction writer and frequent tweeter (@daniellevalore). With 12.1 K tweets and 1,929 followers as of March 2019, clearly her wit regularly wins new followers, many of whom are also writers and editors. An example is her tweet from March 21, 2019, where she writes about buying clothes for the Associated Writing Programs (AWP) national conference: "Usually when I buy something, the internet tries to sell me ten more of the exact same thing, but I bought serious professional clothes to up my AWP game, and now all my ads are David Bowie sweaters and sequins and animal prints like who do I think I'm trying to kid." She also is savvy about sharing links and photos. One of the strengths of Twitter is how easy it is to share media—so easy that there is an old joke about how the main purpose of the internet is to make it easier to share cat photos. Evans shares this tradition and captions the photo of her cat on the mantel in front of a ship picture with "Who's going to tell her that being a pirate involves water?" Indeed. Such personal glimpses of daily life are endearing and work well to create an appealing online persona.

Of course, this blend of personal and professional can veer over to the professional side, too. For example, many academicians in the Ozarks keep in touch with other scholars in their field through Facebook, Instagram, and Twitter. My own Twitter feed began in April 2008, almost at the beginning of Twitter. Many of my 869 followers are scholars in my field of rhetoric and composition as well as followers in my other field, creative writing. I have 4,782 tweets, follow 869 people and have 673 followers. Given that professional focus, it is not unusual that I have a high number of retweets, which are tweets that I forward as part of my Twitter

Sharing cats by tweet.
Image by Danielle Evans, courtesy of Lanette Cadle.

stream, usually something about a colleague's article or a piece of news (such as my March 17, 2019 retweet about the Lady Bears' tournament success). What is surprising though, despite my intent to use Twitter primarily for professional development, is the success of purely personal tweets, measured by retweets and how many times the tweet is favorited. One example of this would be my July 11, 2018 tweet, "Since I was up early anyway, I tried the 6 a.m. Early Morning Vinyasa. It really energized the day, but I wonder how long that will last. Check again in 12 hours? #yogadiary ." That was one of my #yogadiary tweets.

Cat photos remain a crowd pleaser too, as shown by the response to my December 23, 2019 tweet about a poem publication in *WSQ* (*Women's Studies Quarterly*): "Sasha and Mango apply cat rhetoric to the @WSQjournal Protest issue. Can you eat it? No. Can you play with it? No. Humans will feel differently though, and I'm thrilled to have my poem 'Staring at the Future' in this issue." I can't claim to be a Twitter giant, and none of the Ozarks Twitterers, myself included, currently have the kind of following shared by celebrity Twitterers Wheaton and Gaiman; but all have respectable followings and a clear online identity marked by a savvy interaction of blogs, Facebook, YouTube, and Instagram, all curated and promoted via Twitter.

Of course, Twitter is not the only social media out there that Ozarkians excel at. Sometimes the pictures (or videos) do the talking. In that case, Instagram is far more visual than Twitter or even Facebook. Unlike YouTube and blogs, it is not able to be monetized, but successful Instagrammers whose feeds are focused can attract sponsorships that can be extremely lucrative. Additionally, success on Instagram famously can lead to a book contract, as evidenced by the two bestselling poetry books by Rupi Kaur, who began as a slam poet and was (and still is) an "instapoet" on Instagram. Others sell commercial services and promote busi-

nesses via Instagram, and sometimes the line between business and hobby can be blurry. For example, one local instagrammer, Daley Gold-Swan, centers her Instagram feed on wellness, fitness, and travel and has been approached about sponsorships. Her feed is typical of ones that have a clearly defined focus and features a narrative shown through well-thought-out photos. Tim Shelburn is another Ozarkian making a name for himself on Instagram. A food blogger and self-described Backyard Grillmaster, Shelburn makes meat the centerpiece of his Instagram feed with grilled steak, shrimp, and chicken photos that are enough to make a vegan swear off kale. He also features specific brands for charcoal, ovens, and grill, making it probable that if he doesn't currently have sponsorships, he soon will.

The history of the Ozarks shows that writers, artists, and other innovators thrive here. Current and future social media use can only make communicating one's ideas and art easier. In the future, it's possible that trendsetting, news, and other fresh ideas will increasingly originate from places like Springfield, Nixa, Cabool, or Blue Eye—just as it does from New York City, Dallas, or Los Angeles. With the ease of access and communication made possible by social media, anyone can literally be a star on YouTube, Instagram, or Twitter—and thus influence the world.

References

Anderson, Monica and Andrew Perrin. 2017. "Tech Adoption Climbs Among Older Adults." *Pew Research Center: Internet and Technology.* http://www.pewinternet.org/2017/05/17/tech-adoption-climbs-among-older-adults/ Accessed January 31, 2017.

Cadle, Lanette. 2018. "Shadow or the Self: Neil Gaiman and Social Media." *The Comics Work of Neil Gaiman: In Darkness, In Light, and In Shadow,* Eds. Joseph Michael Sommers and Kyle Evelet. U of

Mississippi Press.

Daley Gold-Swan. @daleyfitt. Instagram home page. https://www.instagram.com/daleyfitt/

Jenkins, Jeff. @theskinnyguru. Twitter home page. https://twitter.com/theskinnyguru

Jenkins, Sarah. @jenksie. Twitter home page. https://twitter.com/jenksie

Serfaty, Viviane. 2004. *The Mirror and the Veil: An Overview of American Online Diaries and Blogs.* Amsterdam: Rodolfi.

Shelburn, Tim. @tshelburn. Instagram home page. https://www.instagram.com/tshelburn/

Raine, Lee and Janna Anderson. 2017. "The Fate of Online Trust in the Next Decade" *Pew Research Center: Internet and Technology.* http://www.pewinternet.org/2017/08/10/the-fate-of-online-trust-in-the-next-decade/ Accessed January 31, 2017.

van Dijck, José. 2004. "Composing the Self: Of Diaries and Lifelogs." *The Fibreculture Journal*, FCJ-012. no. 3, 2004.

Wayback Machine. http://web.archive.org

Wheaton, Wil. WIL WHEATON dot NET. http://wil-wheaton.net

III. The Twenty-First Century and Beyond

Silica granule, scanning electron microscope (630x magnification). *Courtesy of Austin O'Reilly, Dynamic DNA Laboratories, Inc.*

A Different Perspective on Life:
A Cell Biologist Looks Through the Microscope
Paul L. Durham

Recently I was asked what I do as a profession. Without much thought I responded, "I'm a cell and molecular neuroscientist and director for the Center for Biomedical and Life Sciences at Missouri State University." What I should probably have said to prevent the blank stare was, "I'm a biology professor," which typically elicits the response, "Oh, so you study animals—that's cool." I don't think too much about what it means to be a scientist. There really aren't many of us in the Ozarks. Unfortunately, most of the time the word "mad" is included in our titles, since Hollywood has done a good job making money on movies portraying (and promoting) the "mad scientist" character type. My personal feeling is that these films were produced by students who didn't do so well in their high school or college science courses. In truth, I am a biological scientist who studies how nerve cells get excited and then send pain signals to your brain. I work to discover ways to quiet the nerve cells involved in causing migraine, trigeminal neuralgia, and jaw pain—known as temporomandibular disorders or TMD, mistakenly referred to as TMJ (which is actually the name of the joint).

Being a scientist takes perseverance and a deep desire to want to know the answers to important questions; however, I've found that being a scientist also requires humility and a sense of humor, given how little we really know and understand about the natural world and even about the functioning of our own cells. As a cell biologist, I want to understand how cells perform their myriad tasks; to do this, I have to use the lab instruments available to me, including—and especially—the varieties of microscopes. The naked eye is a powerful instrument in itself, but it needs help peering into the smallest structures of life. The cells that I study are so small, hundreds of them can fit on the sharpened tip of a pin; the pinhead can hold many thousands. Fortunately, each innovation in microscopy has multiplied our vision: fivefold at first, soon tenfold, then more than a hundredfold, now many thousands of times more than before. English scientist Robert Hooke (1635-1703) was among the first to describe the "little worlds" or *microcosmi* revealed by means of the earliest microscopes; the etchings in

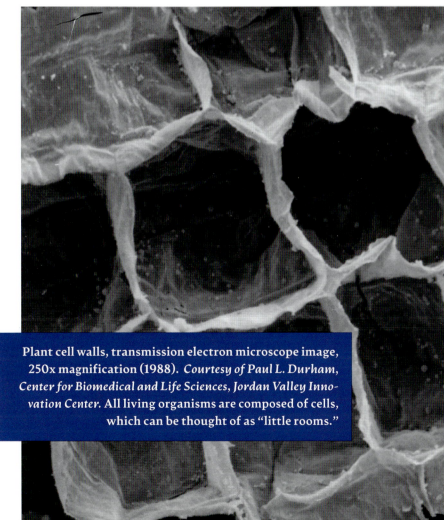

Plant cell walls, transmission electron microscope image, 250x magnification (1988). *Courtesy of Paul L. Durham, Center for Biomedical and Life Sciences, Jordan Valley Innovation Center.* All living organisms are composed of cells, which can be thought of as "little rooms."

his *Micrographia* (1665) remain monuments of early technoscience and are admired still for their artistry and accuracy. Aided by a single-lens candle-lit microscope, Hooke's drawing of the plant material cork reinvented our understanding of biological structure: He is credited with using the word "cell" to describe what appeared to be "little rooms" or *cellulae*, similar to those inhabited by monks. In comparison, here's an image that I captured early in my career, using a scanning electron microscope (SEM) far more powerful than Hooke's old instrument: Clearly it illustrates why the term "cell" is used in biology.

In my laboratory today, I study the activity of cells with the aid of a fluorescent microscope, which allows me to see changes in the level of proteins; these are the molecules that "perform the work" in a cell. To give you some sense of the challenge and excitement of understanding how cells perform their myriad tasks, here's an analogy from the game of baseball. In this model, proteins are the players, each assigned specific functions that allow the team as a whole to perform efficiently. Now, pretend that you have to describe to a total novice the essence of baseball in all of its rules and strategies, but you can only show that person single pictures (snapshots) of the dynamic, ever-changing game. How many pictures would it take to ensure that someone unfamiliar with the game of baseball understands the overall goal and how each individual player contributes to the team effort? In a typical major league game, there are nine position players on the field at any one time, along with several umpires. Honestly, is there a single person alive who knows *all* the rules that govern major league baseball? As complex as the game of baseball is for the average fan to understand, in our world of cells, the game of life is infinitely more complicated; and that's because there are literally thousands of players (remember, we're calling them proteins) in each cell.

Adding to the complexity of our analogy, let's say

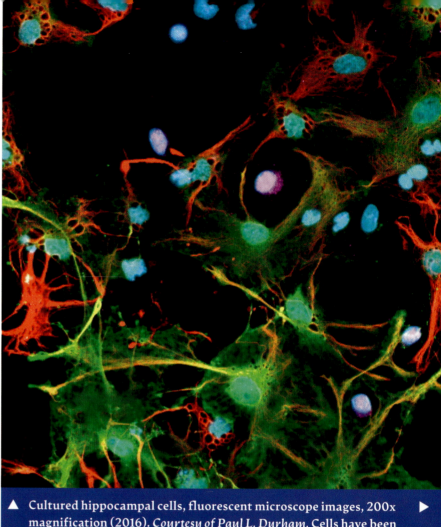

▲ Cultured hippocampal cells, fluorescent microscope images, 200x magnification (2016). *Courtesy of Paul L. Durham.* Cells have been stained with fluorescent dyes to visualize internal cellular structures. Information gained from these images allows us literally to see changes in response to a neurological disease or in response to a novel drug. ▶

that you've been successful in helping your friend understand the rules and teamwork involved in the game of baseball; in which case, you've managed to describe *how one single cell functions in your body*. There are, however, several hundred different kinds of cells in the human body. So, now you have to explain how each cell is uniquely different from other cells—which

186

is like explaining how baseball is different from basketball, cricket, hockey, soccer, football, etc. And all of these teams (cells) with their specialized players (proteins) are using the same field—basically, you! And they're all playing at the same time, though they have to coordinate their games—that is, their cellular activities. Hence, all the different cells of the body are all talking to each other to coordinate their functions to maintain your health and allow you to perform the myriad activities involved in your daily life. This is the challenge that cell biologists face every day: By observation and experiment, we have learned many of the actions and functions involved in biological life at the cellular level. We're in better shape than that novice, whom you tried to teach baseball. But, going back to the early history of microbiology, *we scientists were the novices* trying to learn "the rules of the game" by mere snapshots. And we're still learning. To state

the obvious, our world of cells is very complicated. But that is what makes studying them so exciting: For a cell biologist, discovering a new function of a protein is like discovering a new planet—or *microcosmos*, as Robert Hooke might have termed it.

Much has changed in the thirty years that I have been photographing cells. Back in 1984, when I first became a researcher at the University of Iowa, there was not a computer in every lab and I did not own a personal computer. In fact, I did not even own an electric typewriter but used the manual typewriter that my mom had used in the 1960s. To capture images of cells, I made use of a manual-focus microscope and F200 and F400 Kodak black-and-white film. Sometimes I used color; either way, the film had to be developed. I'd take the photos based on manual settings for aperture and exposure time and then have someone from a local camera shop pick up the film and develop it within twenty-four hours (there was an upcharge for faster processing). I still remember the excitement of holding the roll of negatives up to the light to see if any of the images were usable. The process was incredibly laborious, since you could spend hours taking photographs of cells only to discover that the exposure time was not correct and the images were either too bright or too dark. After viewing the film negatives and determining whether there were some useful images, I'd then spend hours in the darkroom perfecting the art of "burning and dodging" to get the textbook 4" x 6" print for publication.

Thankfully, my days of having to capture images of cells manually were short-lived. Today, our laboratories in Temple Hall and in the JVIC Center for Biomedical and Life Sciences are equipped with state-of-the-art digital microscopes; these allow us to capture amazing images and greatly aid in our understanding of how cells in our nervous system function to cause disease and respond to therapies.

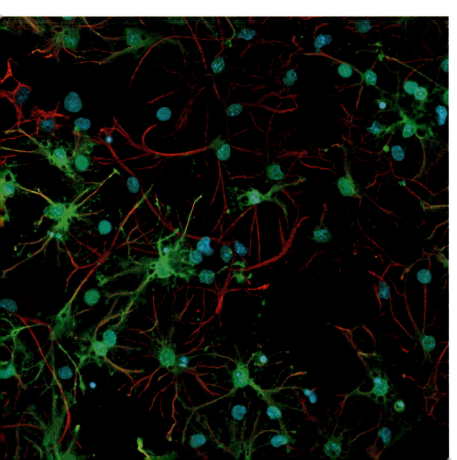

Similar to scientists in other disciplines, I am, first and foremost, an adventure seeker who is driven to discover and learn something new each day. I want to unlock the mystery of how cells in the nervous system function so that I can better understand the pathways that lead to development of chronic pain and identify novel ways to prevent, manage, and possibly cure the incredible burden—physical, emotional, social, economic—caused by diseases of the head and face. After thirty years of being a trained cell biologist, I still get a natural high from learning something that no else knows and then sharing that information with scientific communities via conferences, lectures, posters, and publications and then on to the general public, oftentimes with help from public media. I hope that these images help you to appreciate the beauty and mystery of the incredible cellular world that I have the privilege of studying each day.

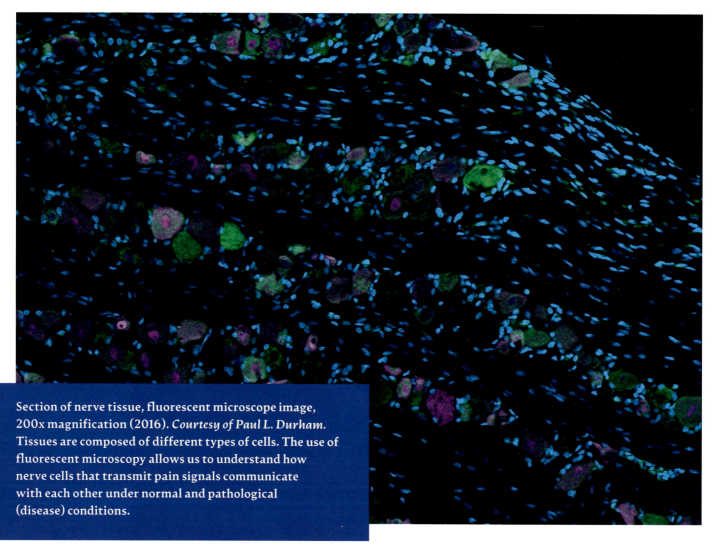

Section of nerve tissue, fluorescent microscope image, 200x magnification (2016). *Courtesy of Paul L. Durham.* Tissues are composed of different types of cells. The use of fluorescent microscopy allows us to understand how nerve cells that transmit pain signals communicate with each other under normal and pathological (disease) conditions.

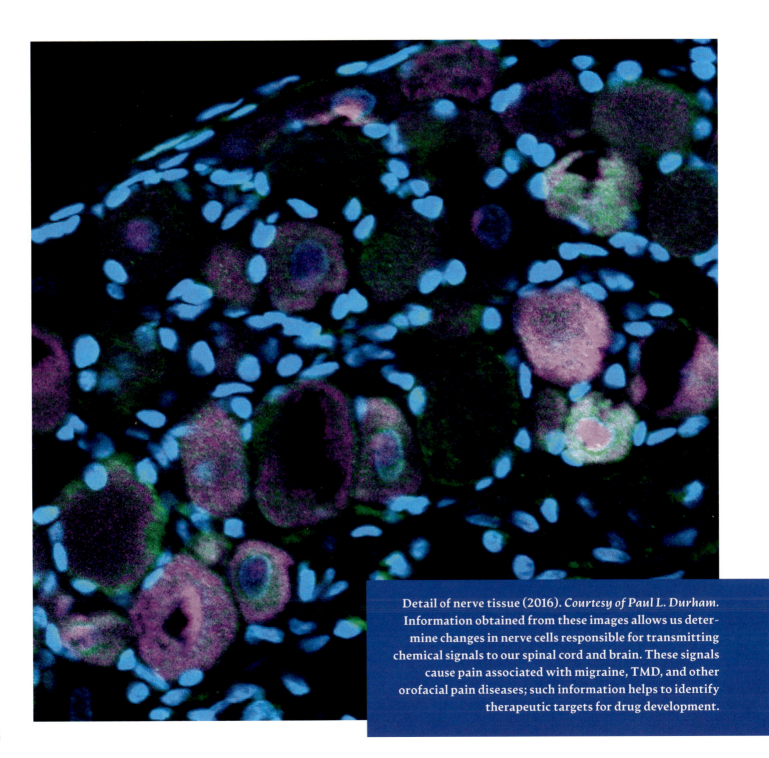

Detail of nerve tissue (2016). *Courtesy of Paul L. Durham.* Information obtained from these images allows us determine changes in nerve cells responsible for transmitting chemical signals to our spinal cord and brain. These signals cause pain associated with migraine, TMD, and other orofacial pain diseases; such information helps to identify therapeutic targets for drug development.

On Data Generation and/as Digital "Drawing":
Agrarian I/O Studies
B. Colby Jennings

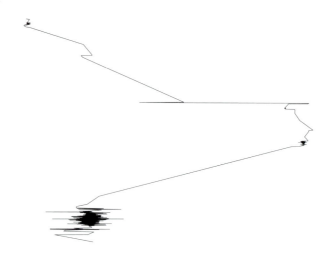

Farm Cat:
5000 GPS location points with path lines.

Farm Cat:
5000 GPS location points with software-enhanced path lines.

As a teacher, researcher, and practitioner in the digital arts, my work dwells in the intersections of art, communication, design, engineering, media, and science. The following body of work utilizes GPS data collected by tracking the movements of various domesticated animals, including livestock, at different times of the year on farms and in fields in Greene and Bates Counties. Featured are sheep (tracked in their movements between two separate fields), cattle (tracked across a field of several acres), a cattle dog (working a smaller acreage on the same farm), and a farm cat (tracked in its movements from the farmhouse to outlying barns).

The equipment used for data-gathering is straightforward in its construction, built around an environment-sensing Arduino Feather Board programmed to receive inputs from a GPS logger running custom firmware; assembled with battery and kept in a weatherproof case, the device is half the size of a deck of playing cards. With the exception of the farm cat, this device was attached to a collar and placed around each animal's neck. The GPS tracking data were then processed using custom software to generate digital images from the input. The resulting output were drawings created by the movements of animals that are commonplace in pastoral landscapes, though rarely (if ever) mentioned in today's recurring conversations over human data generation, data collection, and the resulting uses of those data.

Whether through GPS tracking of mobile devices, the various computers or portals accessed, the collection of web-usage data, or even newly deployed facial recognition software connected to massive arrays of surveillance cameras, people using modern technology generate "drawings" similar to these each day. These "drawings" are hotly debated, raising concerns over privacy and 21st century digital ethics.

When decontextualized—that is, when removed from the animals and the fields from which they were gathered—the information collected by GPS tracking creates visual patterns that take on qualities associated with computer-generated art. Put perhaps simplistically, these images raise a timely question: When does a visual rendering of data cross over from the realm of digital information into the realm of digital art?

This question—combined with an affinity for the pastoral landscape in which I was raised—led me to begin collecting data from animal locations. I saw this as a way to appreciate the data in pure form while nodding to the larger conversations currently embroil-ing our society. When the number of GPS points ran over a few hundred, I needed to develop a method to plot this information other than by hand. I now uti-lize the Java Development Software to create pro-grams that plot the location points, add lines that connect the points, and then stylize the lines for visual effect. Each drawing is titled according to the animal tracked and the number of tracking points plotted.

In labeling these digital images as "drawings," notions of an otherwise traditional medium are stretched and redefined. At its core, this work is quietly positioned to encourage further thought and questions regarding data generation—whether voluntary or involuntary—as well as the means of data collection and how these data may be used. A sampling of agrarian I/O images follows.

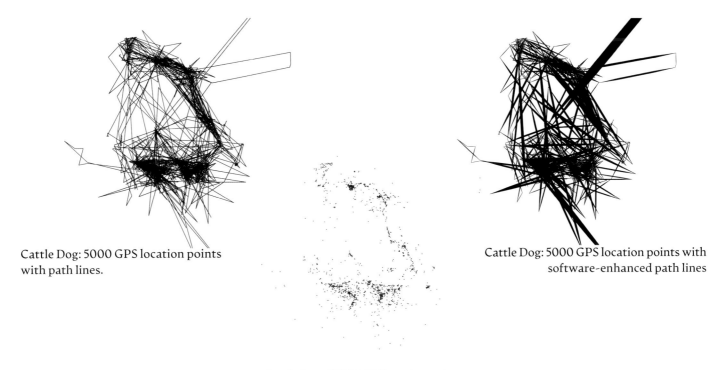

Cattle Dog: 5000 GPS location points with path lines.

Cattle Dog: 5000 GPS location points with software-enhanced path lines

Cattle Dog: 5000 GPS location points.

Sheep: 5000 GPS location points.

Cow in Field: 5000 GPS location points.

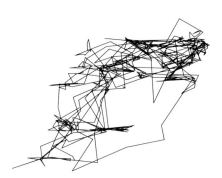

Sheep: 5000 GPS location points
with path lines.

Cow in Field: 5000 GPS location points
with path lines.

Sheep: 5000 GPS location points with
software-enhanced path lines.

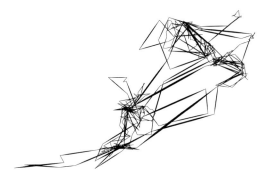

Cow in Field: 5000 GPS location points with
software-enhanced path lines.

A Forum on the Future:

Texts and Interviews of Clifton M. Smart III, Elizabeth J. Rozell, Rachel M. Besara, Thomas E. Tomasi, Lloyd A. Smith, William B. Edgar, Craig A. Meyer, Robert T. Eckels, Frederick Ulam, Martha Cooper, Julius Santos, Phillip "Cloudpiler" Landis, Jacob F. Erwin, and Dan Scott, with written contributions by Christopher J. Craig, Rachel Anderson, Lyle Q. Foster, Susanna B. Reichling, Morris Kille, Jr., Hanna Moellenhoff, and Jennifer Probst

Transcribed and/or edited by Allee Armitage, Tita French Baumlin, Sarah Crain, Jasmine Crawford, Hannah Fox, Cara Hawks, Arylle Kathcart, Jacob Miles, Hanna Moellenhoff, Austin Sams, Grace Sullentrup, Nicholas Stoll, Emma Sullivan, and Jesse Walker-McGraw, with James S. Baumlin

Our forum deserves some explaining. The idea derives from a book. In the final pages of *Inventology: How We Dream Up Things that Change the World* (Boston, MA: Houghton Mifflin, 2016), Pagan Kennedy describes the internet as a living, expanding rhizome, with decentralized open-sourced data-mining as its mode of health and self-protection. As she envisions it, the internet democratizes research, taking research and development out of corporate-academic labs and placing it in the hands—rather, the laptops—of the people. If Kennedy's right, then we need to rethink research—the who, what, how, and why of it. Though many of us teach it and practice it, we can declare that R&D is not for a course grade or a career: *It's for life*. So Kennedy suggests:

> Of course, the centralized, Edison-style R&D system continues to be crucial to the way we keep billions of people fed, clothed, and warm. *But we are already beginning to see the enormous potential of the transparent, decentralized, open, and chaotic R&D system that belongs to no one and lives everywhere.* My hope is that it will evolve into something like a global immune system—that we can reinvent invention itself and put the designed

environment in the hands of the many rather than the few. Nature uses multiple strategies rather than a monoculture to solve problems. The pathogens that threaten us constantly mutate, changing their tactics, making discoveries, hitting on breakthroughs, finding ways to disarm our drugs. Nature's research and development is based on billions of experiments. It is a resilient, decentralized, and wildly diverse scheme. Cancer cells defeat us simply with their numbers and the trial-and-error ingenuity with which they evolve. How to fight back? By recruiting millions of people to try billions of experiments and thus mount a random sweep through the unknown that might unlock the secrets of killer diseases. This system may also be our best hope for surviving the problems that will be unleashed by climate change—from water shortages to mass extinctions to agricultural disasters.

Apple and Google create terrific products that we love. But corporations are not designed to solve social and environmental problems. *We need to evolve another system of invention that mimics the natural world,* so that we become a more robust species, better able to survive and adapt. We need an R&D system like the system that protects our own bodies—open, obstreperous, and resilient. It should gain strength from every attack on it. And this system must be intimately connected to the pain receptors rather than cut off from them. *The people who suffer from the failures of technology— the Lead Users, the whistleblowers, the sick, the disadvantaged, the poor—are the ones who are best able to diagnose our problems. They are—or should be—at the center of R&D.* (p. 223; emphasis added)

Arguably, the health (and even, perhaps, the survival) of our species rests on our collective capacity to invent our way out of the messes—economic, ecological, technological, political, cultural, spiritual, etc.—that we find ourselves in today.

A reporter by trade, Kennedy interviewed a hundred-plus inventors, innovators, and entrepreneurs, discovering patterns in their success. A dozen aphorisms follow, deriving from her work. In our own interviews of local entrepreneurs and campus-community leaders, my team of MSU Honors College writing students found confirmation of her claims.

Successful thinkers, writers, researchers, planners, practitioners, leaders and policymakers possess or pursue many of the following:

1. *They anticipate needs in emerging disciplines and technologies and seek to be "cutting edge."* Kennedy (p. 106):

"We shape our tools, and thereafter they shape us," the media scholar John Culkin once observed. The most ingenious technologies teach us to form new habits. And once you've acquired that habit, the technology is a part of you, and you can't imagine living without it.... [T]echnologies don't need to be slick or fully realized to catch on. Instead, they must open up a crack in the world,

like Alice's looking glass opening a door into Wonderland; we pass through these machines as if they were portals to the future. The[y] are predictions made out of wire, glass, and molded plastic. They let us taste, feel, and smell what could be, which is a sort of sorcery.

2. *In exploring present conditions, they learn "to see differently."* Kennedy (p. 61):

One of the most important parts of the inventing process involves an unusual ability to visualize an imaginary world—to "prototype an idea by constructing it in the mind's eye."

3. *They practice "active imagination" as seeing beyond the horizon of present circumstances.* Kennedy (p. 136):

Look around the room where you're sitting right now and try to "fast-forward" it by fifteen years. To do this properly, you have to ask yourself innumerable questions: Is the air outside still breathable? How do people communicate now? Does America still exist? And on and on.

4. *They engage in elaborate "thought experiments."* Kennedy (p. 128):

[As] a teenager at a progressive school, Albert Einstein learned the principles of mental experimentation, and at the age of sixteen, he started to wonder what it would be like to zoom alongside a beam of light. As an adult, of course, he famously used his imagination to work out his ideas in a landscape he outfitted with hurtling trains, locked trunks, stopwatches, elevators, and beetles. Nikola Tesla, the visionary inventor, also used a lavishly outfitted mental laboratory.

5. *They project possible futures through various means, including narrative.* Kennedy (p. 136):

> You will notice that as you begin working out the details of your future, *you are telling a story.* The R&D lab inside our minds has a highly narrative quality to it. If you're inventing a futuristic machine, you have to tell a story about the people who will use it: Where do they live? What are they worried about? What do they desire?

6. *They take risks, accepting failure as part of the process of invention and, in exploring possible futures, they imagine best-case and worst-case scenarios.* Kennedy (p. 35):

> [Gary Klein] has devised a method—he calls it the "pre-mortem"—to sharpen the imagination, using it as a tool to predict disaster. He asks executives to use their minds to time-travel into the future and then look back at the plan that they're about to put into action. As Klein explains, "A typical pre-mortem begins after the team has been briefed on the plan. The leader starts the exercise by informing everyone that the project has failed spectacularly. Over the next few minutes those in the room independently write down every reason they can think of for the failure—especially the kinds of things they ordinarily wouldn't mention as potential problems."

7. *They look for problems, weaknesses, and blind spots in current disciplines and technologies.* Kennedy (p. 160):

> Every discipline shares the same library of solutions and also the same blind spots. That's why it can be so powerful to tap the knowledge of someone outside the field—that person will come equipped with a different model of reality, a different set of tools, and her own library of solutions.

8. *They learn by close observation of their fellows, seeking out mentors and role models.* Kennedy (p. xii):

> It's crucial that we find out what people actually *do* as they invent things. What are they doing in their minds and with their hands? We need a new field of study—call it Inventology—to answer that question.

9. *They look to cross-fertilize systems and approaches through "hybridization" of disciplines.* Kennedy (p. 164):

> The French term *déformation professionnelle* wonderfully sums up the way our jobs shape our minds; the term implies that every profession comes with its own worldview. If you're a structural engineer, you'll glance at a rotten beam in the foundation of a house and see trouble. If you're a microbiologist, the rotten wood seems like a wonderful place to harvest fungi…. Every form of expertise includes a belief system about what is possible or impossible, what is pointless and what is essential. The winners of problem-solving challenges seem to be relatively free of *déformation professionnelle.* They're interested in connecting, cross-pollinating, and zigzagging.

10. *They don't wait for others to solve the world's problems; instead, they work for the benefit of others and look to solve specific problems for specific communities.* Kennedy (p. 21):

> Some engineers and designers … embrace the practice of ethnography, spending weeks or months among a community in an effort to gain a deep understanding of local problems.

11. *They think "outside the box," looking for collaborations across professions and disciplines.* Kennedy (p. 11):

If you want to understand a problem, *ask a community.* When lots of people experience the same kind of pain or frustration, they generate an enormous amount of information about unsolved problems.

12. *They "mine for information," using web resources.* Kennedy (p. 14):

> The Internet ... is a "negative space," because it belongs to no one and is continuously revamped by everyone. It affords an entirely new ecosystem—a virtual universe—that nourishes the weeds and wildflowers spawned by millions of minds; it's like a vast R&D lab where we all share our own experiments and benefit from the work of others.

Such are the recipes of successful thinkers, writers, researchers, planners, practitioners, leaders and policymakers. In building our "Forum on the Future," we emulated Kennedy, asking members of the community—MSU students past and present, professors and professionals on- and off-campus—to give their recipes for success, as well as their predictions for the near future. Several wrote their own texts; most chose to be interviewed. Some spoke or wrote of their neighborhoods and the prospects for growth; some noted the role of technologies in healthcare and quality of life; several sounded the alarm over technologies and their health impact; several raised questions over ecology and quality of life; and several, being academicians, described an evolving university. Do not expect consensus. Expect, instead, a healthy diversity of viewpoints.

On behalf of my fellow interviewers and co-editors, I thank contributors for the stories and predictions that follow.

—*James S. Baumlin*

Technology Has Leveled the Playing Field

Throughout the 20th century, businesses favored major coastal and metropolitan areas. In the 21st century, location still matters—but not in the same way. It's geography that will put Springfield on the business map: not because its metro region is densely populated or located on a river or coast, but because the Ozarks offers an environment, and a lifestyle, and a community ethos attractive to new generations of entrepreneurs. The future of Springfield and the region will be bright and prosperous, and the Ozarks will continue to grow with the rapidly changing times.

As recently as five years ago, if you wanted to work in a specific industry—say, fashion or technology—there was no cogent alternative to living in or near Los Angeles, New York City, or San Francisco. You lived where other like-minded individuals resided: That was the obvious and unavoidable first step. "Big city life" seemed preordained, necessary to acquire the connections required of almost any business wishing to scale at a significant pace. Now, with advancements in technology and the "rise of the rest," this is no longer the case. The proliferation of technology and ubiquity of social media platforms has produced a level of connectivity previously unknown in human history. As a result, the customer, the distributer, the blogger and/or influencer are all just a click or swipe away. When you consider the rate at which internet speeds and connectivity are expanding, you're presented with a potent recipe for disruption of the traditional business paradigm.

This new paradigm reaches from the individual to the major corporation (and everything in between). As a result, we will see cities like Springfield and metro regions throughout the Midwest reap the benefits of a more discerning corporate culture. In the future, even a sizable company founded on the coast or in a large city may not be able to scale at the desired rate;

the available workforce may prove inadequate or the sheer cost of running a business there may prove too high. Increasingly, companies will turn to the Midwest for purely logistical reasons: cost of living, workforce availability, work ethic, available land, office space, etc.

Technology offers options that heretofore didn't exist. Springfield and the Ozarks will benefit for years to come, in ways that will be as drastic, as far-reaching and, ultimately, as unpredictable as technology itself and the world as a whole. What remains clear to me are the intrinsic and ineffable qualities that place this region and its residents above the rest. A spirit of innovation. A willingness and capability to adapt. A sympathetic and encouraging community that genuinely wants everyone to succeed. A knowledgeable support system and access to resources on par with any long-established business hub. Technology has leveled the playing field and Springfield, Missouri is ready to play ball.

—*Rachel Anderson, Director of the eFactory (part of the Idea Commons, MSU downtown campus)*

Renovating, and Living, Downtown

I was born and raised in Springfield, but I left for five years to get an architectural degree at Kansas University. My career as an architect began roughly in 1984, and I quickly realized that my income was going to be tied to the amount of time I spent staring at a computer screen and drawing. After five years of sore wrists and blurry eyes, I wanted to branch out. I felt that real estate was a natural companion to architecture building renovation. Real estate in downtown Springfield, which I have always had an interest in, was going for such an inexpensive price back then, it was easy to get into renovation work.

I started Jericho Development as a partnership in 1996, bought out my partner in 1997, and have partnered with my wife ever since. The second floor of the 333 South Jefferson Avenue building (our first renovation project) has an office space that my wife and I occupied. Every night, when we came out of the office to head home, we looked at the building that is now 400 Place. It was boarded up and falling in on itself; it was probably the most difficult building in downtown. There have been other difficult buildings, but this one had to be put back into place, structurally. That's what led me to take on 400 Place. We spent eight months putting the building back into place before we could spend two years renovating. It was a very difficult project, but it was difficult coming home from my architectural office every evening and seeing this boarded-up building in need of someone with the ability to do the work. I always tell people that I'm a member of the "More Guts than Brains Club." My wife and I have lived at 400 Place for fourteen years. We designed a beautiful apartment for ourselves that is, frankly, as big as the house we came from. There's a certain simplicity to apartment life. We have the luxury to be bored. My wife and I put three or four thousand miles on our cars, combined, in an entire year. Being downtown, we love being able to head out the front door and then decide which restaurant we want to go to. We can walk to the movie theatre, and that is just nice.

Overall, I would give downtown Springfield a grade of B. I've been in downtown for nearly thirty years and am aware of the overbuilding of living quarters. We've lost a lot of the vibrant daytime activity. Most people who live downtown must leave downtown to go to their jobs. I would like to see it get some vibrancy back with office use (and that would move the grade from a B to an A- or A). Then, we'd have the daytime clientele that would balance the evening and nighttime. More offices downtown would allow people who live here to walk to their job. I think that that would better support the storefront businesses, because they'd have somebody here in the daytime and evening. Now, I'm working part-time for another design firm. I walk three blocks over to their office on South Street, and that's what I wish two thousand other people did.

Downtown is becoming the hotel hub for Springfield, and that's bringing a lot of people downtown. They'll go to the restaurants and they'll shop at the storefront businesses. If we're not able to quickly attract office users, then the next best option is to develop downtown as a hospitality center. Having three to five flagship hotels downtown would go a long way to accomplishing that.

It is hard to predict the future. I have observed that about every five years there's a cycle in downtown. It's like the greater economy. There are upswings and downswings. In the downswings, maybe you'll see some businesses close. We could possibly be in that downswing now: Not that I've seen many businesses closing, but downtown is missing an energy level or vibe. I believe the next upswing could be the biggest upswing that we've seen. There's a huge apartment complex being built, plus the new Hilton. There are so many projects teed up, so you know the next upswing could be an amazing five-year period. These projects will bring along new neighborhood restaurants, clothing stores, and specialty shops. There is potential in all these projects. And we'll see the continued connection of Missouri State's main campus to the downtown campus. It's inevitable.

—*Dan Scott, AIA, MBA and co-founder and owner, Jericho Development Company*

Building Businesses of Color

Big Momma, as she's known around here, brings a passion for great food, coffee, and Southern hospitality to the heart of the Ozarks. She's the pride of historic C-Street, and though you won't see her around very often, you'll always feel her warm, welcoming spirit at Big Momma's. She's for dreamers, teachers, listeners, storytellers, and most of all—friends. Join us at Momma's house where friends become family.

—"Big Momma's Story"
(www.bigmommascoffee.com)

As the owner of Big Momma's Coffee and Espresso Bar on Commercial Street, I'm happy to declare that Springfield is a great city for entrepreneurs, no ifs, ands, or buts about it. I am one of those who constantly remarks on the vibrant entrepreneurial climate. One of the aspects I boast about is the fairly affordable entry point for millennials and people of other demographics to start a business: Startup costs (such as rent, licenses and other fees) are reasonable, the public is supportive, and the trade area is larger than the population size might seem to indicate.

Communities of color have a very interesting history in our fair city and, if current trends continue, a future that should be more robust. The past informs us of the many professionals (doctors, lawyers, dentists, pharmacists), inventors (like Walter "Duck" Majors), and small business folks who, at the turn of the 20th century, made Springfield a regional hub of black trade and entrepreneurship. Newspapers and other records of the time mention some of these early businesses, such as Hardwick Brothers Grocery Store—once the largest in the city. The era of segregation reveals a smaller African-American community, but many businesses thrived, particularly on the North Side, with a few having a national reputation. Catering to Route 66 travelers, Alberta's Hotel (just west of town) was listed in the now well-known *Green Book*, while Graham's Ribs on Chestnut Expressway and Washington Avenue drew patrons from far and wide. Many of these businesses disappeared with full integration, which gave consumers many more choices;

also, younger generations often left the old family businesses, seeking opportunities in other parts of the country.

The 21ˢᵗ century has brought opportunities and a few challenges for entrepreneurship within communities of color. There is a greater focus than ever on how important small businesses are to the general economy and many consumers are aware of the need to shop local. While capital can still be a challenge, there seems to be a broader level of support in securing resources and mentoring.

On an individual level, I have met with a curious blend of learning sprinkled with curiosity over my being a diverse person starting and operating a business. I have given thought to this and attribute some of it to people's lack of familiarity with entrepreneurs of color. Many entrepreneurs of color tend to work in specialized businesses that cater to a specific community, such as barber shops and salons and other forms of personal services. And Springfield has had a history of black barbeque restaurants, so that is somewhat familiar. But opening a coffee shop seems to have put me on a slightly different path. There have been times when customers would assume the college student standing next to me was the proprietor. Or they'd ask if I was the cook. The possibility of me being the owner lay beyond their expectation. Learning that I was the owner would often prompt additional questions, some of which, I think, showed genuine interest while others tried to sort out how a person of color would choose this line of business. Usually, I'd offer that this was an opportunity to fashion a new model of what black entrepreneurship might look like. And I say this because, more than a few times, I have been asked if the store sells barbeque when I have yet to find a coffee shop that had barbeque on the menu. (This is not to say they don't exist, only that I haven't visited one.)

An increasing number of people approach me asking about the business climate in Springfield. There is a growing enthusiasm in the African-American community to start and grow businesses; they remark that seeing other people's examples has helped to fuel the entrepreneur fire in them. And to that I can give an enthusiastic yes: This is the best time to plant and grow a business, because the city is fertile ground for all. Perhaps the more businesses of all types that we get to see in the Springfield area, the more normal it will seem for a person of color to own a coffee shop—or any other business.

Historic C-Street was once a strong business district and it is slowly and surely coming back—and that is good news for all of us. Being an African-American businessman on Commercial Street makes me aware of the diverse communities within a community. And I must admit, there are times when I feel a bit of constraint in the ways that I share publicly the issues of importance to the African-American community. A coffee shop serves as a community hub, so there are times when various political entities ask to stage events, post signs, or use the facility as a meeting or rally point for various causes. There has been more than one occasion when I have wondered whether that event or poster or meeting might offend or create some potential backlash from our customers or the larger community. At some level, I think that all businesses have to consider their public profile; but, for entrepreneurs of color, I think there can be a different sense of the tightrope that has to be navigated.

Still, I consider the overall environment to be positive for entrepreneurs of color and of other diverse backgrounds. In fact, I am fairly excited about recent developments that include additional brick-and-mortar operations owned by diverse folks. City leadership and the general public seem not only welcoming but intentional—proactive, even—in identifying and supporting both non-traditional businesses and diverse businesses. So let's get started planting and growing these businesses, together.

—*Lyle Q. Foster, Assistant Professor of Sociology and Anthropology, Missouri State University*

Baking Healthy, Building Community

I didn't begin my career in baking. Until January of 2008, I had worked in hospitals on the East Coast, assisting in cardiothoracic surgery. During the economic recession, I returned to my hometown Springfield, but no hospital could pay me what I had made back East, so I couldn't find a job.

I had always wanted to learn how to bake, so I began to teach myself new recipes and new techniques. A friend of mine, a master gardener, would bring me different herbs to sample in my baking. That's how my business began: Supplied by my friend, I opened an herb shop at the Commercial Street Farmers' Market. Part of the job was educational, teaching customers how to use herbs in their own kitchens. After a year of working solo, I took on a partner, creating Sisters in Thyme as an herb company. Two years later, we opened up a C-Street storefront.

Soon after we opened Sisters in Thyme, my life partner and business partner, Gina, was diagnosed with celiac disease. For years, I waited for stores and restaurants to take on the gluten-free market and produce foods that people with this medical condition could eat *and enjoy*. Transitioning to a gluten-free diet is not a fad. Celiac disease is real, causing people's bodies to treat wheat products as deadly invaders. In fact, gluten intolerance is becoming a reality for more and more people every day.

I kept waiting and no one did anything about it, so I decided to do something myself and began baking gluten-free breads and desserts, selling them in our bakery-deli restaurant. There is no reason why people with a condition like celiac disease shouldn't be able to eat healthy, delicious food. We serve a lot of food here that our customers don't necessarily know is gluten free; they just know they feel better after eating it. I want people to be able to heal through our food and to leave knowing that they were taken care of. And, you know, people's lives are uplifted by our foods, because they can actually go out to eat knowing that it will be safe. We've had mothers cry, telling us that this was the first time their celiac child truly had a choice in what to eat.

When I began making my baking and cooking about other people's health, that's when I felt the most successful. Right now, we're the region's largest supplier of gluten-free baked goods. That's entrepreneurship with a social mission.

Why C-Street, you ask? When I was a kid growing up in Springfield, it's where we came to shop. This was our mall. There was the denim store, Busy Bean; there were Della's Wigs and Rathbone Ace Hardware. It was a great street then, and I knew it could be great again: I could see the potential. But when I started out, there weren't many open storefronts. Up and down C-Street there was empty building after empty building. I knew that my customers would include the neighborhood's homeless and residents from the Missouri Hotel. So I accepted food stamps. I valued my little neighborhood and was glad to be of service. In fact, when the Hotel closed in 2015, I lost a big part of my business and had to diversify. Catering to more of a sit-down clientele, I moved to a bigger location with more tables and seating.

From my favorite table looking out, I'm "the eyes on the street," you might say. Like other urban neighborhoods, we've had our problems; we've had to make our share of 911 calls. But we C-Streeters share a common purpose and feel safe and at home.

What's in store for the future? When choosing to locate here, I had a feeling that it could become another Beale Street, a historic street in Memphis that started around the same time as Springfield's Commercial Street and has faced similar struggles. I can see us slowly getting there. Commercial street is the most diverse street in Springfield! We have Lebanese cuisine, Dutch cuisine, and good ole' American pizza. There is so much culture on this street, and I do see it as a real hub for culture. We are always actively searching for great businesses to come here to make the street better than it already is.

In 2024, I hope to be right here in my shop on Commercial Street, sitting at my favorite table, looking out the window, waving to friends and welcoming them in. The future success of Sisters in Thyme rests in the success of Commercial Street, which rests in our capacity to create community, which rests in a commitment to inclusiveness and diversity—which includes welcoming all our brothers and sisters of all faiths, of all sexualities, and of all economic standings. Our business will continue to cater to the needs of those with dietary restrictions and provide them with wholesome, delicious options.

—Martha "Marty" Cooper, baker and proprietor,
Sisters in Thyme on Historic C-Street

Emerging Technologies in Health

My story begins with a sickness, an environmental toxicity that took years for local physicians to diagnose. And when the diagnosis at last came, my symptoms had cascaded to the point seemingly of no return. Without going into details, I was told that my medical condition—tied to an overtaxed autoimmune system gone haywire—"had no cure," "rarely improved," and "typically got worse." Lucky me.

If medicine back then "had no cure," then, by golly, I would research the condition on my own. I took to the web and searched. And I did find a treatment protocol, one that lay outside of the "American mainstream" medical paradigm, involving Eastern acupuncture and naturalist homeopathy. It required many dozens of treatments over a year's time, for which I had to travel out of state. Spending much of my time on the road, I lost a year of my life. But I got my health back, and my life back.

That was a decade ago. I suspect that many practitioners today would scratch their heads when presented with my old symptoms, but there are some in the local medical community who would figure it out fairly quickly and make the right referrals. For medical understanding has evolved along with technology: The acupuncture treatments that were done "by hand" a decade ago can now be delivered by cutting-edge computer programs that are available locally. Though I occasionally come into contact with chemicals that throw my autonomic nervous system into a tizzy, my health nowadays is maintained *by a computer program* housed in a local clinic specializing in modalities of "energy medicine." That term may be unfamiliar to some; I predict that it will become commonplace in time.

"I sing the body electric," wrote old Walt Whitman. I sing along with him. Our bodies are organizations of energy; we're bioelectric machines—true dynamos. Every major clinic has its sonograms and CT scans and MRIs, illuminating the body from the inside. More recent is the mainstreaming of computer programs that treat the body—bones, muscles, systems, organs and all, the brain included—energetically. There's the Low Energy Neurofeedback System (LENS), Pulsed Electromagnetic Field Therapy (PEMF), and Scalar Wave—brainchild of the late great Nikola Tesla—among other programs, systems, and devices. I make no claims for their efficacy and leave it to readers to look these up. The web offers lots of information, pro and con, on technologies such as these. Even as I write, I note on my calendar an afternoon appointment for K-laser treatment on my ankle (a sports injury). God bless the machine and those who know how to use it!

In 2024, I imagine myself sitting in a lounge chair in a coffee house on South Street sipping a good coffee while writing on the interrelations between health and the environment. I'll be looking for ways to include "the city" in our collective explorations of ecology, given that most of us live in urban settings with urban lifestyles. I'd hope that our lives will be less stressful, but I doubt it. Our world is saturated with chemical compounds of human manufacture whose long-term effects remain unstudied: We can't help but breathe in and ingest them and take them in through our skin. There's a lot of toxicity in the air, water, and soil, in our homes and workplaces, on our streets, in our relationships, on the internet and in the airwaves. Those environmental stressors won't be going away.

We'll continue to get sick—likely in new ways— and continue to look for cures. We'll look to our local healthcare professionals, though an increasing number of us will "do homework," self-researching options and alternatives. The old "passive," obedient patient will evolve into an informed consumer, someone who wants a say in his or her health decisions. Medical technologies will keep up (I hope) with the medical conditions that we'll face in the 21st century. And our practitioners will keep up (I hope) with emerging science. Then as now, information will be key; finding that information will be key; and communicating it will be key.

—Julius Santos is the pen name of a freelance writer living in Springfield

Therapy by Technology

Mr. Santos, whom I've known for some years, asked me to dialogue with his contribution, since I "practice" what he "preaches" with respect to frequency therapy. So, here's a story. In 2003, when I was halfway through chiropractic college, I heard a talk by a chiropractor with the U.S. Postal Service Pro Cycling Team. U.S. Postal won the Tour de France that year. Anyone with knowledge of that hyper-competitive sport knows that cyclists are always looking for an edge (which explains the incidents of doping that ultimately

rocked cycling, taking down the legendary Lance Armstrong). And that edge includes both performance and recovery—recovery from injury as quickly as possible, addressing not just healing of tissue but reduction in pain and inflammation. That year, the pro cyclist's "healing edge" came from laser therapy. In effect, the cyclists were Lead Users in a technology that would evolve from sports medicine into general medical use.

That talk got my attention. When I began my private practice, I bought my first dual-frequency ten-watt laser. Following the published protocols, I treated many patients for tissue healing, swelling reduction, and pain mitigation. And then I discovered something: The laser gave relief to people with pain from shingles and from postherpetic neuralgia or "post-shingles syndrome." I questioned the manufacturer and found that there was no established protocol, but that other doctors had seen similar results. So we pitched in and worked with the manufacturer, establishing a protocol for shingles.

Having helped with the protocol, I applied it to a gentleman in his nineties whose postherpetic neuralgia presented as a constant moderate pain in the left-side rib cage and frequent stabbing pain under his armpit that would bring him literally to his knees. After three treatments of the laser protocol, he reported no more stabbing pain under the armpit. We released him after fifteen treatments, when he reported only occasional mild, generalized pain. On the day we released him, he admitted to me that he had been considering suicide, because the pain had been so bad. That conversation put me on the path of finding newer and better ways to heal with frequency.

Let me tell one more story about a technology mentioned above, whose application surprised me. The Low Energy Neurofeedback System (LENS) is a passive therapy, in which a patient is connected to an EEG through a few wire leads applied to the scalp.

Many other neurofeedback systems require an active engagement by the patient, who, watching a computer monitor, tries to manipulate screen activity by means of his or her brain activity. With these, the patient needs to be old enough (and calm enough) to perform the required mental "task." Our clinic's LENS operator was posed with a unique challenge when asked to work with a two-year-old who had suffered brain trauma from shaken baby syndrome. She was in foster care and her symptoms included sleep disturbance (she was able to sleep only forty-five minutes at a time), delayed development in speech, right-sided motor dysfunction, and severe attention-deficit disorder. The results after the first session were unremarkable, but after the second session the girl's eyes made contact with and stayed focused on the therapist. This *was* remarkable. Her speech and physical therapist noticed her calmness and requested an appointment with her medical doctor, thinking she might be sick. When put to bed that evening, she slept through the night. The judge overseeing her case had authorized the LENS therapy and now wanted a letter outlining all that had happened. A few months later, her therapist reported improvement in speech and in her right-side physical deficit, predicting that she would catch up to her peers in six months or less.

Though I'd be considered a Lead User in these and other frequency therapies, I look forward to the time when modalities such as these become mainstream.

—*Morris Kille, Jr., D.C.*

Science and Food Supply

By 2024, the planet's ever-increasing population will be fed by GMOs—genetically modified foods. The future of food production is already here. Though GMO food production remains a matter of public debate, it is an inevitability that the world will come, in time, to accept.

As a child, I read about people suffering through genocide, war, and hunger. My tender heart wanted to save the world by any means necessary. Growing up on a small farm in Southern Illinois, my family was never anti-GMO. My father grew GMO crops on our farm. I listened to him talk about anti-GMO campaigns but didn't understand the issues. As I got older, I learned more about the world and its problems. When I started thinking about career paths, my choices were greatly impacted by my upbringing: With farming in my blood, my interests turned to plant biology. As an MSU undergraduate, I found myself in a lab doing research in genetics. Initial projects ranged from plant genome sequencing to the effects of abiotic stress. It was my good fortune to be accepted into an internship program—a prestigious one at the Donald Danforth Plant Science Center—where I first worked with GMO plants. I was impressed by the center's efforts to develop crops that could grow well in harsh environments with minimum pesticide or herbicide use while increasing yields and maintaining a high nutritional content. While these GMOs would not fix all the problems my young mind had vowed to solve, I knew that these crops, once difficult to grow, could aid whole countries in feeding their own people.

I have just graduated with a B.S. in Biology and, in the fall of 2019, I will be starting my Ph.D. in molecular plant biology. I will be continuing my lifelong goal of helping the world, one research project at a time. I do take science as a vocation—a calling—and I am hopeful that my work in the development of GMOs will make a difference.

I think back to my parents' conversations around the kitchen table. More than I could have understood as a child, I realize now that "farming for the future" involves both science *and* public policy, the latter informed largely by popular-media perceptions.

Popular culture shows an increasing distrust of science and scientists; in a world driven by technology and specialized expertise, this is a dangerous affair. People need to eat, and they need to trust what they are eating; they need to rely on the people who grow their food; and the growers need to rely on the science underlying their work. Corporations, research institutes, and university labs are not "enemies of the people" in providing high-yield, pest-resistant, nutritious foods in abundance. I acknowledge that many people today mistrust GMOS; since childhood, I have listened to their arguments. Now I would ask a favor of them—that they pause and listen calmly to me. Again, the scientist must become an educator and communicator. I embrace these roles. I want to show people that our goals are as noble as developing vaccines or digging wells. In the future, I hope to show people what science can really do. In so doing, I'll be fulfilling my lifelong goal.

—*Jennifer Probst, doctoral candidate in Biology and MSU alum (B.S. 2018)*

Brain Science

I began my college education in the field of music. During my third year, I was working as a nurse's assistant in an Alzheimer's unit, though the patients I cared for suffered from a wide range of conditions: Alzheimer's disease, dementia, gunshot wounds to the head inducing amnesia, and much more. I experienced all of this around the time that my psychology introductory course was focusing on brain science and its intersection with psychology. After completing an undergraduate degree in psychology, I went on to earn my master's and doctorate. It wasn't until I was working on my thesis that I become increasingly interested in the neurophysiology of psychology. It's what led me to becoming a practicing neuropsychologist.

The entire field has exploded in technological advances over the years. Everything was very simple when I was entering the field. The devices we used to record brain activity were slow and imprecise. Thinking back to my doctoral dissertation, I was one of the few people with a personal computer, and now computers are used for everything. This, of course, parallels the advancements within my field. The same technology we used back when I was a graduate student has evolved to be more precise and detailed; my role as a neuropsychologist has evolved. The technology at our disposal has become incredibly detailed and accurate, to the point that we can identify the direction of neural firing.

In a way, I have subspecialized in analyzing physiological data of the brain and psychological data simultaneously. New therapies have emerged. I have recently become interested in neuromodulation, which studies non-invasive methods used to change brain activity. Using one method, we are able to alter brain activity using transcranial direct current stimulation. This term is commonly associated with electroconvulsive therapy. Electroconvulsive therapy administers electrical currents strong enough to force neurons into firing in new patterns. In contrast, neuromodulation administers a small electrical current (most of which is shunted away from the brain via skin tissue, bone, cerebral fluid, and other factors), which alters the excitability of neurons and—in contrast with electroconvulsive therapy—simply encourages a shift in firing patterns instead of forcing them. A second form of neuromodulation that does not require the induction of energy is neurofeedback. In this therapy, instantaneous information about brain activity is displayed to a patient, who is tasked with training the brain so that it reflects a normative model for someone of their demographics. Research is currently being done on light as a form of neuromodulation. It turns out that certain frequencies of light that penetrate the skull affect the functioning of the mitochondria, metabolism, and excitability of neurons.

Within the foreseeable future, I'm expecting

technology to advance so that neuropsychologists are able simultaneously to perform brain imaging while patients are taking tests. That isn't possible with the way we currently use MRI technology and administer psychological evaluations. And, of course, all evaluations will soon be computerized. Results will be more accurate and instantly scored.

—*Dr. Frederick Ulam, Neuropsychologist at Burrell Health Center, Springfield*

Computerizing Creativity

I began my studies in college with dreams of becoming an opera singer. I studied music education as I was fulfilling a degree in vocal performance. Thankfully, I had an education degree as a backup plan. My voice on stage was never quite loud enough to perform professionally. The downfall of my future in performance gave rise to my studying the voice in my undergraduate research. I wanted to know everything about it: how it works and, most importantly, what makes it louder.

While I was studying speech science, a professor of mine shared his theories on how computers could understand speech. This was what sparked my interest in computer science. He offered me a job in his computer speech recognition company, where I worked for ten years as a computer scientist. My work there consisted of studying computer speech recognition and possibilities for future applications of speech recognition programs.

It was from this research in computer speech recognition that I then became interested in machine learning. This is a subset of artificial intelligence, in which computers use patterns of input data to complete tasks without explicit instructions. One experiment I did with a past student was to develop a program that created blues music using probability analysis. The computer used the probability of any one note following any other note in current blues music to create new melodies. In fact, those who listened to our machine-generated music could not distinguish it from "real" music.

David Cope, a well-established professor, author, and composer, famously developed this technology to compose music. Stuck in a "composer's block," Cope created a computer program that was inspired by existing music to produce new music. Similar to my blues experiment, his program rearranged motifs of well-known pieces of Mozart and Beethoven, for example, to write music in the style of these historical favorites.

However, Cope's research has given rise to heated philosophical debates over the question, "What is creativity?" Some opposed to Cope's method of producing music claim that the product of his design cannot be claimed as "creative." They claim that merely reorganizing previous work is not true creativity. Cope bites back to argue that his program is using the same creative process of famous composers, as they, too, simply rearrange notes and melodies to create new masterpieces. Would you say that Mozart and Beethoven weren't creative? All in all, if a computer creates music that touches you in a way "real" music would, it doesn't matter. I believe that art and meaning is in the mind of the listener.

—*Lloyd A. Smith, Ph.D., Professor of Computer Science, Missouri State University*

Virtual Reality: A New Way to See the World

Virtual reality is here to stay.

But what exactly is virtual reality? Most of us remember the iconic red plastic View-Master: Popular in the 1960s and beyond, the View-Master is a toy stereoscope whose interchangeable cardboard "reels" presented a series of paired color slides. We'd press a lever and the reel's "split-eye" 3D images came into view. Reel after reel, the View-Master "immersed" us in strange still-life worlds, occupying us for hours. Now, think of the View-Master, but on steroids. A virtual reality headset like the HTC Vive or the Oculus Rift uses a similar "split-eye" technique to simulate depth; at the same time, it supercharges the sensory experience by integrating high-resolution computer graphics, high-fidelity stereo sound, and haptic sensitivity (touch and motion) that immerse the user in a self-contained sensory world within which he or she can explore and interact.

The commodification of virtual reality (VR) took off when the gaming community embraced 3D audiovisual technology as the "next step" in home entertainment. VR precursors like *The Sims* (2000) and *Second Life* (2003) allowed access to an alternate computer-simulated reality, but the user never really transcended the awareness that he was playing a game. In *The Sims*, the player was a god-like overseer of the simulation, controlling characters from a distance like chess pieces. Even in *Second Life*, although the player was "in" the world, he existed only as an avatar—seeing not through his own eyes, but rather being observed as a separate persona (and with a measure of emotional detachment) from the eyes of his human counterpart and his vantage point outside of the game. (Toggling to "mouseview" in *Second Life* did allow the player to scroll down into the "intraocular" visual perspective of his avatar—as if experiencing the external world through the avatar's physical body and sensorium—but the process was clunky and more than a little bit disorienting!)

In Springfield, the public—kids and adults—can pay to try out VR gaming at Andy B's Restaurant and Entertainment in their open-air arena. And at least two local businesses, Self-Interactive and Splitverse, are involved in developing VR programs for business use. (Self-Interactive is working with Cox Health to produce training simulations for their hospital staff, and Splitverse recently teamed up with Oculus and LIV to produce the *Beat Saber* announcement trailer for the new Oculus Quest all-in-one virtual reality system.) In 2016, as the result of a Meyer Library Summer Innovation Grant proposal by staff member Janelle Johnson, Missouri State University began providing virtual reality (VR) experiences free of charge to students, staff, and faculty on its Springfield campus. In the fall of 2019, MSU Libraries launched a one-credit-hour "badging" course, giving students an opportunity to explore the emerging "immersive technologies" and to evaluate their potential use within students' respective fields of study.

Meyer Library uses the HTC Vive system in its Virtual Reality Room. Two Google Cardboard units are also available for checkout by members of the Missouri State community. Students in Art 110 (New Media) have been using Tilt Brush, Google's 3D paint environment, for the past two years to create art projects. Tilt Brush is a fairly sophisticated virtual reality program, but intuitive enough that even a beginner can produce some nice effects.

Such is the "present state of VR technologies" in Springfield; as for its future, some predictions follow.

Gaming and Recreation: The primary motivating force in the early stages of VR development came from the gaming and recreation industries. While the educational merits of a shoot-'em-up game or zombie apocalypse simulation may not be immediately evident, the technological advances—creating realistic blood splatter, for instance, or enhanced motion-tracking abilities—benefit all user applications, educational as well as entertainment.

Two of the more popular games in the Missouri State VR Lab are *Fancy Skiing* and *No Limits 2 Roller Coaster Simulation*, both of which fight the balance between a fun and realistic experience versus an unpleasantly nauseating nightmare (people *have* lost their balance and fallen over while playing the downhill skiing game). Other games, like *Fantastic Contraption*, *Human: Fall Flat*, and *Gadgeteer*, can be helpful in exploring basic physics principles of cause-and-effect. Another visually stunning program, *Walden*, takes players to Walden Pond to relive Thoreau's first year in the woods. And *Beat Saber*? Well, it's just plain fun!

Education/Problem Solving: Working through simulated environments, virtual reality can encourage "no risk" exploration and experimentation. *Universe Sandbox* is a space simulator that is both visually dazzling and thought provoking. (What does it take to make a planet habitable?) Springfield's software development studio, Self-Interactive, has created numerous VR simulations for corporate training—including simulation of a meatball-making machine: Participants can shrink themselves down to the size of a meatball and ride the conveyer belt, getting a bird's-eye view of the machine's internal components.

Universal Design/Universal Access: Is virtual reality the great equalizer in education? Already it has proven its unique abilities to transcend barriers and restrictions—restrictions, for example, of money, time, logistics, and physical limitations. Students can travel back in time and experience the sinking of the Titanic or fight for survival on a World War II German U-boat. Students can explore the solar system, examine marine life at the bottom of the ocean, or travel through the circulatory system of the human body.

Some of the programs are still in crude stages; the resolution in Google Earth, for instance, can be disappointing in the Street View mode, but the chance to fly around the world like a bird and visit new locations or revisit old memories is hard to beat.

One complaint about virtual reality systems—at least the high-end ones like HTC Vive and Oculus Rift—remains the cost. But, as with the personal computer, improvements in the technology keep reducing the price and the size of systems. Missouri State University, like other schools across the country, is embracing the new technology and providing free access to its students, staff, and faculty. Student and faculty projects are incorporating the design elements in large and small ways, and cross-disciplinary collaborations are strengthening the overall educational experience in the Ozarks.

A new generation of VR-savvy consumers and producers is on the horizon, and the future is looking positive—excitingly so.

—*Susanna B. Reichling, Library Associate in Music & Media at Missouri State University*

The Emerging Library

I've noticed recently that my strolls through Meyer Library don't leave me thinking about the books. But how is this so? Hundreds of thousands of books, the most tangible source of humankind's accumulation of knowledge, are housed here; yet my thoughts aren't about the books. The university is the world's most suitable place for acquiring knowledge, and books are the very embodiment of this ideal; yet my thoughts aren't about the books. I wonder, though, if my dissonance is justified. The archiving of resources has been the first function of libraries for centuries; but is this still the case? I cannot help but think that the essence of the college library is changing.

The resources are meaningless without students to use them, so shouldn't the library be more devoted to the students? And, with the continuing emergence of public access—the public's right to view publicly funded scholarship—the library will certainly have to adapt in various ways. The tension I feel between the traditional library and the emerging library, then, is a good tension. The future of Meyer Library excites me.

As the needs of the library's users evolve, so must the initiatives of the library. A student population in need of high-powered computers is not well served by a library focused only on archiving resources. Meyer Library will be devoted to the continual assessment of and fulfillment of students' actual needs. This will create a non-traditional library, but this deviation from tradition will be for the better. As a university populated primarily by commuter students needing a productive space to work throughout the day, Meyer Library seeks to provide abundant and high-quality study areas. Resource collections will be relocated and stored compactly, and different types of furniture will be used. Since most MSU students are not wealthy, Meyer Library will continue fighting for open-educational resources. Practically, this means an increase in the number of textbooks in circulation from the library. And the fight for open-educational resources is intricately connected to another pressing concern: public access of resources.

Public access of resources will, albeit slowly, become the dominant system of resource-sharing in the United States. This change has been in progress for some years, but it is a messy change. Corporate publishers, who profit mainly from selling their copyrighted materials to colleges, must change their business models significantly. Additionally, the concept of scholars gaining reputation through publishing only with esteemed publishers—so-called tier-one journals, university presses, etc.—will change, because public access will make information widely available regardless of publisher. Since resources will no longer be found in one library only but can be accessed by anyone at any location, libraries will not need to fight for rare and prestigious materials. Resources will be found and accessed primarily through databases, so physical volumes will become of less importance. And libraries will need to use archiving resources more effectively: Having more resources means a growing need for search engines and finding aids. Meyer Library will continue to advocate for public access and work to encourage its use among students and faculty.

My work is made worthwhile by the broad implications of these changes. Students with improved study space will be more productive and satisfied in their work. Faculty and students with public access to the latest research will themselves be able to produce quality research. Even ordinary citizens, those with no professional academic training, will have access to (and be able to experience) the inherent joy of discovery and learning. The future of the library is not dedicated simply to the storing of books. It is a future of meaningful change.

—*Rachel M. Besara, Associate Dean of Libraries,*
Missouri State University

Dealing with "Big Content"

Right now, the universe of web-based content is undergoing its greatest expansion ever. We don't know how it's going to play out. Week by week, the sheer amount of content is growing exponentially. For instance, some of the most worthwhile thinking is taking place in the world of blogs, on almost any topic you could think of. This has changed the way we've understood vetting for the last hundred years: The public, sometimes brutally, does the vetting. This happens throughout social media as a whole. Content is being produced and vetted with every post, comment, like, and reaction.

How do we deal with the vast increase in data? As a species, the human being faces a series of limits or problems. One is cognitive. Considering all the content that exists, there is a certain threshold of exposure that any one person can experience. It's been growing as a problem in the last fifteen to twenty years, especially. The second limit is one of our own mortality—simply, we don't live that long. But content does. Content can outlive human beings; it can live for millennia. We don't. And then there's the "big world" problem: Literally, it *is* a big world, even with global transportation and communication. We need centralized places where we can store and access global content. The "big world" problem is just as bad, if not worse, in a massive digital world. Huge search engines will not always compensate. Once people get content, the fourth limitation is their struggle with it. This is the mediation problem: Google and its ilk have convinced people that they can do anything for themselves but, in reality, people don't understand how to find content and use it to make meaning. People's comfort with just "surfing" content is an expanding ignorance.

Enter the librarian, whose job is to help overcome these inherent human limitations. We select key materials, describe them, and organize them so that people can navigate the human cognitive limit. We make legal arrangements to make content available to combat the time limit. We develop places—libraries—to maintain content in a central location to make the big world smaller. And we provide reference, research, and consultation services to guide the public in their content searches, providing necessary mediation.

As the sheer volume of content is growing, the question and challenge becomes: How do we meet people where they are and still practice the traditional aims and values of librarianship? Perhaps our main task is to remain uncomfortable with the content. It is our responsibility to make it available to people, no matter how ideas of society shift. We can offer the ability to navigate things people aren't familiar with and not comfortable with. And it's our responsibility to be permanently uncomfortable, making material available that we don't necessarily agree with.

Ironically, too, the role of the physical library remains. People have thought that, as digital content evolved, physical libraries would cease to exist. That hasn't happened. The world of content has become ubiquitous; it envelops us. Paradoxically, people need a place to retreat from data-overload, shutting it out so they can think; and they come to libraries to do that.

The role of the physical place has flipped. For most of history, the world was information poor and the library held that information. Now the world is information rich and the library offers refuge, a means of calming the constant bombardment of "big content," which can mean big data (i.e., numbers) but also an overload of texts, images, sounds, and multimedia.

—*William B. Edgar, Ph.D., Associate Professor and Reference Librarian, Missouri State University*

Becoming Smarter Than Our Smartphones

When I was still getting my education, we were taught that the brain and its structure is permanent—which is absolutely not true. Neuroplasticity—the changing of the brain in terms of its experience—is very real. Our use of technology is changing the way our brain works, affecting the structure of the brain at a microscopic level. Our understanding of these changes is limited, but it's happening. In this respect, the power of technology is a double-edged sword. It can be harnessed and used for benefit as we better understand how the brain works and how to use technology to interface with the brain; but, in the meantime, it's changing the brain in unpredictable ways.

The byproducts of technology can be toxic. Psychologically, we need to teach people to be actively aware of the consequences of their actions. We also need to be smarter than our phones—to be sustainable with ourselves. We have recognized that blue light emitted from smartphones alters the neurohormone function of the brain; it affects our regulation of the day-night cycle. Luckily, we have found ways to curb that effect, but it is going to take a lot more than that. We need to train our basic behavior to put the phone down and turn it off. Smartphones are addictive, with many of the same effects as in addictive substance abuse. Many of the same neurochemical changes take place, and manufacturers are aware of that. It encourages me that some youth are aware of the problems with technology. They're aware that they've have been used as pawns—manipulated through this great social experiment. I see them changing their own behavior and taking control themselves, which is very encouraging.

We also need to be aware of technology in terms of environmental sustainability. In some ways, we have rushed head-forward in developing all these electronic devices with little or no thought to the ecological consequences. We get a new phone every year or two. What happens to all the phones that are discarded? A lot of them may be recycled, but not all. Where do these go? All these phones are electronic and have to be powered. Where is that energy coming from? Are we looking into clean ways to power the electronic devices that are being developed? I think that we are going to need to catch up with ourselves and develop clean, renewable ways to power our devices.

I think that we're at a point in time when certain emergent technologies demand government regulation. Look at artificial intelligence in self-driving cars. Certain critical principles have to be kept in mind by their programmers. A self-driving vehicle has to have ethical decisions built into its programming, since choices with profound ethical consequences will need to be made. For example, if you're driving and a dog and a human are crossing the road from opposite sides, is the program going to prioritize the dog or the human? All these ethical choices have to be built in. The government's involvement is critical. Companies can't regulate themselves in this regard.

—*Dr. Frederick Ulam, Neuropsychologist at Burrell Health Center, Springfield*

Preparing for the Graduating Class of 2024

The demographics of our faculty, staff, and students will change in the next five years. As a country and as a state we are getting more diverse ethnically and demographically, so I would think that the university will have both a more diverse student population as well as a more diverse faculty and staff, and all will be positive developments, in my opinion.

In 2024, MSU students are likely to be less affluent than our current students, because of the jobs that are being created: About two-thirds of these jobs are going to require a post-high school credential, and we are living in a state now in which about 40% of the population has a post-high school credential. So, more people will have to go to college. At that point we won't be taking the top 30% of high school students but, instead, the top 60% of a class. How will we maintain a quality curriculum, and will we expect the same high quality of performance of these students? They may be first-generation college students; they may need more support; they may have more financial challenges. All this raises further questions: How do you teach? What kind of support programs do you need? Does it change your scholarship programs?

Right now, 84% of MSU students come from Missouri. I would think that in five years, this will change, and the change will have a positive effect on the quality of education we offer. I would hope that, within the next five years, the U.S. will become again a more meaningful player on the world stage, such that more international students will again be coming to the United States for their education. In the past, that has been a great way for American students to have relationships with all sorts of people from different backgrounds—people they might otherwise not have the opportunity to meet if everybody in the classroom is from Missouri.

—*Clifton M. Smart III, President, Missouri State University*

The Evolving University

The demographics of America have shifted, creating pressure on universities. I am at the tail end of the Baby Boomer generation; now we are entering the baby bust. This has led to fewer traditional college students attending universities, including in Missouri. The effects of Americans having fewer children trickle down, so universities are bracing for a dire enrollment decline. Competition among colleges has increased. To survive downturns in enrollment, Missouri State is going to need to market more aggressively. Another way that Missouri State can withstand declining enrollment pressures is by recruitment of international students; however, the number of exchange students has declined in the past few years. Finding ways to attract more international students could seriously improve MSU's enrollment.

Another large shift in the world of higher education is the change in attitudes regarding jobs. In the average life today, people are pursuing six to seven different careers. These career changes could be an opportunity for MSU to evolve, since people will need life-long training and education. In the future, colleges may be continuing-education centers, not just schools for eighteen-to-twenty-two-year-olds. These future students will be interested in making sure the school acts responsibly. We're already seeing requirements on schools to be more accountable to students, faculty, legislators, and the population as a whole. People want to know where higher ed is spending their money and whether or not schools are being wasteful in terms of finances or degree programs. In addition, students want to know whether attending a certain college will help them get jobs and whether they will be able to finish their degree within four years. In the future, this interest in accountability will only increase. Consumers of higher education are becoming more and more sophisticated, so MSU will have to prove that coming here is a good investment.

Predictions for the future envision a higher education landscape that is drastically different from the modern one. Some smaller liberal arts schools have had to close, and smaller universities have been absorbed by larger ones. Students and faculty may be able to move from institution to institution, with students taking some courses from one school and others from another. Professors will function more as entrepreneurs, and students may seek out specific teachers for specific classes. If this happens, new organizations will form to give credentials to faculty and others will develop systems to bank students' credits. The overall number of faculty members may decrease. In addition, they could be divided into research faculty and teaching faculty, allowing them to devote more time to a specific area. In this environment, regulation may go down and academia may be governed by economic principles. Instead of spending money on buildings and facilities, institutions will invest their resources on teaching and learning.

In the future, technology will become an even larger part of education than already exists. Current technology will become antiquated very quickly. Online classes will become more common and advanced. Students may no longer attend many seated, face-to-face classes, and technology such as the virtual reality "classroom of the future" at Stanford and the Harvard will become more common. In addition, soft skills will be emphasized, and education will be individualized. All this change is inevitable. Faculty must evolve or be left behind.

—*Elizabeth J. Rozell, Associate Dean and the Director of the MBA Program, Missouri State University*

Finding a Vocation

After I completed my associate's degree in general education, I wanted to gain independence. I started working in a car parts factory to support myself; however, I was miserable to the point that I felt enslaved. I realized how meaningless work is when the main goal is making money. I became introspective about what career would satisfy my need for fulfillment. The decision to become an English teacher came from my love of literature and of helping others. So I got my bachelor's degree in English literature. Now, as an MSU graduate student teaching multiple English courses, I am so grateful to have a job where I can go home at the end of the day feeling like I benefited others. My favorite part of teaching is seeing the "light bulb" go off when explaining material to a student. I have, for example, one insecure student who's a very talented writer. Whenever I give her feedback on her papers and whenever I tell her how well she's doing, I get to see her confidence rise that much more.

My perspective on the future, as an English teacher, is both optimistic and pessimistic on the change technology is making in culture and in the classroom. Ebooks, for example, are cheaper and more environmentally friendly. We will steer away from the bulky textbooks we have now and, for the majority, we'll be getting our information from the internet. But there's a potential mortality of knowledge in this transfer to technology. Books definitely are becoming a casualty. I think we are playing a dangerous game here. If we put all this stock in technology and something happens, we lose it all.

Inevitably, technology can cause distractions in the classroom and that can be hard to handle for the teacher and, often, for other students (when one or several others are absorbed in personal texting during class). Technology has also made discourse—at least in the classroom—somewhat diminished, because students who are absorbed in texting or blogging aren't used to talking to the students next to them, let alone to the entire class.

I encourage my students to think critically, to examine and defend their points of view well. We also dissect pieces of literature, such as George Orwell's *1984*, looking at how social justice was being represented at that time and at how it is being represented now in issues of technology, sustainability, gender, race, and much more. In contemplating the future, it is important to discuss political and social issues occurring now.

—*Jacob Erwin, graduate teaching assistant in English, Missouri State University*

Accommodating Differences

I am on the cusp of having a stutter and being a fluent speaker. In other words, I am someone who communicates by using circumlocution (i.e., the navigation around a single word or idea by using other words to avoid stuttering). While in speech therapy, I became aware of this ability to avoid my dysfluencies by working around them—I became aware of my ability to adapt. This process requires continual mindfulness in communication, including the shortening of words like "disability studies" to "dis-studies." This adjustment allows me to appear more fluent. However, it was not until I was in the process of completing my dissertation in composition/rhetoric and attended a conference that I became interested in disability studies. I attended a panel of disability scholars and realized my dysfluent speech was an opportunity to write about rhetoric, composition, and stuttering.

From within disability studies, technology, such as computers and cell phones, offers many avenues to communicate. It has connected us globally to the rest of the world and to each other. Likewise, many resources have been provided in terms of disability accommodation. For example, *AccessNow*, an application for cell phones, allows users to input data about disability accommodations at restaurants and other public venues. This allows those with disabilities to make better decisions before travel. Within our current political climate, however, these accommodations and awareness of differences can be seen as a detriment or even an affront to our freedoms. I think this misrepresentation of strength is something people need to talk about, along with the suppression of perceived weakness.

These evolving dynamics increase our collective need to grapple with power structures and their impact. As humans, we want to be part of the group that is better, so in being a part of a weaker group, one suffers from exclusion. In elementary school, I was told by a teacher that I was stupid—which, I hope, is not true. Yet, the perception of my stutter by others is that it stems from a position of weakness rather than normalcy and certainly not from advantage or added ability. (I might ask, can *you* circumlocute like I can?) Our society tends to glorify the experiences of those with disabilities—as if completing normal or daily tasks is something extraordinary. While the experience of a disabled person might be extraordinary, the tendency to draw on weakness to demonstrate that perspective is damaging to our culture and our understanding of (dis)ability. I think the basis of who we are as humans is the same and ability is only a version of difference.

In the future, I expect that disability will continue to be seen through a normalized lens. That normalized thinking divides groups into abled and disabled, which hinders our ability to evolve as a culture. The fear that those with differing abilities might be discarded is very real. But if those with differing abilities can be seen as aiding the greater good, we can collectively move beyond ableist thinking. If we can stop equating disability and weakness, we will gain a new way of understanding ability. At this point in my teaching, I have learned that new perspectives and changing

understanding, through the acknowledgment and acceptance of those other perspectives, is critical to learning. To learn from those othered perspectives, we must be consciously open to dialogue so we can, then, learn from another's unique perspective. It is those differences that will allow humanity to grow from where we are now.

—Craig A. Meyer, Ph.D., Assistant Professor of English at Texas A&M University - Kingsville and MSU alum (M.A. 2008)

Bear POWER

Following the Americans with Disabilities Act (1990), the early 1990s brought innovations in assistive technology that began to have major impact in education, employment, and other aspects of life. These emerging devices or systems included screen readers for computers, scanners that convert print to various forms of digital access, and, of course, the internet; these helped to place those with disabilities on even footing with their peers on campus and in the community. Just a few years ago, libraries on and off campus were not accessible to many with disabilities; now, anyone with a smart phone with voice-over features can access virtually everything digitally.

While the digital age has clearly changed the landscape in higher education with respect to accessibility, the changing laws and regulations have certainly been a challenge. Laws or technology alone will not change the hearts of people, nor do they create the mindset of inclusiveness. A shared vision for universal design, coupled with an understanding of the role of digital material, will eliminate barriers and help facilitate access to the educational process and vital community resources. A number of the elements of universal design are subtle and easy to implement.

In January 2019, Missouri State University adopted Blackboard Ally, a product that provides faculty with real-time feedback about the level of accessibility of digital material being used in classes. In the faculty dashboard, each digitized item is shown with an icon resembling a fuel gauge: Material with high accessibility registers "in the green," while low accessibility registers "in the red." The goal is to help faculty "raise the gauge" on their instructional material. In effect, we just pulled the Band-Aid off and sent the signal to everyone that *accessibility is a process*. There will be much support for those who will be making changes to their digital course material, and their efforts will help to make our campus a welcoming place for everyone. The system can also help us look across the university to gauge what progress we are making in overall accessibility. MSU has enjoyed a great reputation for doing the right thing; our websites, for example, have been recognized as being the best among comparable universities for their level of accessibility.

As stated, the iPhone has likely become one of the great equalizers as an evolving technology. Voice-Over is a feature under accessibility settings in every iPhone. This immediately enables those with reading disabilities or visual impairments to access the same information available to all students. Audio description and closed captioning have also dramatically changed how those with disabilities participate in typical recreation activities on campus and in the community. Most movie theaters now offer headsets that provide ways of improving access to the latest releases. Amazon Prime and Netflix originals all include audio description and forms of closed captioning. There are many forms of TV and movies that simply would not be understood without this technology, which has become commonplace. I recently went with some friends to see the movie, *The Green Book*. While the dialogue was rich, I would not have understood some of the key messages in the movie without audio description.

The real opportunity is the extent that faculty, students, and community leaders embrace those with disabilities, digital technologies notwithstanding. The inaugural class of Bear POWER students started in January 2019, and truly has become a clear example of how far we have come, providing access to higher education and job readiness for those with intellectual and developmental disabilities. As the MSU website reads, "Bear POWER (Promoting Opportunities for Work, Education and Resilience) at Missouri State University is a two-year, five-semester inclusive college program for individuals with intellectual disabilities. The program was created to offer a post-secondary education opportunity to students with Intellectual and Developmental Disabilities (IDD) transitioning out of high school."

Bear POWER became a shared vision among faculty, staff, and students from across the university and is having an impact on our regular admitted MSU students as they strive for cultural competence—one of the pillars of our Public Affairs mission. Again, laws and technology will not change the hearts of people; rather, if we can embrace the notion that everyone can learn, then everyone can enjoy access to experiences that will enhance quality of life and promote true cultural competence within our communities.

—*Christopher J. Craig, Deputy Provost,*
Missouri State University

Conserving Energy, Saving Money

Upon arriving at Missouri State, I noticed inefficiencies in the way we operated the heating and cooling of our buildings: We were restrained by our budget to make corrections the conventional way. At a higher education facilities management conference during the mid-1990s, I learned of a new concept that was not yet embraced by a lot of people. It's called performance contracting. We were getting a revenue stream to pay our utility bills, but I knew that we could redirect parts of the funding by conserving energy. Performance contracting consists of paying for energy improvements through the savings generated. To implement performance contracting, it required that the university take out a $5.5 million loan. It took me over a year and a half of meetings with the administration and MSU Board of Governors to gain approval for this loan.

We installed an EMS—Energy Management System—throughout the major buildings on campus. Similar to programmable thermostats in homes, this EMS allowed us to schedule systematic heating and cooling, instead of maintaining a constant flow of energy. It was a major improvement that saved us close to a million dollars a year.

Another performance contract project I was involved in was the second phase of what had become the MSU Energy Program. It began after dinner with my former mentor from the Navy. At the time, he worked for an international mechanical engineering firm and was interested in getting involved with higher education. Following a consultation with them at MSU, the project was implemented. We created a design that allowed us to remove the inefficient air conditioning units and replace them with a bigger, more efficient system. The sharing of the chilled water produced by the larger units was another relatively new concept. I learned we could connect and cool major buildings through the improved technology of underground composite piping. For a while, the campus looked like a giant mole had burrowed under it while we were installing the new piping; however, once completed, the two phases combined for considerable savings. Over a fifteen-year period, the MSU Energy Program avoided $34 million in energy costs, with over $15 million in actual savings to the university. By 2021, the loan should be paid off and MSU will be saving over $2 million in actual savings a year.

There were many major developments in technology that allowed for these sustainability opportunities. Scheduling software systems (enabled by the advancements in computers) allowed temperature control for occupant comfort in a major space only while occupied, subsequently tracking the efficiency of each department. Developments in lighting—the transition from incandescent to compact fluorescent to LEDs—were also important. All these developments and installations cost money, but that is the beauty of promoting energy efficiency—I could promote an idea and explain how money was going to be saved instead of how much money was going to be spent.

While maintaining efficient buildings, the dominant goal has been to create a comfortable learning environment for the students. Missouri State became the energy program that others emulated and wanted to hear about. From Texas to Manitoba, I was receiving interest about energy savings; that was very gratifying. Energy efficiency became a commitment: a Public Affairs issue. MSU and its close partnership with City Utilities became a model for the city of Springfield.

—Robert T. Eckels, Facilities Management Director (retired), Missouri State University

Land and Stewardship

I am a medicine man whose vocation and mission, as passed down through generations of the Ahkehkt clan, is to heal the individual, family, community, society, and planet. Though the indigenous technologies of healing are traditional, they are continuously adaptive to conditions at hand, relying always on the gifts of the Earth: water, soil, stones, plants, animals, air, heat, cold—all gifts, but one gift. The Sehaptin cosmos is a unity, wherein the strength of one element sustains the rest, while weakness in one diminishes the rest. A Eurocentric worldview tends to dissect and analyze and compartmentalize, whereas the indigenous worldview seeks unity and synergy and synthesis. Science and religion are one: both gifts, but one gift.

I have been asked how I view the present state of things. The medicine man's technology rests in knowledge, discernment, and decision making. Without these, tools are useless. People come to me with their illnesses, but the illnesses are not theirs alone: They're signs of exhaustion and toxicity in the water, soil, plants, animals, air. We humans are a microcosm of the planet: Having wasted its resources and destroyed its health, we find ourselves wasting away. We live longer than generations before us, but do we live better? The Eurocentric habit is to exploit nature, depleting its resources. In using up the land, we are using up the future. Look at the soil. Once a region of native forests and prairies, southwest Missouri was transformed into farms, orchards, and ranches. One hundred years ago, Missouri fed itself. Now, almost all our food is shipped in. If we relied on the land in its current depleted state, we'd starve.

We have a way forward, which the land itself teaches if we're willing to look, listen, and learn. The cosmos is a unified, living creation. Left alone, the soil will regenerate; but one hundred years of human exploitation might take one thousand years to restore at nature's pace. The soil needs our help in regenerating; and we best serve the soil, not by behaving as chemists, but by learning to behave like the millennial forests and prairies that built layer upon layer of topsoil. The technology of proper stewardship consists of a shovel, a rake, and a wheelbarrow. My current homestead was soil-poor when I first took possession of it, showing a quarter inch of tilth depth over most of the acreage. After four years, the soil depth is sixteen inches in places and rising. Yes, I grow my own herbs and vegetables, organically. My family takes care of the soil, and the soil gifts us in return.

I have been asked about the planet's future. The language of the Sehaptin people speaks the unity of the cosmos; and, with every loss of a species of plant or insect or animal, we have to relearn our language. Every loss of species diminishes the whole of life and the language of life. Creation is a song. But ours is a diminished song, with fewer words and fewer notes. Today, the indigenous people sing a song of mourning. With proper stewardship, we can bring some measure of healing to the land. We can add notes to the song and sing it in a more hopeful key. As we do so, we shall restore to the planet its original, recuperative powers. And we'll be restoring our own health, as well.

—*Phillip "Cloudpiler" Landis, Ranking Elder and Native American Practitioner, Sehaptin Culture*

Sounding the Alarm over the Damage Done to Our Planet

Though not specific to the Ozarks, changes on a global scale may certainly reach us here. Such developments include environmental changes like global warming, mass migrations of species, and rising ocean levels. Our planet has already begun to shift towards something less recognizable. For example, there was a species of rodent endemic to a very small island in the South Pacific. Ocean levels have risen so much that at high tides, the island is covered by water. There were four hundred of these rodents, and now they're all gone. Coastal cities like New York, Tokyo, and London will be underwater one day; this forces not only human populations to move inland but wildlife as well. Animals that aren't adapted to those new environments will attempt to outcompete each other for resources, and some may be fatally defeated. If you've ever wanted to see the Great Barrier Reef or a glacier, now's the time to do it before they're gone.

The second problem is the relaxation of the current U.S. administration on environmental policies. "Who needs the EPA?" they ask. "Why should we spend money to save endangered species?" Those in the White House are taking a *laissez-faire* approach to the industrial sector, allowing big businesses to regulate themselves. Two years ago the effects of climate change weren't as recognizable; now, people are wondering why there are so many violent storms and droughts. I like to use the analogy of Jenga with my intro to biology students. When you pull out one of those wooden blocks, maybe nothing happens. Once a block vital to the integrity of the tower is removed, the whole thing collapses. We're playing this game with the future of our planet.

Here's a third problem. There is currently more distrust in scientists than there ever has been. Lay people seem to have this split mentality that I find extremely odd. When faced with these pressing issues, the general public doesn't tend to worry because they reason that scientists will always find a solution before the world falls apart. At the same time, they staunchly deny the warnings of the very same scientists. They put so much faith in the scientific community to save them but concurrently reject the validity of their claims. The wooden blocks are like species in a natural community. We are one of those species, and we depend on the others in ways we don't always recognize.

Our priorities are entirely misplaced. The most important thing for our future may not be that I get to eat all the beef I want. We could eat the grain fed to the beef instead and live quite well while reducing the toll on the planet. Do I really need strawberries in the supermarket at wintertime? It costs thousands of gallons of gas to ship these fruits grown in warmer climates when I could easily grow them in my backyard and can them for winter. We think that going back to the way things were is such a hardship. In reality, our current way of living is self-centered and destructive to the planet, unlike the simpler lives led by people throughout most of human history.

For people to make any real change happen, they must be convinced. Simply telling them about the damage we are doing to the planet obviously won't work; our situation must become bad enough that it's self-evident. Fear is the greatest motivator. A healthy fear of the future is likely to bring about the most reform.

—*Thomas E. Tomasi, Ph.D., Professor of Biology, Missouri State University*

Grieving for the Planet

"Do you ever think about climate change?" I asked a family friend recently, someone my grandmother's age: "Does it worry you?"

"No," she replied: "I don't have much time left, so what do I have to worry about?"

I've been doing research on the subject of ecology and surveying local attitudes; hence, my questions. I asked another family friend, someone my father's age. Compared to the first person's apathy, his response was downright militant. "It's all fearmongering and lies," he said: "Don't even talk to me about it." One of my mother's friends was next. "I do think about climate change, occasionally," she said: "It makes me sad, though I don't worry about it. I do worry for young people today, who have their whole lives ahead of them—and I worry for their children."

My own brother and sisters were next. "Mostly every day," said one sister: "How can you *not* think about climate change, given its presence on social media?"

"I think about it multiple times a day," said my second sister, again pointing to social media: "It's inescapable."

My brother's response was more dramatic: The looming fear of climate change fills him "with existential dread and pessimism." It's a fear that he lives with daily.

I asked friends my age the same questions. Their words echoed my brother's. "It's hopeless," one said. "I see no future," said another: "There's no way to fix what we've destroyed."

I find it hard to reconcile these responses, ranging from a grandparent's indifference to a parent's denial to a brother's dread. All are problematic in their way. As a psychology major who aspires to clinical practice, I shall be caring for the emotional health of clients for whom climate change will be more than a fearful prospect: It will be their reality—our reality. My reality. Psychology is already recording the effects: As K. Hayes writes in *Climate Change and Mental Health* (2018), the sheer scope of the crisis "can incite despair and hopelessness, as actions to address the 'wicked problem' of climate change seem intangible or insignificant in comparison to the scale and magnitude of the threats." And I suspect that there's something beyond despair lurking in the feelings that my siblings and friends express.

There's a term for this something: *ecological grief.* As A. Willox writes in "Climate Change as the Work of Mourning" (2012), ecological grief strikes people who, "around the globe ... have experienced or are currently experiencing grief and mourning responses to changes in their environment or due to the deaths of non-human entities or understand the need to grieve for non-humans." This grief, despair, sense of helplessness, and hopelessness in younger generations is not going anywhere anytime soon. Climate change will continue to take its physical, economic, and emotional toll.

So, I have new questions to ask. How do we deal with the ecological grief that has become increasingly present, even invasive, in our lives? How can we restore hope? It's not just the environment that needs healing. Helping humanity cope with these emerging mental health issues could be world-changing, even world-saving, if we do it right.

—*Hanna Moellenhoff, Honors College Psychology major, Missouri State University*

On the Artistry of Science:

Nano-Imaging by Dynamic DNA Laboratories, Inc.
Austin O'Reilly and Rhy Norton; introduction by James S. Baumlin

Science ... is a way of "seeing." Given its explicit late modern hyper-visualism, this is more than mere metaphor. There remains, deep within science, a belief that seeing is believing. The question is one of how one can see. And the answer is: One sees *through, with, and by means of instruments*.

—Don Ihde, *Expanding Hermeneutics: Visualism in Science* (1999)[164]

If you're looking for a successful Ozarks startup, look no further than to Dynamic DNA Laboratories, founded in 2015 by Springfield's own Austin O'Reilly. Partnering locally with MSU's Jordan Valley Innovation Center and contracting with clinics and industries nationwide, Dynamic DNA is a direct-to-consumer genetic testing lab specializing in molecular, cell, and microbiology services (https://dynamicdnalabs.com/research/capabilities/). Genetic testing and biomedical research are its bread and butter, but a third service—microscopic artwork—sets Dynamic DNA apart.

Joined by his lab manager, Rhy Norton, and others on the Dynamic DNA team, O'Reilly uses cutting-edge technology to magnify our human powers of sight many thousand-fold, reinventing our perceptions of the natural world. An album of microscopic artwork follows. A couple of these pieces may be recognizable and a few more guessed at; others, surely, will surprise.

---

164. Quoted from David M. Kaplan. *Readings in the Philosophy of Technology* (Lanham, MD: Rowman & Littlefield, 2009), p. 518.

Dynamic DNA Lab's scope room with scanning electron microscope.

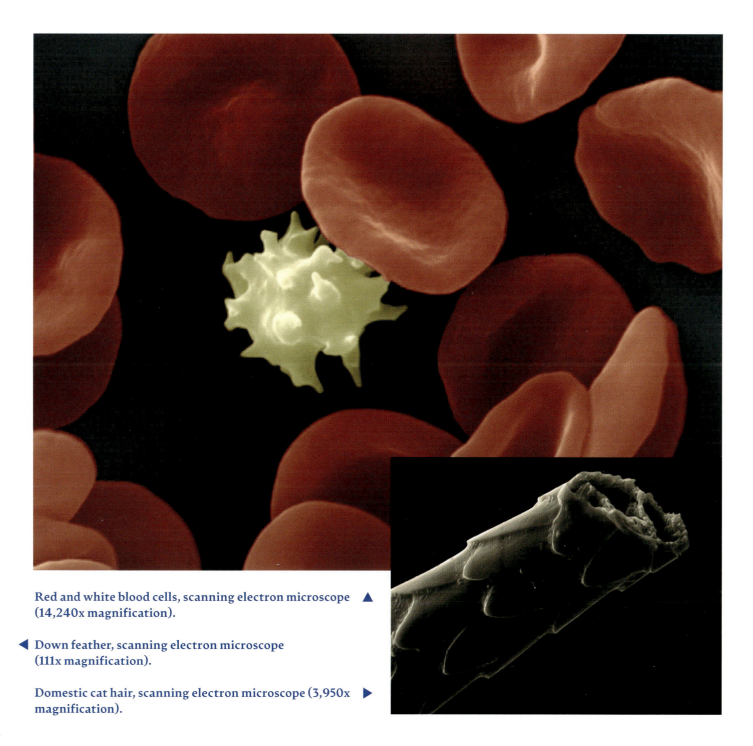

Red and white blood cells, scanning electron microscope ▲
(14,240x magnification).

◄ Down feather, scanning electron microscope
(111x magnification).

Domestic cat hair, scanning electron microscope (3,950x ▶
magnification).

221

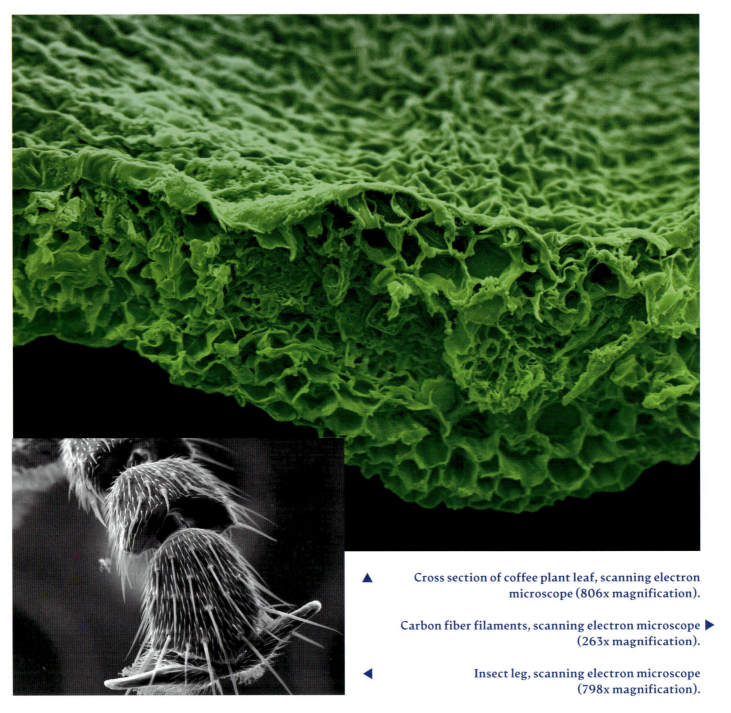

▲ Cross section of coffee plant leaf, scanning electron microscope (806x magnification).

Carbon fiber filaments, scanning electron microscope ▶ (263x magnification).

◀ Insect leg, scanning electron microscope (798x magnification).

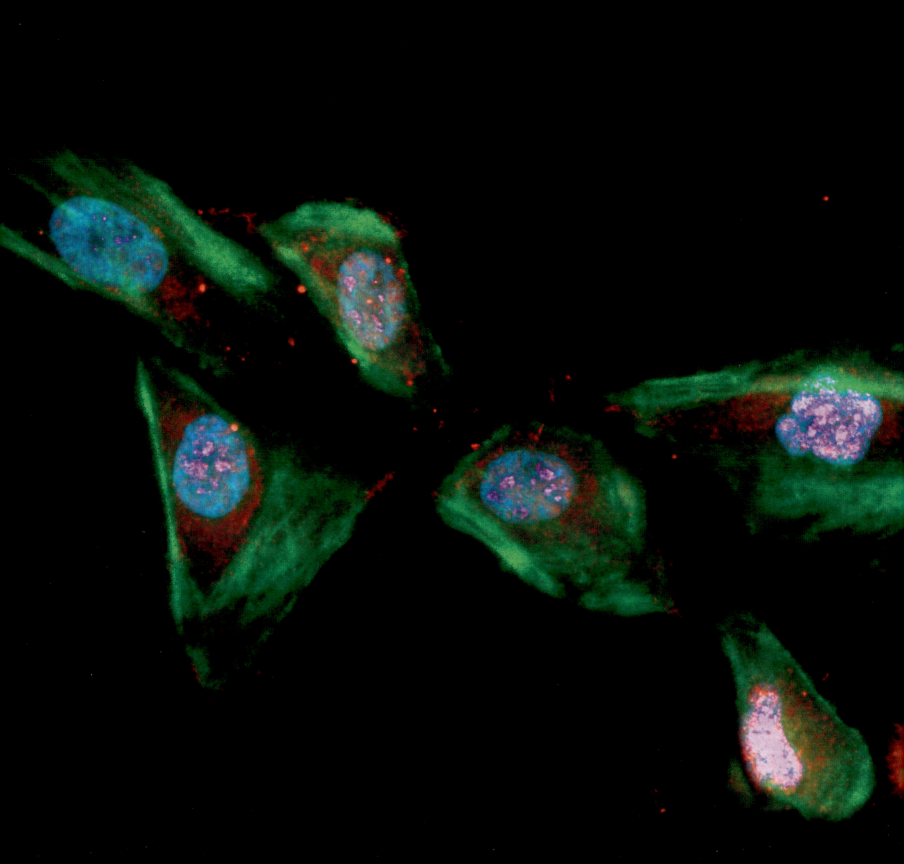

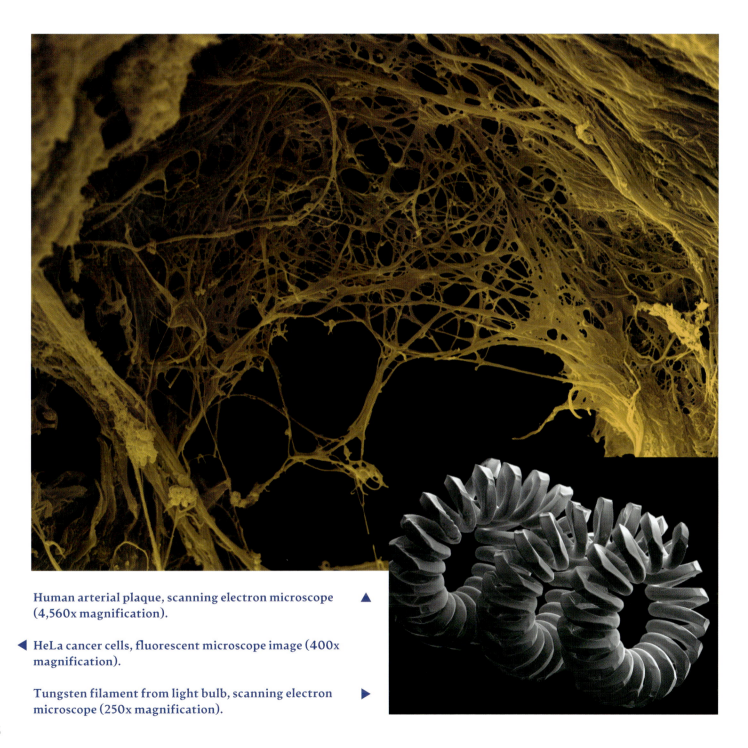

Human arterial plaque, scanning electron microscope (4,560x magnification). ▲

◀ HeLa cancer cells, fluorescent microscope image (400x magnification).

Tungsten filament from light bulb, scanning electron microscope (250x magnification). ▶

225

Computing in the Ozarks and Beyond:
A Tech-Family Perspective
Kenneth R. Vollmar and Nick Vollmar

I possess a device, in my pocket, that is capable of accessing the entirety of information known to man. I use it to look at pictures of cats and get in arguments with strangers.

> —Redditor's response to the question, "If someone from the 1950s suddenly appeared today, what would be the most difficult thing to explain to them about life today?"

The Ozarks has not been noted for its advances in computing, but the region is well equipped for improvement through education. In 2015, Arkansas became the first state to require that every public and charter high school offer computer science courses (Lapowski, 2015). Beginning in 2019, Missouri has added computer science to its menu of technical courses required for high school graduation ("Missouri Governor," 2018). Within and around the boundary of the Ozarks are at least twelve universities with flourishing undergraduate degree programs in computing and three with doctoral programs.

This essay will speculate on advances to future computing that result from changes to network, software, and hardware. The essay concludes by considering some technical and societal issues of advances in computation and communication. The authors—a father-and-son duo—provide a complementary look at the state of computing and its near future: Ken from the point of view of technology, and Nick from the point of view of commercial software development.

Father Goes First

Every discussion of computer technology should begin with a recognition of the astonishing rate of performance improvement within the relatively short history of electronic computing (Routley, 2017). As a rough example, the Cray-2 was the fastest supercomputer in the world in 1985, with peak performance of 1.9 GFLOPS (GFLOP is 10^9 floating point operations per second) for a price of $32 million. In comparison, an Apple iPhone 5, released 2012, had *three times* the performance (6 GFLOPS) for a price of $749. At this writing, the iPhone 5 has been surpassed in its turn: Now obsolete (that is, unsupported), it sells for about $70 on eBay.

Since about 1970, computer performance has been fairly accurately modelled by Moore's Law, which holds that the density of fundamental integrated circuit components doubles every two years. This exponential rate of improvement cannot be expected to continue indefinitely simply by reducing the size of transistor etching onto silicon: Today's integrated circuits are approaching physical, molecular limitations where etching becomes impossible. The end of this means of achieving performance makes it a challenge to predict future developments in computers.

The distinctions between "personal" and "technical" computing should be made, because of the differing perspectives of groups of users (although there is a nontrivial overlap between the two). We use the term "personal" to describe users of computing for primarily mobile, social, communication, and informational applications. For instance, smartphones may be used for on-the-go informational services and communications with friends and coworkers, but may not be as well-suited to inputting large amounts of text or simultaneous display of several windows of information. Several characteristics of network, software, and hardware support this type of computing and the ease of its use. For instance, widespread distribution of cellular transmission towers and Wi-Fi hotspots support wireless access, software design makes updates seamless and virtually unnoticed, and hardware design enables device functionality in a portable package.

We use "technical" to describe users of computing for primarily business or scientific analysis, finance, research, and other technical purposes. This type of computing requires high capacities and capabilities for data manipulation, display, transmission, and storage of amounts of data that were unthinkable only a few years ago. For instance, an issue in current technical computing is that the amount of data is too large to be contained on an "off-the-shelf" computer, leading to development of new algorithms that process data in manageable subsets rather than as a single whole.

The first aspect of computer technology that we consider regarding future updates is the system of networks that support communication between computers and people. A list of significant milestones in networks follows (dates are approximate, given variations in accessibility of the technologies):

1985: Personal computer network connectivity is over phone lines, limited by speed to small files such as emails and bare-bones websites.

1990-2000: Fiber-optic cables are installed, enabling transmission of cable TV and (formerly) broadcast TV signals, "hardwired" internet, and telephone.

2000-2010: Widespread accessible distribution of Wi-Fi routers used to connect computers and devices to the internet at a range of about 100 yards (100 meters).

2000: Introduction of Bluetooth technology, which connects devices directly to each other, at a range of about 10 yards (10 meters).

2020: 5G Wi-Fi—high speed, low range.

At this writing, "wired" internet in the U.S. seems fairly well established in urban and suburban areas. (There exist occurrences and valid concerns of a classist "digital divide" of access to computing and connectivity by all people, including persons with disabilities. This issue will continue indefinitely.) Internet access in rural areas is available through the same mechanism (with some cost) as cable TV: individual subscription to commercial service and a small satellite dish.

It's clear that network connectivity is essential to provide continuous access to internet sites. Commerce, finance, education, social media, health care, and entertainment all rely upon near-instantaneous communication. The next upgrade to network technology, beginning about 2019, will be the 5G (Fifth Generation) wireless networks (Goldman, 2018):

> [M]uch of the 5G network will travel over super-high-frequency airwaves. These higher frequencies bring faster speeds and more bandwidth. But they can't travel through walls, windows, or rooftops, and they get considerably weaker over long distances. That means wireless companies will need to install thousands—perhaps millions—of miniature cell towers on top of lamp posts, on the side of buildings and inside homes.

That is, a 5G signal will require infrastructure installation of 5G antennas, whose maximum range is well under 100 meters with no obstructing walls, etc. Computer users have fairly recently witnessed and financed construction and installation of Wi-Fi networks, and 5G requires a similar round of installation, construction, and infrastructure for each user.

The second aspect of computer technology is software, whose capabilities are increasing in exciting and sobering new ways. The terms under which proprietary software may be used is Digital Rights Management (DRM). Significant milestones in software follow (again, dates are approximate):

1970-2000: Personal computer software is generally single-machine (software is resident on the hard disk of the local machine). This is a common software distribution model: The computer user is the permanent owner of the purchased software. This model requires frequent user action to update new versions (e.g., to patch newly discovered bugs, new features, upgrades, etc.).

2000-2010: An increasingly used software distribution model is "right to use" under a fixed-length-term license. The software is downloaded and resides on a local machine, with the term of use determined by date.

2016-2017: The complexity of software is testing the limitations of humans to understand and verify. Go is a complex board game, significantly more difficult than chess. A computer program "AlphaGo," developed by a Google team, defeated a human "Go" master (Hutson, 2017). A vastly improved program, "AlphaGo Zero," defeated the AlphaGo program 100 games to zero. It is commonplace for artificial intelligence programs to produce results which are demonstrably, quantifiably successful and accurate, but whose techniques and inner mechanisms have not been rigorously proven (Monroe, 2018).

2010-2020: A new software distribution model is SaaS (Software as a Service). In the SaaS model, software does not reside on the user's computer, but instead is accessed and used on a web page. This software usage model relies upon and is made possible only by continuous and reliable network connection.

Several successful software products have been developed or maintained by large teams requiring significant financial investment (social media, Wikipedia, Google products, etc.). And yet, open-source software for common tasks—word processing, spreadsheet, etc.—has the capability to replace Microsoft Office as a standard. We are left, then, to ask: What is the future of small teams or "new players," niche-market software, or multiple developments of essentially the same software product?

The final aspect of computer technology is hardware that executes computation or communication at a "personal" level and at a "technical" or industrial level. These two categories are necessary, because phones and tablets are certainly computers but are used in significantly different manner than are "technical" computers. Significant milestones in hardware at the personal level follow (dates again are approximate):

Cellular phones have evolved from the early 1990's novelty to an essential, multi-purpose personal tool.

1990-2005: Fixed-keyboard phones.

2000: Text messaging.

2005: Camera in phone.

2007: Apple iPhone.

2007-present: Increased memory, resolution, available apps.

The primary design consideration of "personal" computers (phones or tablets) is their small size (pocket or backpack), being generally used as communication and entertainment devices that do not require significant text input. As a result, phones and tablets have only small keyboards, typically implemented as bounded areas on a touch screen. (Peripheral, larger keyboards are a popular addition to tablet computers.)

The primary design consideration of "technical" computers (office or accessible server) is computation performance. Whether used for number-crunching or for traditional clerical tasks, technical computers are selected on the basis of cost performance. Considerations and techniques for maintenance of cost performance are new designs of computer architecture and assembly language sets, integrated circuit density, and effective and efficient use of cores.

It's Nick's Turn

The first computer I remember typing on was a Compaq running Microsoft Windows 95. I was four or five years old—it was around 1997—and I was starting up Busytown, a game based on the children's books written by Richard Scarry. I could insert the CD, but Windows wouldn't auto-run the game after you put the CD in the drive. Instead, you had to execute two DOS commands at the command prompt. My father (and co-author, Ken Vollmar) had written these commands on a sticky-note and made sure I knew how to get to the command prompt and type them in. I remember the emptiness of the black screen of the command prompt, to which I would add one line of mysterious pale text. For me, there was no meaning in the words I was typing. It was just the sequence of characters that you had to press before playing *Busytown*.

The memory of running *Busytown* calls several things to my mind. In the first place, it shows the low degree of expectation that users have for software: We will tolerate bugs in software that we would not in appliances or clothing. I am also reminded that software development aesthetics has not changed very much in the past twenty years. Even now, at my work on software considered to be state-of-the-art, I spend much of my time typing in a fixed-width font against a black background, either editing files or issuing commands at a prompt fundamentally the same as the

one I used then. And, lastly, the memory of starting the game segues into the memory of the game itself: How simple it was, with its lovingly shaded pixels, the little world map, various single-screen levels, and primitive sound effects. When I think about *Busytown*, I think about how much more, in terms of features and content, we users expect from the world of software today.

It's not that anything Earth-shattering has been invented since the days when I played *Busytown*. In my view, the most significant technology in computing is the *idea* of an internet; and the particular network that we call "the internet" was already widespread by 1997. Advances in computing technology have allowed developers to write more and more intensely complicated applications—whether they are computation-intensive, network-intensive, or both. What has changed, within my particular field, is that software developers have had more opportunity collectively *to adapt to the internet*. On this subject, as in many other ways, my father has gone before me admirably: He's explained *why* the advances in computing technology have been possible, so I'll touch on what I think these advances mean to software developers such as myself.

To begin with, I draw a distinction between application and platform development. The internet has democratized application development by allowing virtual hosting, or PaaS (Platform as a Service). This decouples, as never before, the producer and consumer of software. As a consumer, you can download and use software on demand, without buying a physical CD or having one mailed to you. Meanwhile, as a producer, you don't need to pay up front for the printing or shipping. Applications are (in principle) cheap to develop, and applications developers can (in principle) operate anywhere. They can have their pick of the litter when it comes to programming languages, database APIs, and virtualization technologies. Meanwhile, the internet has tended to centralize platform development. This happens in much the same way that the internet (and the telephone and telegraph, its predecessor technologies) has tended to centralize every other related industry: by making central coordination cheaper and more effective. If an applications developer wants to run a large service in "the cloud," one's options are limited to a few big players: e.g., Google, Amazon, Tencent, Alibaba, and Microsoft. (As an example of this discrepancy, I can name several companies that have software teams based in the Ozarks—Great Southern, O'Reilly, Expedia—but I don't know of any big PaaS provider that operates a datacenter in the area. The closest I can think of is the Google datacenter in Council Bluffs, Iowa.)

As I see it, the biggest crisis facing software developers is also the oldest one: *how to get paid*. The internet began its life as a research/academic network viewed as a message board or, more expansively, as a library: Everything was free, and user activity was basically untracked. As I started my career, this model was starting to see "growing pains," since businesses needed to train consumers that they can/should pay for content.

Since that time, the music, TV entertainment, and film industries have largely succeeded in getting people to pay for website services. But news, books, and magazines have not similarly succeeded in receiving payment for services—possibly because of the original internet-as-library metaphor, with its expectation of free access to content. (So far, the news-related website approach to monetizing content is the use of advertising networks or "paywalls.")

This leads to a further, more troublesome strategy of web-based profiteering. There's a saying that has gained currency: "If you're not paying for it, that means *you're* the product." The fact that networks have more bandwidth and that storage is cheaper allows for a massive telemetry: *Every input and output can be recorded and transmitted ... somewhere ... for any purpose.*

It's well known that smart TVs of every brand will "phone home" with data from their screens (Maheshwari, 2018). Meanwhile, on the internet, essentially the same thing is happening with your browsing activity. On any large consumer website, just open your browser's network viewer and observe all the outgoing requests to ad networks and god-knows-where-else. All of this information can be transmitted every time you perform any action; and it can be stored anywhere, for a theoretically unbounded amount of time—although, given the state of PaaS, the data will probably be stored in a datacenter belonging to one of a few big players: See the point made above. This is not in the realm of "tinfoil hat" paranoia: By our browsing activities, our clicks and keystrokes provide data *on ourselves* that become marketable commodities. With this recognition, we have reached a watershed moment in our culture: *Where do we go from here?*

Father and Son Together

To bring our viewpoints together, we consider a few examples of the increasing presence of computation and communication in society. Here, then, are some predictions on the near future of computing.

- At the time of this writing, cybersecurity is an area of high attention. Too often, malicious users are able to detect and exploit computer security flaws and obtain sensitive data; and, also too often, data security is merely reactive, providing redress to victims of data disclosure and identity theft. However, reactive steps do not address the software and hardware systemic roots of cyber insecurity data loss. It's likely that data security will be addressed by *development and adoption of protocols for encryption and blockchain-based verification*, providing reliable personal identification and a way of establishing personal identity.

- *Virtual reality displays will soon be standard*, with two-dimensional displays becoming obsolete for graphics. Already there are inexpensive headsets to hold phones for binocular, three-dimensional projection. For example, 5G networks support continuous connectivity, uninterrupted by movement. In the near future, as software is designed and completed for those displays, every phone-sized device will be able to provide 3D displays for gaming, education, training, and entertainment.
- Custom sensors and input devices to be used with off-the-shelf computers will enable all imaginable types of interface to a computer. *These will contribute to computer accessibility* and wider use by people with disabilities or preferences.
- *Different expectations of privacy of personal information and anonymity will evolve*, in response to new forms of data collection and retention. Examples of these include facial recognition in a crowd; widespread and warrantless license plate scanning and retrieval; collection and maintenance of DNA databases for identification of individuals using DNA of near relatives; and voluntary surrender of health data access to insurance providers in exchange for coverage.
- We shall see an increasing number of arenas in which *computers have superior performance to humans and may be expected to replace humans.* For several decades, computers have outperformed humans in arenas such as diagnosis of diseases; but their success has been downplayed in favor of "more trusted" human physicians. Recent demonstrations of computer-based superiority in some arenas such as games of high intelligence (i.e., Go), facial recognition, and shoplifting prevention (Metz, 2017) have publicized areas in which the quality of computer-based decisions can compete with human-based decisions.

- Internet of Things (IoT) is the term for objects with independent sensors and data-producing capability, connected to some monitoring and reacting computer (NIST, 2019). For instance, IoT includes wearable fitness monitors, agricultural conditions sensors, and smart refrigerators. IoT devices use some type of wireless communication with a "host" computer; but at present, this communication too often uses an unsecured form of transmission, one easily observed and compromised. This has serious consequences in devices such as medical implants and automobiles. *The prevalence and criticality of IoT devices will force the development and adoption of encrypted, secure communication standards.*
- *Increasing prevalence of software in a browser (instead of resident on the user's computer) will have significant impact on user experience and developer economic model.* First, the user's computer becomes primarily a display node for internet-delivered content, with reduced requirement for standalone computational capacity. Next, the software developer's task is simplified: The developer does not need to write software to be executed internally on the user's computer, but rather to be executed on some computer of the developer's choosing and then *displayed* on the user's computer. Finally, the reduced requirement for standalone computational capacity means that end-user computers can be less powerful, which means lighter (battery weight) and cheaper (processor capacity).

The history of computer development has been characterized by alternating advances in networks, software, and hardware, with each improvement in one area serving to support and provide opportunity for growth in the other two. At this time, avenues for research and improvement are clear in each of these three aspects. Our reaction to this partial list of predicted developments is an anticipation of their arrival: We look forward to the advances and potential of these products and technologies.

References

Goldman, David. 2018. "What is 5G?" *CNN Business*, February 25, 2019. Retrieved 25 March 2019. <https://www.cnn.com/2019/02/25/tech/what-is-5g/index.html>.

Hutson, Matthew. "This computer program can beat humans at Go—with no human instruction." *Science Magazine*, October 18, 2017. Retrieved 25 March 2019. <https://www.sciencemag.org/news/2017/10/computer-program-can-beat-humans-go-no-human-instruction>.

"If someone from the 1950s suddenly appeared today, what would be the most difficult thing to explain to them about life today?" *AskReddit*. Retrieved 25 March 2019. <https://www.reddit.com/r/AskReddit/comments/15yaap/if_someone_from_the_1950s_suddenly_appeared_today/>.

Lapowski, Issie. 2015. "So, Arkansas Is Leading the Learn to Code Movement." *Wired: Business*, March 20, 2015. Retrieved 25 March 2019. <https://www.wired.com/2015/03/arkansas-computer-science/>.

Maheshwari, Sapna. 2018. "How Smart TVs in Millions of U.S. Homes Track More Than What's On Tonight." *The New York Times*, July 5, 2018. Retrieved 25 March 2019. <https://www.nytimes.com/2018/07/05/business/media/tv-viewer-tracking.html>.

"Missouri Governor Signs Bill on High School Computer Science." 2018. *The Joplin Globe*, October 30, 2018. Retrieved 25 March 2019. <https://www.joplinglobe.com/news/missouri-governor-signs-bill-on-high-school-computer-science/article_65699e7c-dd0d-11e8-b3ce-a74d9e42a8b8.html>.

Metz, Rachel. 2017. "I Tried Shoplifting in a Store Without Cashiers and Here's What Happened. Checkout Systems Are Going Autonomous." *MIT Technology Review*, September 6, 2017. Retrieved 25 March 2019. <https://www.technologyreview.com/s/608765/i-tried-shoplifting-in-a-store-without-cashiers-and-heres-what-happened/>.

Monroe, Don. 2018. "AI, Explain Yourself." *Communications of the ACM*, November 2018, vol. 61 no. 11, pp. 11-13. Retrieved 25 March 2019. <https://cacm.acm.org/magazines/2018/11/232193-ai-explain-yourself/fulltext>.

<NIST. 2019. "What Is the Internet of Things (IoT) and How Can We Secure It?" National Institute of Standards and Technology. Retrieved 25 March 2019. <https://www.nist.gov/topics/internet-things-iot>.

Routley, Nick. 2017. "Visualizing the Trillion-Fold Increase in Computing Power." *Visual Capitalist*, November 4, 2017. Retrieved 25 March 2019. <https://www.visualcapitalist.com/visualizing-trillion-fold-increase-computing-power/>.

Ends of Antecedents:
Some Views of Natural Color and Form
Photography by Chris Barnhart

1.

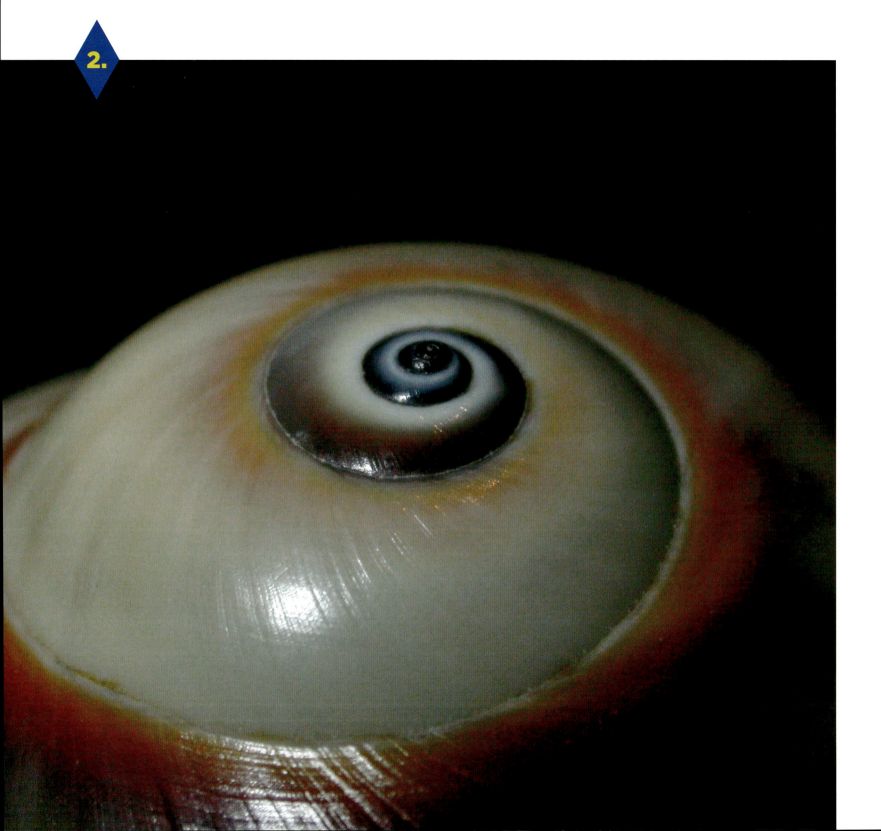

2.

3.

4.

6.

7.

8.

9.

14.

Images

1. Cherry gum (2017).
The resin exuded by cherry trees for defense anticipated chewing gum and gummy bears.

2. The logarithmic shell-spiral of a moon snail (2004).
"Seek not for ends, but for antecedents." —D'Arcy Thompson

3. Lichen on stone in the Boston Mountains (2009).
Lichens are teams of algae and fungus, combining their talents to colonize bare rock.

4. Colonies of crustose lichens (2009).
In a two-dimensional surface world, competition creates sharp boundaries between species.

5. Decomposers hard at work on an Ozark forest floor (2009).
"Into every empty corner, into all forgotten things and nooks, Nature struggles to pour life, pouring life into the dead, life into life itself." —Henry Beston

6. Pie pan after the dishwasher (2018).
Oxide layers create the prismatic effects, and the streaks result from patterned flow around holes and dents.

7. Frost framing a window in January (2014).
The crystalline forms of frozen water result from cohesion of the tetrahedral molecules—a wonderful balance of rule and randomness.

8. Water grass and reflected sky (2017).
One of the photographer's favorites, promising worlds above and below.

9. Carp at Fellows Lake (2009).
An iconic and voracious invasive species, though introduced intentionally from Asia as sport fish.

10. Caterpillar of small-eyed sphinx moth, on wild cherry stained by leaf-spot fungus (2013).
The unseen artists in this composition are the birds that overlook well-camouflaged meals, generation after generation.

11. Water striders skating on the surface of a pond (2012).
"If there is magic on this planet, it is contained in water." —Loren Eisley

12. Eggs of pipevine swallowtail butterfly (2013).
All of its life stages contain toxins from its host plant, and the adult swallowtail is mimicked by other species for protection from predators.

13. Pumpkin spider (2017).
This colorful form of the marbled orb-weaver is inconspicuous among fall foliage and berries.

14. Polyphemus moth (2006).
A native Missourian, its dramatic "eyes" may startle potential predators when the moth raises its wings.

Open(ing) Access:
On Building a Digital Commons at Missouri State University
Rachel Besara

The Ozarks region is relatively rural, with many of its institutions founded comparatively recently. (For example, the region's Fourth District Normal School—state-funded precursor to MSU—was established in 1905.) And the terrain made travel challenging. Accordingly, access to research information, and research data, was rare. In the past, scholarly publications (particularly academic journals) were to be found almost exclusively on the shelves of colleges and universities, with only a few, mainly students and academic researchers, having complete, easy access. The general public was welcome; but the infrastructure required to hold large collections of print publications was such that, in practice, only the university community and exceptionally driven members of the public could take advantage of the scholarly literature. With the poor roads and rugged terrain (even into the 1960s, many rural roads were dirt and gravel), this meant that, historically, even fewer residents of the region were able to take advantage of published scholarship.

With the advent of the internet and online journals, one might think that the problem of access has been solved once and for all. This, unfortunately, is not (yet) the case, given concurrent changes that have taken place in scholarly publishing. Schools that once subvented scholarly journals underwent rounds of budget cuts; given rising print costs, scholarly organizations could no longer pay for their journals by membership dues alone. Commercial companies, such as Elsevier, Wiley, and Springer, bought out many of these scholarly associations' journals, making them available in print and online. But, once in possession of these journals, the commercial companies began to charge high fees, inflating yearly far beyond the cost

of living, and putting the online literature available outside of the library behind a high paywall. With the Ozarks being a poorer region of the country, this has kept access beyond the means of many, even if they know how to find the literature and have the technical means to access and download the titles.

Despite widespread computer use/storage, raw research data for investigation have been even harder to get. The data used for scientific or other research projects remained locked on paper, available only on the hard drives of a lab or work computer or, perhaps, available in the cloud locked behind one or two accounts. While the researchers gathering data recognized the value of its information, it was not easy for them to share their work in any way other than distilled into a journal article, report, or book; and even these remained relatively difficult to access, requiring travel to a library and at least a basic knowledge of how to search for the desired information. Despite the advent of the internet, the size, complexity, and infrastructure of data sets made them difficult to distribute and to be decoded by others. This was complicated by the fact that, due to the historical limitations of data distribution, researchers rarely even considered how to manage and describe the data they produced in ways that would make them discoverable and usable by others. At best, this might happen in a limited way if, at the end of one's career, a researcher's papers were donated to an archive; but, most often, these items would not be able to be found or used unless the searcher already knew of the work that had been done in the past. Furthermore, the data searcher would have to be able to travel to wherever the archive was located and spend time sifting through the paper documents or old computer files. This meant that a great deal of potential knowledge was lost.

This is changing, however, for the Ozarks and the nation. There is now a growing expectation that publicly funded research data—such as those produced by Missouri State University, the University of Arkansas, the Missouri Institute of Science and Technology, and other institutions—will be made publicly available and open to all. The infrastructure is now in place for this to be done, not just in theory but in practice. The following paragraphs will discuss the key drivers of many of these changes, as well as the role that Missouri State University is playing in opening up public access to scholarly publications and research data, here in the Ozarks and beyond.

A key value of libraries is making information as available as possible to the communities they serve and to the public at large. Within the last thirty years, with the rise of the internet, there has been a much stronger push to make information openly available, openly accessible, to all. To understand why this did not become standard practice when the infrastructure first allowed it, the traditional cycle in disseminating scholarly information needs briefly to be discussed. Traditionally, a scholar or researcher at a university would do research, which would be distilled into a paper. This paper would then be sent to a publisher, often sponsored by a scholarly or professional association, whose editor would review the paper, send it to other experts in the field for evaluation and improvement (all on a voluntary basis); and then, if the paper was deemed worthy, it would be published. The scholarly or professional organization was most often run at relatively low profit margin, and, to limit liability, the standard practice was for authors to sign over the copyright to their work for free in exchange for publication. The publication would then be sold at reasonable cost (to encourage the widest dissemination), most often to academic libraries which, in turn, would make it widely discoverable and accessible, enabling

further creation of new knowledge. The performance of an academic researcher and, often, one's continued employment would be determined by the number and quality of publications produced (and by the reputation of the presses/journals one's work was published in), all of which would be decided by this process.

But when commercial, for-profit presses bought out the scholarly presses, the price of journals began to increase. Libraries and universities had no choice but to pay the hyper-inflating prices in order to stay competitive by making sure their students, instructors, and researchers had access to up-to-date information.

Not surprisingly, many university faculty and librarians found this new model problematic. They answered by joining the open access movement. Participants in the open access movement would do what they could to make the publications from their research as widely accessible as possible. For example, a mathematician might post a preprint version of a paper in a discipline-based repository, such as arXiv (Cornell, 2019), in order to make that paper widely available. While this is laudable, many publishing companies would try to suppress practices of this type, being detrimental to their business model. Or, an author might publish in an open access journal— for example, one supported by the Public Library of Science (Public, 2019)—which doesn't charge readers to access its content. Unfortunately, since these will be relatively new venues, they might not have earned the same reputation as established journals. These challenges were (and still are) compounded by the performance review model for faculty, which remains bound to the review/publication/press reputation cycle; and, since an increasing number of established journals has been bought out by commercial publishers, a university's faculty—its junior faculty, especially—remain under significant pressure to conform to the model of the major commercial publishers.

A major breakthrough occurred in 2013, when the White House Office of Science and Technology

Typing at the keyboard: computing in Meyer Library. *MSU Photographic Services.*

established guidelines that federally funded science be made publicly available (Holdren, 2013). This does not just cover the traditional journal publication that sums up a phase of a project; it covers the gathered data that are analyzed to form the conclusions as well. Failure to comply will lead to denial of future public funding. This created a challenge for all publishers of scientific research. If what they were publishing was required to be publicly accessible, how could they profit? Who would buy publicly available information? This has led to a scramble for different business models, and the outcome from that standpoint is still unclear. Many of the publishers are now trying to get payment in advance of publication through author page fees, where an author has to pay the publisher a set fee per page published, rather than charging the university through library subscription fees after publication.

Missouri State University, through its Libraries, has addressed this issue and contributes to the university's Public Affairs mission by building a digital commons, called BearWorks. Established in 2016 (Peters, 2017), BearWorks creates "a record of pub-

lication, research, and scholarship at Missouri State University" (Missouri, 2019). While this started with the gathering and entering of past research in theses and dissertations, BearWorks also provides the university with an avenue to meet the requirements for public access of federally funded research. Authors can deposit their articles and research data in Bear-Works, where anyone in the world can access them for free—unlike the licensing terms restricting so many of the resources to which the Libraries now subscribe, since these allow the public access only if one is within Meyer Library. This means that someone living in Shannon County can access the same resource as a student living in the Blair-Shannon House on the MSU campus, and this is something that was not possible before. It is as simple as the author checking and working with the publisher to ensure that a copy of the research article or other work can be placed in BearWorks.

The challenges facing the deposit of research data are more substantial than those facing standard research outputs, such as journal publications. But these challenges lead to the current cutting edge of the open access movement. From the beginning of a research project, the data involved must be gathered, recorded, documented, and described in such a way that others can use it. While this is an achievable goal, it poses greater difficulty because researchers, in many cases, have not had to document their data for future use by others. In situations involving the use of human subjects in research, the way forward becomes even more challenging: What are the ethics of sharing human subject data long-term with the public? How does it change the practices going forward? Many of the social science fields are just beginning to tackle these issues, which biology and medicine have been struggling with over the last five years.

BearWorks allows MSU to provide ethical, sustainable leadership in accessing new knowledge, in the form of publications or data created by its faculty and students. It opens up access to new perspectives to the larger community. It allows for feedback and interaction from the community to the researchers and opens up possibilities for new types of collaboration between the university and the community. For example, new approaches to community-based science are now possible because the community can see and can contribute to data-gathering in MSU-led research. This could lower the barriers to undertaking "citizen science" movement projects, where scientists, data managers, and citizens collaborate to uncover new knowledge across the traditional boundaries dividing the university from the broader community. These initiatives could lead to a far greater understanding of the Ozarks region's environment, cultures, and ecologies. The impact of BearWorks as a digital commons of open Ozarks data has yet to be measured, but the potential is enormous.

Works Cited

Cornell. (n.d.). *ArXiv*. Retrieved from <https://arxiv.org/>.

Holdren, J. (2013, February 22). Increasing access to the results of federally funded scientific research. Retrieved from <https://obamawhitehouse.archives.gov/sites/default/files/microsites/ostp/ostp_public_access_memo_2013.pdf>.

Peters, T. A. (2017, June 1). *Bearworks Institutional Repository*. Retrieved from <https://libnotes.missouristate.edu/2017/06/bearworks-institutional-repository/>.

Public Library of Science. (n.d.). *PLoS*. Retrieved from <https://www.plos.org/>.

Missouri State University. (n.d.). *BearWorks*. Retrieved from <https://bearworks.missouristate.edu/>.

The Digital Auto de Fe of 1601 Project:
Modeling Cultural Competence and Global Research Collaboration in the Virtual Reality Classroom of the Future

John F. Chuchiak IV, Antonio Rodríguez Alcalá, Justin Duncan, Argelia Segovia Liga, Dulce Martínez Roldán, María del Carmen Rodríguez Viesca, Hans B. Erickson, María Fernanda Barrón, Wendy Arcos, Andrea Flores Navarrete, Ledis Molina, Michaela Šimonová, and Sarah Powell

> "Which is better? To believe and say you do not believe, or not to believe and say you believe?"
> —Words of Mariana Núñez de Carvajal, Crypto-Jewish woman sentenced to be burned at the stake after an *auto de fe* in Mexico City (March 25, 1601)

Missouri State University and the MSU Honors College—in conjunction with our international partner, *Universidad Anáhuac Mayab* in Mérida, Yucatán, México—are using new digital technology in virtual and augmented reality to make global connections and contributions to research and pedagogy. From exploring molecular models in three dimensions to traveling back in time to early 17th century Mexico, Missouri State University and its Meyer Library stand on the cutting edge of the digital revolution in collaborative interdisciplinary research in virtual and augmented reality.

In 2017, the Meyer Library created a dedicated space for the development and beta testing of virtual and augmented reality technology. The digital tools for this revolution in interactive classroom technology are found in the Library's *Achievement Studio: Promoting Interdisciplinary Research and Education* (ASPIRE), which serves as a workspace for faculty, international exchange partners, and students to pursue projects in interdisciplinary working groups.

Within the new fields of digital humanities, the use of virtual reality simulations and interdisciplinary re-creations of historical Virtual Worlds offer powerful tools for active leaning. These new technologies give access to the world at large, expanding students' experiences of cultural, historical, and religious difference. By bringing faculty, students, and international researchers together collaboratively in global projects of research, such re-creations help model for all involved a respect for multiple perspectives and cultures—the sort of respect that underlies cultural competence, one of the pillars of Missouri State's Public Affairs mission.

Committed to open access and digital humanities research, the *Digital Auto de Fe of 1601* gives the Ozarks region access to an entire Virtual World of colonial Mexico. Using the UNREAL® virtual reality software developed by Epic Games, Inc., the project re-creates the Mexican Inquisition's 1601 *auto de fe* as a historical simulation, featuring an interactive high-fidelity videogame-like visualization in a high-resolution format. Through an open access website (https://www.autodefeinnewspain1601.com), the project aims, additionally, to make the Virtual World simulation available to all without cost. The *Digital Auto de Fe* can be used in middle and secondary schools, in college classrooms, and by humanities researchers interested in the complex and polemical history of the Spanish Inquisition.

The ultimate goal is to develop a virtual research environment for the study of the public performance of the Mexican Inquisition's celebrations of the *auto de fe*.

The *auto de fe*, or "act of faith" in English, served as the most elaborate public spectacle in what was otherwise the most private and secretive action of an Inquisitorial Tribunal. Although most previous scholars have identified the *auto de fe* as ostensibly a form of religious ritual, more recent scholarship has begun to understand that the Spanish Inquisition's *auto de fe* ceremonies served not only religious, but also political, cultural, and didactic purposes.[165] Combining the politics of both the secular and the religious and imbuing its ceremony with hierarchical and political messages—messages that concerned the nature and structure of social and racial hierarchies—the Inquisitorial *auto de fe* served to warn the Catholic faithful of the dangers of heresy. It also served to delineate the proper hierarchical social and cultural spaces of what the Catholic Church and the Spanish Crown believed were the natural order of Spanish colonial society.[166]

The innovative use of digital technology enables a multidisciplinary re-creation of the setting, sounds, sights, and events related to the public celebration of one of the better documented *autos de fe* in New Spain: the *auto de fe general* of 1601. To achieve this, the project combines the interdisciplinary skills of historians, costume designers, historical architects, illustrators, computer programmers, and digital designers.

Drawing on the available visual and textual primary source records as well as on archaeological evidence, it utilizes software for architectural modeling and acoustic simulation. The result is a reconstruction, as accurate as possible, of the setting, events, and public pageantry of this awe-inspiring event.

Creating a Virtual World of 17th Century Mexico City to Study the Mexican Inquisition and its Lived Human Experiences

> We understand that the past did not happen in 2D and that it cannot be effectively studied or taught as a series of disconnected static images.
> —Donald H. Sanders, "Why Do Virtual Heritage?" (2008)

Using historical simulations and virtual reality to teach is not a new concept; however, it is only now emerging as a viable way to teach history.[167] Recently, groups have used historical documents to create simulations, immersive environments, and virtual worlds that serve to provide historically accurate information to students and to draw interest.[168] The idea for

---

165. For the major historiography on the *Auto de Fe*, see Francisco Bethencourt, "The Auto de Fe: Ritual and Imagery," *Journal of Warburg and Courtald Institutes* Vol. 55 (1992): 155-168; Alejandro Caneque, "Theater of Power: Writing and Representing the Auto de Fe in Colonial Mexico," *The Americas* vol. 52 no. 3 (1996): 321-343; and Maureen Flynn, "Mimesis of the Last Judgment," *Sixteenth-Century Journal* vol. 22 no. 2 (1991): 281-297.

166. For the best discussion of the symbolism and significance of public spectacle in colonial Mexico, see Linda Curcio Nagy, *Great Festivals of Colonial Mexico City: Performing Power and Identity* (Albuquerque, NM: University of New Mexico Press, 2004).

167. See Jeremiah B. McCall, *Gaming the Past: Using Video Games to Teach Secondary History* (New York: Routledge, 2011). The epigraph above cites Donald H. Sanders, "Why Do Virtual Heritage?" in *Archaeology: A Publication of the Archaeological Institute of America* (March 13, 2008), retrieved June 1, 2019 (https://archive.archaeology.org/online/features/virtualheritage/).

168. For just a few of the recent similar projects, see the following:

Pox in the City (http://loki.stockton.edu/~games/PoxFinal/Pox.html): *Pox* is a digital roleplaying game in the history of Medicine.
Virtual Paul's Cross Project: A Digital Re-Creation of John Donne's Gunpowder Day Sermon, London1622 (https://vpcp.chass.ncsu.edu/): This VR project recreates the experience of hearing John Donne, the English Dean of St. Paul's Cathedral, deliver his sermon commemorating the failed Gunpowder Plot (November 5, 1622) in the Cathedral courtyard in London.

the *Digital Auto de Fe* came out of a master's thesis by Justin Duncan. Duncan's thesis focused on the spatial representation of power by the Inquisition.[169] The project has attempted to answer several historical questions that seem simple but are very difficult to assess if only the methods of traditional humanities research and textual analysis are used. Only by re-creating the events, scenes, sights, and sounds of the *auto de fe* held in Mexico City on March 25, 1601 in a real-time 3D virtual world can the viewer (student/scholar) come to appreciate the frightening process of organized public terror created by an Inquisitorial *auto de fe*.

To date, most efforts in the re-creation of what scholars have termed Virtual Worlds or Virtual Cultural Heritage have aimed at accurate representations of historic structures, cultural objects, or artifacts.[170] In most historical uses of virtual reality technology, little attention has been paid to how human actors and human institutions interacted with the built environment. Similarly, little time is spent in examining how the human aspects of daily life shaped the cultural heritage or built environments under study. The virtual reconstruction of the life of the buildings, objects, or artifacts and their "human story" have remained intangible for the most part, though these life stories and human aspects of the (re)built historical environments are the "'intangible heritage' to which contemporary people can actually relate."[171]

Digital historical models of buildings and spaces offer only a glimpse at one aspect of the past—a snapshot in time—albeit a glimpse with some sense of precision, given the use of new technologies in combination with historical archival and archaeological and architectural methods of accurate reconstruction. The human usage of the spaces of the built environments of the past, and the human attitudes and cultural traditions which occurred in relationship to or within these built historical structures, are far more difficult to re-create than the physical manifestations of historic buildings, cities, states, etc. The human element of historical actors of the past and of their interactions with the historically reconstructed space remains a gap in current research. As scholars have lamented, these so-called Virtual Heritage Environments or Virtual Worlds "suffer from the lack of 'thematic interactivity' due to the limited cultural content and engaging modules largely used in photorealistic video gaming systems."[172]

The first phase of our joint international research project, *The Digital Auto de Fe*, has sought to integrate the human aspect of the real lives and experiences of people who encountered the repressive apparatus of the Mexican Inquisition—whether as accused heretics, as officials of the Holy Office, or as spectators (drawn from the general public) at a major public event of punishment known as an *auto de fe*. By examining the spatial nature of the *auto de fe* and the dis-

Virtual Harlem (https://www.evl.uic.edu/aej/papers/cga-harlem.pdf): This VR project "lets students experience the Harlem Renaissance of the 1920s and 1930s as a cultural field trip," allowing a single-player avatar to move freely around, giving an immersive experience of the city streets and sights.

Romelab (http://hvwc.etc.ucla.edu/): UCLA's *Romelab* is a multidisciplinary research group whose work uses the physical and virtual city of Rome in studying the interrelationship between historical phenomena and the spaces and places of the ancient city.

169. Justin Duncan, "Performing Theaters of Power: The Holy Office of the Inquisition's General Autos de Fe in Spain and Spanish America and the Visual and Physical Representation of Inquisitorial Power, 1481-1736" (2014), M.A. Thesis, Missouri State University (https://bearworks.missouristate.edu/theses/1170).

170. See Mohamed Gamal Abdelmonem, Gehan Selim, Sabah Mushatat, and Abdulaziz Almogren, "Virtual Platforms for Heritage Preservation in the Middle East: The Case of Medieval Cairo," *Archnet-IJAR* vol. 11 no. 3 (2017): 28-41.

171. Abdelmonem et al., "Virtual Platforms for Heritage Preservation," p. 28.

172. Abdelmonem et al., "Virtual Platforms for Heritage Preservation," pp. 28-29.

tribution and use of public space in early 17th century Mexico City, this project has taken what some have called the "spatial turn" in the digital humanities.[173] Thus, this phase of the project focuses on the interactions of historical personages with the built environment of the 17th century Palace of the Mexican Inquisition; it focuses as well on the relationships of these historical actors with the functions of the institution of the Inquisition, and on their interactions and experiences within, outside, and around the re-created ritual, cultural, and judicial space of an Inquisitorial Palace.

Modeling Cultural Competence and Teaching Empathy in a Digital Humanities Context

One of the primary goals of this project is to emphasize the relevance of humanistic and historical scholarship on religious intolerance in the past to contemporary debates over modern issues of religious and racial persecution. By examining the nature of religious intolerance and persecution through the story of one young Jewish woman's ordeal and forced participation in the *auto de fe* of 1601, we explore ways of creating empathy in modern audiences, encouraging tolerance and mutual understanding through historical simulation: By this means, we aim to counter the recent, increase in anti-Semitism and other alarming trends of religious intolerance.

As we have noted, recent scholarship has come to understand the political, cultural, and didactic purposes of the Inquisition's *auto de fe* ceremonies. Our challenge, therefore, is to re-imagine how its sermons and public sentences, being social and political as well as religious gatherings, functioned to bring together church, state, and people for punishment, instruction, inspiration, and the creation of a common cultural

identity. The project will provide detailed information about 17th century architecture, dress, religious symbolism, and common processionary procedure of the time period, all of which will enhance our current knowledge of the human experience of life in 17th century New Spain.

Another major goal of this project is to demystify the institution of the Inquisition. The project will provide access to a vast amount of information about the structure, organization, and day-to-day activities of the Inquisition—information that will be made available to the general public for the first time. The project also re-creates in detail the major buildings and architectural features of the streets along the processional route. First among these historical re-creations of 17th century Mexico City is the virtual reality reconstruction of the Mexican Inquisition Palace and its developmental stages: By bringing the architecture of Mexico City in 1601 alive for the interactive viewer, it offers both students and scholars the rare opportunity to experience a major 17th century city in its splendor. A team of architectural historians have helped with the re-creation and design of historically accurate buildings and built environments, offering an intensively researched focus on the utility and usage-flows of these buildings by real historical actors.

To create empathy and encourage tolerance—two important aspects of humanistic studies of the past—the *Digital Auto de Fe* project attempts to design and implement several interactive and 3D digital re-creations that visually and interactively portray for the scholar, teacher, and student the human experiences, pains, shame, and public punishments related to these acts of religious intolerance.

The basic research questions that the project team hopes to address with this re-creation of the Virtual World of Mexico City in 1601 are as follows:

---

173. For a good discussion of the "spatial turn" in digital humanities, see Richard White, "What Is Spatial History?" (Stanford University Spatial History Project, 2010).

What would a penitent have actually seen and experienced during the procession of an *auto de fe*?

With an avatar-style approach, the viewer (student/scholar) will be given a direct point-of-view access to the experience of any one of the actual historical actors who participated in the *auto de fe* of 1601.

Placing the viewer "inside" the persona of a convicted heretic will help create empathy and better understanding and a personal connection to the past.

Dialogues, conversations, and speeches can be experienced in either 17th century Spanish or in English translation, with virtually accurate acoustically re-mastered sounds, music, and other visual and audio stimuli that might have been experienced by a spectator of the event.

These types of virtual visualizations and re-creations can also help highlight the painful nature of anti-Semitism, highlighting the problems and pains involved with religious and racial and ethnic intolerance—problems very much at the center of the human condition even today.

How would a lower-class or mixed-race *casta* resident of Mexico City have perceived and experienced the event of the *auto de fe*?

What would the view of various spectators have been, based on their varying positions, social class, and/or racial caste?

How and in what way would the Inquisitors and highest-ranking members of the religious and political elite manifest their power through the spatial creation and manipulation of height, position, and religious and political symbolism?

By viewing the staging and event from the visage and point of view of an Inquisitor, the scholar and student can come to understand issues relating to the spatial representation of power, and also understand the hierarchically stratified nature of the society in New Spain.

By modeling and analyzing these and other research questions, the *Digital Auto de Fe* will serve as a useful tool in examining issues of gender, race, class, status, and political position in colonial Latin American society. With further applications and usefulness beyond the virtual re-creation of the *auto de fe* itself, this project will offer the viewer the chance to delve deeper into the society, culture, and race relations of colonial Mexico City at the turn of the 17th century.

Encountering Culture in a Virtual World: Teaching Race, Caste, and Class through Virtual Simulations

Another goal of the *Digital Auto de Fe* is to portray the relationships among clothing, social status, and caste identity. The dress and costumes of the period represented the power and authority that individuals held within their social and racial position.[174] Each type of dress and accessory held a specific meaning, portraying either the status or power of the wearer or the lack thereof. Many groups of people participated in the ceremony, from the poor to the wealthy and powerful.

---

174. Royal sumptuary laws prohibited certain *castas* from wearing various types of textiles. For an example of the role of the Inquisition in policing these laws, see Martha Sandoval Villegas, "Indecencia, vanidad y derroche en algunos trajes novohispanos de fines del siglo XVII: Conceptualización del mal a través de la indumentaria," in Erik Velásquez García (ed.), *Estética del mal, memorias del coloquio Internacional de Historia del Arte* (UNAM, 2013),pp. 49-83.

The Viceroy of New Spain, Gaspar de Zuñiga y Acevedo (1596?). *Courtesy of Museo Nacional de Historia, Mexico City.*

Clothing and costumes served as an essential means in Mexican society of distinguishing social groups from one another. Therefore, not only will the project re-create the dress of the time period, but there will be an array of information on the specific symbolism of the clothing and designs used by the characters (see Fig. 1).

Figure 1. The Viceroy of New Spain, Gaspar de Zuñiga y Acevedo. *Character created by MSU Student, Ledis Molina.*

The project aims to show era-specific clothing for each participant in the ceremony, as well as the clothing of the general populace that witnessed the event along the street. Each character is fully interactive and their dress, race, caste, and social status are explained in detail. The racial and social makeup of Mexico City in 1601 will be portrayed proportionately, based on the available census and population documents known as *padrones*. In this manner, the relative number and ethnic identities of characters and bystanders will represent an approximated view of the varied racial and *casta* makeup of Mexico City in the early 17th century (see Figs. 2-3).

Illustration and design work involving several of the major characters used in the Virtual World have been mocked up by graphic artists Dave Gibbon,

Figure 2. Concept art for the initial facial design of the African Slave, Juan Mozambique, assistant of the chief jailor of the Mexican Inquisition. *Concept art by Michaela Šimonová, digital conversion by Ledis Molina.*

Figure 3. African Slave, Juan Mozambique. *Concept art by Michaela Šimonová, digital conversion by Ledis Molina.* Mozambique was tried for illegally taking secret notes to the prisoners and processed in the *auto de fe* of 1601.

Ledis Molina, and Andrea Flores Navarrete. Michaela Šimonová and a separate team from the Comenius University of Bratislava, working in conjunction with MSU Honors student Sarah Powell and other student artists, are currently aiding in the creation of more concept art; designs for further digital characters are in various stages of development.

Spanish language transcription, translation, and analysis of the original archival primary sources of the Mexican Inquisition form the core of the historical documentation. Dr. Argelia Segovia Liga leads the team of historians and students who are currently accessing, analyzing, and transcribing materials from one of the largest and most complete surviving archives of any Inquisition Tribunal, the Mexican Inquisition's surviving documentation from the *Archivo General de la Nación* in Mexico City. A significant number of original sources are also found in private libraries and museum collections in the United States, such as in the Conway Collection at the Helmrich Center for American Research (part of the Gilcrease Museum and Collections in Tulsa, Oklahoma); taken together, these offer a very intimate and minutely documented look at the past of this repressive institution and its historical actors, officials and, in many cases, its victims. One set of documents (to be discussed in a separate digital publication in Bear-Works Digital Commons) will be the trial transcripts of the case against Guillermo Enríquez, a Flemish sailor and one-time privateer.

Studying Human Interactions with the Built Space: Virtual Reality Re-Creation of the Palace of the Mexican Inquisition

In the execution of these goals, the *Digital Auto de Fe* focused in its first phase on the central *traza* or grid plan of the 17th century Mexican capital city in general and, more specifically, on the plaza of Santo Domingo,

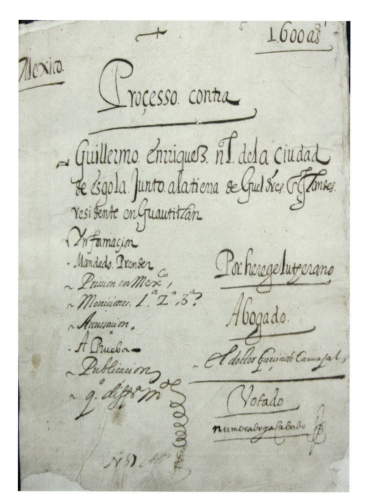

"Inquisition trial against Guillermo Enríquez, native of Flanders for heresy" (1600). *Conway Collection of the Helmrich Center for American Research, Gilcrease Museum.*

whose centrally located palace complex once held the Tribunal of the Holy Office of the Inquisition (see Fig. 4). In studying more than just the built environment, this project has investigated and incorporated numerous historical, cultural, archaeological, and architectural methods, sources, and interpretations, in order to offer a historically supported virtual re-creation of the cultural heritage of 17th century Mexico City.

Figure 4 : Plaza de Santo Domingo in Mexico City, with view toward the Palace of the Mexican Inquisition as it appeared *ca.* 1655. *Virtual re-creation by Dulce Martinez Roldán.*
Interactive visualization can be seen here: https://kuula.co/post/7PHQJ

In 2016, after extensive preliminary work on the themes and initial digital character designs, an opportunity arose in the MSU Honors College to expand its international partnerships with the *Universidad Anáhuac Mayab*. A specialized research exchange program created in 2017 between the two institutions launched the second phase of this project under the co-direction of Dr. Antonio Rodriguez Alcalá (professor of Architecture and Virtual Cultural Heritage reconstruction at *Anáhuac Mayab*), who now serves as the project's chief architectural consultant, and Dr. John F. Chuchiak IV (MSU professor of Colonial Latin American History), who serves as chief historical consultant in conjunction with historian and Springfield Public School teacher, Justin Duncan.

With an international collaborative research agreement in place, the MSU Honors College and the School of Architecture at *Anáhuac Mayab* began a fruitful research and student exchange program

focusing on expansion of the project's second phase. Incorporating at this stage intensive research by students of architecture from Mérida, Mexico, and Honors College students from MSU in the fields of history, language, linguistics, art and design, and several other disciplines, this interdisciplinary international working group began its re-creation of the 17th century Palace of the Mexican Inquisition. As the project has developed, a much larger international interdisciplinary research team evolved: Following the Fulbright Research Fellowship at MSU of Dr. Milan Kováč (professor of Ethnology and Cultural Anthropology at Comenius University), students and concept artists from Comenius University of Bratislava, Slovakia have added their own skills, talents, and expertise, with Dr. Kováč's students joining the various research teams.

The symbiosis of research fields and disciplinary methods is being applied in a research project whose main purpose is to bring back to life some of the key

Figure 5. Interior central patio of the Mexican Inquisition Palace (*ca. 1655*). *Virtual re-creation by Dulce Martinez Roldán.* Interactive visualization can be seen here: https://kuula.co/post/7PH9j

elements in the development of the institution known as the Tribunal of the Holy Office of the Inquisition. Through the combination of (and interactions between) the humanities and information technologies, this second phase has focused on the virtual reconstruction of two of its most representative spaces: the Second Audience Chamber of the Mexican Inquisition Tribunal, and the Secret Archive and Library Room of the Holy Office. The richness of the historical subject and its surviving documents allows the integration of research outcomes from several disciplines (using historical documentation, virtual architectural and artistic reconstruction, virtual

museum spaces, and the study of the evolution of the built environment) all in the same project (see Fig. 5).

Due to the nature of the available information and surviving inventories (which contain descriptions of equipment, furniture, provisions of elements, the flow of historical human actors, and their user-flows within the palace structure), it has been possible to create a historically documented and visually enriched Virtual World. The integration of the digital technology and its methods became the next step, which consisted of moving this 2D historical documentation into the digital realm of three-dimensionality. That is, the reconstruction phase began with the use of

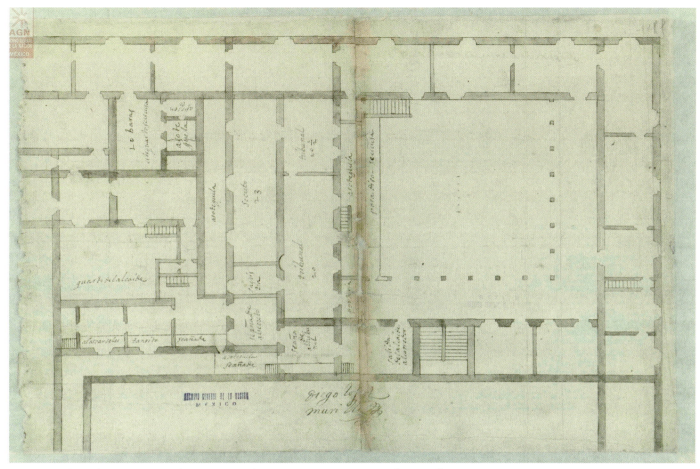

Map and plan of the interior of the Palace of the Mexican Inquisition (*ca.* 1655).
Archivo General de la Nación, Mexico City.

two-dimensional drawing software to re-create the architectural plans of the Inquisitorial spaces within the Mexican Inquisition Palace; this reconstruction was based on a 1655 plan of the architect Diego López Murillo, which exists in the collections of the National Archives in Mexico City.

Subsequently, the team modeled the interiors based on the architectural typologies of the time, using standard types of masonry walls, wooden coffered ceilings, and wooden doors, among other aspects of the built space. Within the model, a rigorous system of notation was maintained, leading to a uniform system of codes to document evidence and the sources of historical, architectural, and archaeological information used in the historical reconstructions.

The placement phase of the integration of art assets of the Virtual World was rigidly regulated by historical documentation, which included detailed inventories, descriptions, and visitation records of the Mexican Tribunal, all of which indicated with great

Figure 6. Preliminary version of the Sala del Secreto, or the Secret Archives of the Mexican Inquisition. *Virtual re-creation by Antonio Rodríguez Alcalá, with material and cultural objects designed by María del Carmen Rodríguez Viesca.* Interactive visualization can be seen here: https://kuula.co/post/7PHzs

precision the location of each official and their equipment and accoutrements, as well as the placement of their furniture, cultural materials, etc. Environmental elements, such as the placement of Inquisition trial files on the shelves of the Secret Archive, were incorporated by taking into account the characters involved and historical descriptions of the layout of the Secret Archives of the Inquisition (see fig. 6).

The environmental conditions were also replicated with care: Since many of the interior rooms were dark interior spaces without windows, they remained totally occluded from natural lighting and required the creation and placement of candles, lamps, and other historically accurate materials and means of lighting; these provided the rebuilt space with the physical and ambient characteristics of the actual surroundings (see Fig. 7). The privacy and secrecy demanded in the audience chambers of the Mexican Inquisition required the use of re-created lighting based on candles and other torches mounted on chandeliers which, when incorporated, offered a more realistic re-creation, impressing upon the viewer the fear and terror that a suspect might experience when brought into one of the smaller, dimly lit audience chambers of the Inquisition.

The contrast of the darkness of the Inquisition Tribunal's interior chambers with the light and open patio plan of the main entrance areas (and even with

Figure 7. Preliminary version of the Sala de la Audiencia "de los retratos" of the Mexican Inquisition, with ambient lighting as the chambers would have appeared in the 17th century. *Virtual re-creation by Antonio Rodriguez Alcalá, with material and cultural objects designed by María del Carmen Rodríguez Viesca.*
Interactive visualization can be seen here: https://kuula.co/post/7PH9j

the patio of the secret prison section of the palace) is stark, serving as a reminder that Inquisitorial imprisonment was meant more for holding prisoners for the duration of their trials than for long-term imprisonment as a form of punishment (see Fig. 8).

The Pedagogical and Ultimate Research Results of the Project

In targeting its audience, the *Digital Auto de Fe* aims to attract more than just scholars and advanced researchers. The principal goal is to educate the public about the Inquisition's *auto de fe,* as well as to illustrate aspects of life in colonial Mexico. Professors and

teachers worldwide teach courses on the Inquisition, and this project seeks to increase the instructors' and students' knowledge through Virtual World re-creation. The technology allows teachers to show the simulation in class to a whole group, to make it an individual class assignment, or to assign the simulation to be watched and interacted with at home. In addition, resources will be provided for teachers and students to assess understanding and learning objectives.

As a Virtual World of Mexico City in 1601, the project offers advanced scholars the ability to engage with the simulation as a research tool. Scholars of both the Inquisition and of colonial Mexico will find in the

Figure 8. Interior courtyard patio of the secret prisons of the Mexican Inquisition.
Virtual re-creation by Dulce Martinez Roldán.
Interactive visualization can be seen here: https://kuula.co/post/7lK49

materials and reconstructions of the built environment, as well as in the representation of the social,
racial and ethnic backgrounds and costumes of the
characters, a wealth of information for research purposes. The linked primary sources, images, maps, and
other historical documents and archaeological artifacts will offer advanced scholars a virtual museum
filled with materials both textual and physical to work
with and utilize in their research and pedagogy.

The Once and Future Ozarks:
Sometimes Things Change So Much They Stay the Same
Dale Freeman

The Futurist is in an enviable position. He talks confidently of things to be 100, 200, 300 years from now, knowing full-and-damn-well that neither he nor the people who hear him will ever live long enough to prove him wrong.

—The Ozarker, 1989

The times they are a-changin' in the hamlets, villages, and towns across the Ozarks.[175] Profound statement. It could have been written a century ago, or today, or a century from now. Or on and on until Armageddon, which a few residents of the Ozarks firmly believe will start at eleven o'clock tonight, or by tomorrow afternoon at the latest.

Despite the Ozarkers' legendary reluctance to accept change, the fact is that their beloved hill country—our land of Milk and Honey and Isolated Splendor—has been in a state of flux since God and the Native Americans first walked these shrouded hills and wavy prairies eons ago.

And so it continues as we prepare to enter the 1990s. Like it or not, fight it or not, more changes—dramatic, drastic, even dreaded ones—they are a-comin'.

In case you haven't noticed, things just aren't the same anymore in the smaller communities of our boundaryless region, from Seneca to Sullivan, from Naylor to Neosho, from Hermitage to Huntsville, from Protem to Plato. They never will be again. Only foggy memories and sweet dreams keep alive the hopes that scores of our towns will be as they once were or, more realistically, as we thought they once were.

As the bells ring out to herald the 21st century and beyond, they toll two tunes: "Auld Lang Syne" and "Happy Days Are Here Again." They peal forth their message of cheer and despair. Many of our communities will grow and prosper, some beyond our wildest dreams or fears; many others will suffer; some will barely hang on; some will die; some will merge with near-neighbors for strength; and some new communities will be born, like the Kimberling Citys and the Bella Vistas and the Viburnums of our world today. More profundities.

"Whatever happened to my hometown?" you ask. Hail' the interstate passed us by; the railroad left; the shoe factory shut down; the mine petered out; the bank closed; the doctor retired; the high school got consolidated; the churchhouse burned; the city well dried up; all the young people skeedaddled 'til we didn't have enough kids left to field a ping-pong team or a one-man band; and the old folks they either died off or moved closer to a hospital. Hail' we ain't even got a postmark left any-mores.

—Town Native, 1937, 1989, 2037, 2089

Fictional village aside, what will it *really* be like in the small town of the Ozarks a century from now? Even Nostradamus wouldn't touch that one with a ten-foot crystal ball. Only a Damphool Native Phuturist would tackle it, one of those who still sometimes pines for a return to the small town of our childhood, where Living was easy but Life was hard.

But lest we be found guilty early on of practicing what the logicians call *dicto simpliciter*, an argument

---

175. [First published in 1988, Freeman's article was anthologized in *OzarksWatch* vol. 5 no. 3 (1992): 54-56. Some thirty years later, we revisit its wistful prognostications and have asked that he provide a postscript. —*Eds.*]

or observation based on unqualified or hasty generalization, let me make something opal clear from the outset: It will take several keys—call them complex combinations—to come anywhere close to unlocking the future of almost any Ozarks town.

> Shoot far! I can remember 'way back when Springfield had only six-seven hunnert thousan' people in it.
>
> —Oldtimer, 2099

We turn the keys, we open the door, and what do we see?

The emergence of at least three mini-megalopolises, extensive, heavily populated urban areas which might include any number of cities. The prime candidates:

> Springfield—from Bolivar to the north, near Branson to the south, near Lebanon to the east, Mount Vernon to the west—perpetuating itself as the Queen City of the Ozarks as it moves well past the million mark in population.

> The Tri-State region around Joplin, including Carthage, Pittsburg, Miami, and Neosho.

> And northwest Arkansas, the Rogers-Springdale-Bentonville-Fayetteville area, destined to become the Land of Opportunity's most populous region.

Why the three? All gateways to the great Southwest, they will continue to grow as, or evolve into, *the* primary centers of higher education, health care, communications, transportation, retailing, distribution, sporting events, cultural affairs, services, and industry for the Ozarks. All these are "musts" today and in the future.

And it will come to pass each of the mini-megalopolises will have many moons revolving around the urban suns, oases where bedroom communities, satellite industries, schools, medical facilities, and the inevitable Wal-Marts will spring up and flourish. What becomes of the crossroads country store? It's already a 7-Eleven.

It will be a never-ending battle to prevent the big fish from gobbling up the little fish.

Only strong leadership, inherited or imported, good schools and health care facilities, a bent toward unpopular regional planning, and a helluva lot of luck will prevent those satellite communities from losing their identities, their personalities, and their independence in years to come.

> There is one thing you can say about Springfield: It is geographically located.
>
> —Welcoming Address by Springfield Mayor Otis L. Barbarick to Missouri Realtors Convention, 1950

So, too, geographically located are several Ozarks towns and small cities and resort regions that do not have to be swallowed by the metro monsters of Springfield, Springdale, or Joplin to survive or even thrive.

They are the stand-alones, the isolates, either whose distance from the urban areas or whose uniqueness set them apart and will allow them to endure.

They are the ones with good chances to retain or to build sound economic bases—some brought on by their own free enterprise and ingenuity, some produced by the federal government, some by natural and man-made beauty.

They are the Rolla-Fort Leonard Wood-Lebanons, the Bransons, the Harrisons, the West Plainses, the Poplar Bluffs, the Lakes of the Ozarks regions.

Their future growth, or their future deterioration, however, hinge on many ifs—stricter environmental protections, adequate water supplies, government programs or the withdrawal of government programs, better roads, agribusiness prosperity and, near the top of the list, attractive tourism. Any one or all of the above apply.

In many cases, local leadership (and, yes, political clout) will help determine their existence in the 21st century and beyond.

And who is to say that other factors, say philanthropy or slick entrepreneurship, will not play a strong role in those and other communities? Although the three may disagree, there will always be a Sam Walton or a Don Tyson or a John Q. Hammons somewhere out there to revive or goose or re-invent a community long thought near the grave.

Throughout the Ozarks, where the once heavy lines drawn between town and rural living have been all but obliterated, there also will be certain anomalies, inconsistencies from the norm, that bring about a town's birth or rebirth, its death, or its stagnation.

The anomaly could be produced by a new military base, a new government impoundment or facility, discovery of the Mother Lode, a regional airport, a new freeway, a new state or federal park. It could be produced by luck. Growth and Prosperity by Chance.

> Used to be the only time folks come back home was for Decoration Day or to be buried in the family plot. Nowadays, the pavement's plumb hot with newcomers, old-uns and young-uns, movin' in from Des Moines and Peoria and Wichita and natives comin' back from Dee-troit and Visalia and See-attle.
>
> —Comebacker, 1987

Greta Garbo would have a heckuva time vanting to be alone in the Ozarks. There will be few hiding spots left out there in our boondocks of the future. Move over and make room for more.

The influx of retirees and returnees, come-heres and come-backs, will continue, undoubtedly exceeding the present rapid pace. Possibly hundreds of thousands of them will be invading the region, most of them hunting the good quiet life they've read so much about in Chapter One of their yellowed Book of Nostalgia.

They will continue to re-populate our small towns—for a while. They will be more affluent and better educated than their predecessors. Ninety-eight-point-six percent of them will be white. Many will be valuable and progressive additions to their adopted communities. Some will be dead-headed obstructionists. And most will not settle finally for the more isolated small towns they originally yearned for so much. Eventually, most will be drawn to the larger cities of the area for a variety of reasons—primarily for jobs, health care, higher education, better transportation, entertainment, and security.

The question: Will their tremendous numbers turn their new home bases into the same type of environment they fled?

It is possible.

Sometimes things change so much they become the same.

> I ain't been to the courthouse since the last hangin'.
>
> —Native Ozarker, 1948

Once assured of their important and necessary role in the future of the country, county-seat towns in the Ozarks no longer are immune from change. Soaring costs of government, particularly in currently decentralized areas such as public safety, could lead to future consolidation of programs among counties. Increased budgetary problems plus expanded computerization of records (including assessment, collection of taxes, and election procedures) also could diminish the traditional need and preference for single-county officials.

Those once-sacrosanct capitals of local power without nongovernmental economic bases could suffer. Some county seats most certainly will be sorely wounded right where they have sat so comfortably for so long.

Just had to write and tell you about our simply awesome evening. Brock and I took the Elevated to Heliport then jet-coptered to Eureka Springs in five minutes. We'd already ordered dinner from our telemenu at home so it was waiting for us when we got there. Had a fabulous and inexpensive meal of home-grown lobster (only $375 per, imagine it). After din-din, we hopped the hydrofoil for a quick sprint to the casino at Branson, where Brock didn't waste much time dropping several thou (ha!) on a blackjack game satellited in from Vegas. Then we took the slow way home—a 20-minute ride on the Bullet Train back to Joplin and our own king-sized air bed. Love, Melinda.
—Nightly Telefax to Mom, 2075

Ah, the wonderful and mysterious future. How can we even envision the explosion of new ideas, the new discoveries and challenges that are bound to both startle and change us. But our forefathers were in the same fix. Imagine: George Washington's Farewell Address on world-wide television. Christopher Columbus' walk on the moon. The battered survivors of Wilson's Creek gawking through Springfield's Battlefield Mall. Ridiculous?

Well, you and George and Chris and the brave boys in the Blue and the Grey ain't seen nothin' yet, podner.

And how are our cherished towns and villages in the Ozarks to survive these mysteries to come, these revolutionary blasts of inevitable change?

What is to become of the Bolivars and Buffalos, the Marshfields and Monetts, the Hartvilles and Houstons and Humansvilles, the Loose Creeks and Lamars, the Green Forests and Greenfields, the Avas and Auroras, the Flippins and Fordlands?

Will they do or die? Will they flourish or stagnate? Will they try to whip 'em or seek to join 'em? Fanciful guesses of things to come are easy, but palmistry is not exactly a pure science.

In addition to the usual asterisks barring War*, Pestilence*, Famine*, Flood*, and Fire* that should accompany all predictions, every forecast of the future for Ozarks towns and villages should bear a further disclaimer:

The only certainty is uncertainty; the only constant is change.
—University of Missouri Futures Committee Report, 1980

Yep, friends and neighbors, it's going to be mighty interesting to see what happens in the Ozarks. And yep, it's a dadgummed shame you-all won't be around to tell me about it.

Peace.

A 2019 Postscript

This is supposed to be a solicited addendum to a fake-news piece this Olde Editor wrote some three decades ago about the long-to-be future of our beloved (?) Ozarks. The premise then was that most of us won't be around to enjoy or despise what's-to-be, allowing all those futurists or prognosticators or modern-day Nostradamuses out there to have at it all to their sefs.

Frankly, we have never met a live futurist or soothsayer or whatever you want to call or brand 'em. Scratch that. Frankly, we have met many pseudo-futurists in the past, including those scores of politicians or wannabes or egoists of a different ilk whose crystal-balling can only be described as indescribable.

Which leads us to the only futurist we've ever known. Or envied.

He was Charles Claiborne Williford.

He was a professional futurist, a forecaster.

He was the official U.S. weatherman, based at Springfield's Municipal Airport before it became part of Branson …

And Mr. Williford, which he was never called, was pretty sure what was going to happen in the future— if the future didn't last more than three or four days. Ozarks weather moved rather quickly, you see.

To you newcomers and you youngsters, C.C. Williford probably was the best-known, the most vilified, the most admired, and the most talked-about person in the Ozarks for a good many years in the halcyon days of the mid-1900s. He also was a character, a perfect fit for the region and the times, after he migrated from his native Little Egypt near Cairo, Illinois, where the Ohio merged with the Mississippi.

For what seemed like a half-century, Charlie Williford was The Weatherman, a favorite adopted son of the hill country. His 8:30 a.m., 12 noon, and 4:30 p.m. broadcasts were must-listen-tos, particularly on far-reaching radio station KWTO (Keep Watching The Ozarks). He reached the unreachable, including the cloistered Trappist monks, near Ava, who would gather around the monastery's one radio solely for any weather news that might affect their grape-growing abilities. (The good brothers later took up concrete block-making.)

As a primary news source, C.C. was quoted almost daily in the Springfield newspapers and became an early favorite of reporters, including this one. He was Jolly Cholly most of the time, but he pulled few punches in his forecasts. One of his favorite weather warnings, addressed to area farmers as a front approached: "Get all your stock on high ground, boys. She's gonna be a toad strangler."

He was Ol' C.C. to all. As his popularity grew, he became a noted public speaker, traveling the area to address everything from high school commencement ceremonies (including mine) to community events with dinner on the grounds. He enjoyed parties, especially private ones attracting journalists. Accompanying him to those events was his comely blond wife, Dorothy, who was a talented whistler (yes, a whistler!) and often the parties' unpaid entertainment.

And Cholly certainly became the first, and perhaps the last, U.S. meteorologist to have an elephant named in his honor. The baby pachyderm had been donated to Dickerson Park Zoo, thanks to a Springfield civic club's fundraiser, and its naming was a community-wide popularity contest. The fact that the winning name, Ol' C.C., was given a female was beside the point. The human Ol' C.C. thought the presence of an animal Ol' C.C. was a blast and reportedly became a frequent visitor to the zoo to chum with his unlikely namesake. So …

Why this piece about somebody most of you never heard of? It was merely a thirty-year follow up, manufactured to tell about a good person, a *real* futurist, someone who faced his jury, the public, daily.

If only we had more of them.

Peace.

Droning the Ozarks:
Innovations in 3D Topography
Toby Dogwiler

Valley Water Mill Dam: 3D point cloud showing dam and profile. *Courtesy of Toby Dogwiler, MSU Department of Geography.*

A confluence of technologies—of drones, digital photography, and computer modeling software—are revolutionizing research and teaching in geography and related disciplines. The following are some local examples of drone photography and photogrammetry prepared by faculty and students in the MSU Department of Geography, Geology, and Planning.

Imaging the Valley Water Mill Dam

At the northern edge of Springfield just to the east of N. Farm Rd. 171 lies Valley Water Mill Park. The Watershed Committee of the Ozarks is headquartered by the pond that is created by the dam that is shown in 3D above. This three-dimensional model was gener-

ated through a process called structure-from-motion photogrammetry. A series of overlapping photos taken by drone are combined to create a three-dimensional representation of the landscape. The following is one of seventy-seven orthogonal photos used to construct the model.

In the composite image that follows, the blue squares hovering over the model indicate the location where each photo was taken. The mission planning software that controls the drone ensures that each photo overlaps about 80% with adjacent photos at the top and sides. The resulting three-dimensional model has a resolution of approximately a centimeter per pixel; its accuracy, similarly, lies within a centimeter.

Valley Water Mill Dam: orthogonal photo taken by drone. *Courtesy of Toby Dogwiler, MSU Department of Geography.*

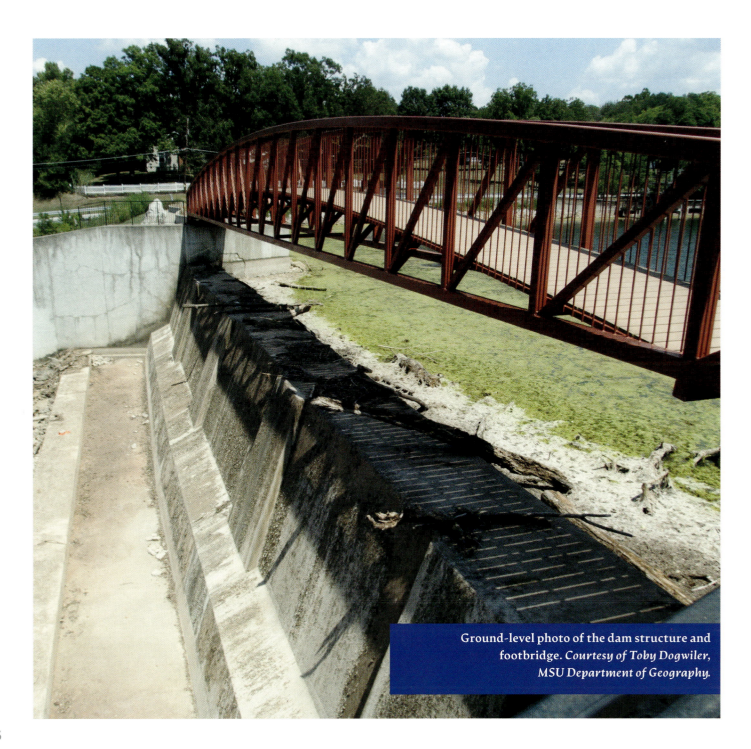

Ground-level photo of the dam structure and footbridge. *Courtesy of Toby Dogwiler, MSU Department of Geography.*

Perspective 30°

faces: 7,960,002 vertices: 3,982,179

Valley Water Mill Dam: 3D mapping software. *Courtesy of Toby Dogwiler, MSU Department of Geography.*

In the "old days" of traditional plane surveying, it would have been impossible to obtain such high-resolution topographic data. Drone-based photogrammetry allows small areas such as this to be mapped rapidly—within an hour or two. For photogrammetry, most time in the field is spent surveying ground control points; these are used to geospatially calibrate the 3D model, ensuring that subsequent model-based measurements and analyses are accurate in terms of distance, area, and volume.

In a photo on the facing page, a student holds a real-time kinematic (RTK) GPS that receives satellite positioning data, while the theodolite (to the student's right) measures variations of distance and elevation in terrain. Through a combination of RTK and theodolite data, the orange triangle at her feet—a ground

control point for calibrating the three-dimensional model—can be accurately georeferenced with respect to latitude, longitude, and elevation. (Again, the accuracy achieved by these technologies falls within a centimeter.)

Surveying Flood Damage on the North Fork of the White River

While the Valley Water Mill Dam proved a useful exercise in technology, this next project proved far larger in scope.

In April 2017, heavy rains led to devastating floods throughout the Ozarks. Some of the worst flooding occurred on the North Fork of the White River in Mark Twain National Forest. Given the dense forest and underbrush, much of this area is difficult to

Drone and triangles used as ground control points. *Courtesy of Toby Dogwiler, MSU Department of Geography.*

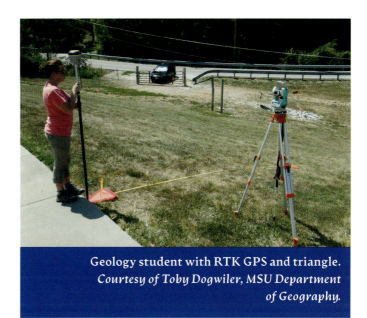

Geology student with RTK GPS and triangle. *Courtesy of Toby Dogwiler, MSU Department of Geography.*

access. In assessing the damage, my colleagues and I gathered data on the river's riparian zone through some old-school on-the-ground observations and surveys combined with drone-based photography and three-dimensional modeling.

As the following photos attest, it's by drone that one appreciates the full extent of damage to the North Fork's once-forested riparian zone.

In the first photo below, note the toppled bridge pier in the stream and the bridge deck removed by floodwaters, deposited several hundred feet downstream. The second photo is an orthogonal image of the same area, with the drone camera pointed straight down, perpendicular to the land surface. The third photo shows the awesome power of floodwaters: Prior to flooding, the forest covered the full zone up to the river's edge.

Clearly, drone-based photography provided the broadest perspective of the flood damage. As well as base maps for our ground-level observations, the drone images also enabled computer-generated mapping and analysis of areas that would have proved too large for field personnel to visit and assess efficiently.

Being a science of observation and measurement, geography has gained some powerful new tools: Already, they are changing the way we see the Ozarks.

Drone photography of flood damage, North Fork of the White River (2017). *Courtesy of Toby Dogwiler, MSU Department of Geography.*

Drone photography of flood damage, North Fork of the White River (2017). *Courtesy of Toby Dogwiler, MSU Department of Geography.*

Drone photography of flood damage, North Fork of the White River (2017). *Courtesy of Toby Dogwiler, MSU Department of Geography.*

Nature Reclaiming Its Own:
Some Campus Images
Photography by Arjo Mitra

Pill bugs or roly-polys congregating under a concrete lip behind the Meyer Library (2019).

Climbing vines reclaiming a wall near the Plaster Student Union (2019).

An insect larva with somewhere to go, keeping to the sidewalk (2019).

Notes on Contributors

Antonio Rodríguez Alcalá is Professor of Architecture at the *Universidad Anáhuac Mayab* in Mérida, Yucatán, México. He has collaborated on works of historical and architectural restoration and conservation of the built cultural heritage of Yucatán and other states throughout Mexico.

Rachel Anderson is Director at the eFactory, an entrepreneurship and business development center. She is the cofounder of Alumni Spaces, a tech startup company. She is also the cofounder of Rosie, an organization that supports, assists, and serves as an advocate network for current and prospective female founders, business owners, and leaders in the Greater Springfield area.

Allee Armitage is an undergraduate major in Anthropology with minors in Marketing and Intercultural Communications. "I am on the MSU Swim and Dive team," she says, "but swimming does not define me."

Chris Barnhart is Distinguished Professor of Biology at Missouri State University. Along with his research in mollusks, his research and teaching interests include ecology and conservation.

James S. Baumlin is Distinguished Professor of English at Missouri State, where he teaches early English literature, critical theory, and writing. He is a founding editor of Moon City Press and edits the OSI Series in Ozarks History and Culture.

Rachel M. Besara is Associate Dean of Library Services at Missouri State University. Her interests include the technologies and policies of open access research. With William B. Edgar and James S. Baumlin, she is co-editor of *Living Ozarks: The Ecology and Culture of a Nature Place* (2018).

Brooks R. Blevins is the Noel Boyd Professor of Ozarks Studies at Missouri State University. He is the author of seven books, including *A History of the Ozarks, Vol. 1: The Old Ozarks* and *A History of the Ozarks, Vol. 2: The Conflicted Ozarks*.

Christopher Bono is a History major at Missouri State University with a minor in Religious Studies. He works to help preserve Springfield's history in MSU Special Collections and Archives.

Loring Bullard is an environmentalist and author who has taught courses in biology at Drury University. He has served as Executive Directive of the Watershed Committee of the Ozarks.

Greg Burris served as Chief Information Officer and Vice President for Missouri State University and Springfield City Manager before founding the Give 5 program and becoming President/CEO of the United Way of the Ozarks. He pole-vaulted for a school without a pole and enjoys French silk chocolate pie, typically alone in a dark corner.

Lanette Cadle is Professor of English at Missouri State University where she teaches rhetoric and writing. She frequently writes about social media and also publishes poetry. Her poetry collection, *The Tethered Ground*, is forthcoming from Woodley Press (Fall 2019).

John F. Chuchiak IV is Director of the Honors College and Director of the Latin American, Caribbean and Hispanic Studies program at Missouri State University. He is Professor of Colonial Latin American History and holder of the Rich and Doris Young Honors College endowed professorship.

Jim Coombs is Associate Professor and the Maps and Geospatial Information Services Librarian at Missouri State University.

Martha Cooper is a Springfield native, specialty baker, and owner of Sisters in Thyme Bakery and Deli on East Commercial Street. She is fiercely supportive of Historic C-Street and the communitarian values—diversity, equity, opportunity, and quality of life—that C-Street affords.

Christopher J. Craig is Deputy Provost at Missouri State University and Professor of Special Education. Dr. Craig teaches graduate research classes for Education majors. His research has focused on assistive technology and policy impacting persons with disabilities.

Sarah Crain is studying Psychology, Dance, and Communication at Missouri State University. She plans to combine these disciplines to bridge science and art in the practice of expressive arts therapy.

Jasmine Crawford is an undergraduate at Missouri State University pursuing a major in Exercise and Movement Science and a minor in Biomedical Sciences. A native of Kansas City, she runs on the MSU Track Team and is a member of the Honors College.

Tom Dicke is Professor of History at Missouri State University, where he has spent much of his career studying the relationship between large and small business. He is somewhat surprised to have lived half his life in the Ozarks.

Toby Dogwiler is Professor and Head of the Department of Geography, Geology, and Planning at Missouri State University. Dr. Dogwiler's research includes applied GIS & 3D imaging, geomorphology, hydroclimatology, and water resources.

Justin Duncan is EBS Special Education Teacher in Springfield Public Schools. He received his M.A. in History at Missouri State University. His M.A. Thesis, "Performing Theaters of Power: The Holy Office of the Inquisition's General Autos de Fe in Spain and Spanish America and the Visual and Physical Representations of Inquisitorial Power, 1481-1736" (2014) inspired the *Digital Auto de Fe of 1601* Project.

Paul L. Durham is Distinguished Professor of Biology and Director of the Center for Biomedical and Life Sciences at Missouri State University. In addition to classroom teaching, Dr. Durham is principal investigator in a multidisciplinary laboratory focused on migraine and other orofacial pain diseases.

Robert T. Eckels is a retired Missouri State University Facilities Management Director (1993-2014), who strove to provide a quality teaching-learning environment for students and faculty.

William B. Edgar is Clinical Associate Professor at Missouri State University, where he teaches information literacy. His research interests include library management and the future of industrial society.

Hans B. Erickson has a B.F.A. in Computer Animation and is currently pursuing his M.S. in Digital Cultural Preservation at Missouri State University. He is a contributor to the *Digital Auto de Fe of 1601* Project.

Jacob F. Erwin is a graduate teaching assistant in English at Missouri State University. His passions include teaching, literature, writing, and working out at the gym.

Dale Freeman is a retired reporter and editor of the *Springfield News-Leader*. Under his editorship, the paper took the following as its motto: "'Tis a privilege to live in the Ozarks."

Lyle Q. Foster is a community entrepreneur, university professor and diversity trainer. On the university level Dr. Foster teaches at Missouri State University in areas of Social Problems, Multicultural Education, and Public Sociology. He is the founder of Tough Talks and supports the MSU Office of Diversity and Inclusion as a faculty trainer.

Hannah Fox is a Missouri State University Psychology major with a Child and Family Development minor who plans a career in school psychology.

She is a tutor at Kumon and an assistant teacher at Grow to Know Preschool.

Tracie Gieselman-Holthaus is Archivist at Missouri State University Libraries' Special Collections and Archives. She is an appreciator of books, photography, art, music, and the Ozarks.

Daniel D. Goering is Assistant Professor of Management at Missouri State University. He teaches and researches at the intersection of psychology and entrepreneurship. He is an entrepreneur himself, living in Springfield while working on his startups in Tokyo, Japan remotely and during the summer .

Joan Hampton-Porter is starting her sixteenth year as Curator at the History Museum on the Square. She curates a collection of over 100,000 artifacts, of which approximately 20,000 are textiles. Museum collections focus on life in Springfield and Greene County from the 1820s through the 1980s, with strengths in business, education, religion, and home life.

Cara Hawks is a student at Missouri State studying Exercise and Movement Science. Upon graduation, she intends to go to physical therapy school.

B. Colby Jennings is Associate Professor of Digital Arts at Missouri State University. He teaches courses in physical computing, creative coding, time-based art, web art, and installation/experience/interaction. His current research focuses on alternative perspectives in collected or repurposed found data, mass communication, and irony. He still tries a head stand occasionally and feels weird speaking in third person.

Arylle Kathcart is majoring in Intercultural Communication and Sociology at Missouri State University. She is involved in MSU Panhellenic policies and culture and aspires to a career in law.

Morris Kille, Jr. graduated from Palmer College of Chiropractic in Davenport, Iowa (2005) and practices in Springfield, Missouri. In 2001, he retired from the U.S. Navy, having specialized in nuclear power; he served aboard three submarines and one surface ship as well as in nuclear maintenance commands.

Michael R. Kromrey is Executive Director of the Watershed Committee of the Ozarks. He is an Ozarks native passionate about sharing nature with people.

Allen D. Kunkel is Associate Vice President for Economic Development and Director of the Jordan Valley Innovation Center at Missouri State University. He is responsible for helping guide the Innovation Center in its mission of supporting businesses concentrating on advanced technology, biotechnology, life sciences, and nanotechnology research and development. Prior to joining Missouri State University, he served as Manager of Regional Development with the Springfield Area Chamber of Commerce.

Phillip "Cloudpiler" Landis is a board-certified naturopath under the aegis of the American Naturopathic Medical Certification and Accreditation Board-District of Columbia. From 2002 to 2012, he served as the Elected Principal Medicine Chief of the Nemenhah Indigenous Traditional Organization; currently, he serves as the Tehk Tiwehkthihmpt or highest-ranking religious leader of that organization.

Argelia Segovia Liga is Instructor of History and Global Studies at Missouri State University. Dr. Liga received her doctorate in History and Heritage of Indigenous Peoples from the Faculty of History and Archaeology at Leiden University, the Netherlands. She is an ethnohistorian and historian of indigenous cultures in Mexico and contributor to the *Digital Auto de Fe of 1601* Project.

Shannon Mawhiney is Digital Archivist for Missouri State University's Special Collections and

Archives. She is also president of the Christian County Historical Society.

Craig A. Meyer is Assistant Professor of English at Texas A&M University - Kingsville. His research interests include histories of rhetoric, composition studies, disability studies, and creative nonfiction. He was one of the founding members of Moon City Press and has worked on several of its publications.

Jacob Miles is a junior majoring in Mathematics at Missouri State and hopes to further his study of number theory in graduate school. Although a lover of math, Jacob thoroughly enjoys studying the humanities as well as playing music.

Arjo Mitra attends Missouri State University, where he studies Philosophy and practices mindfulness. He takes his camera wherever he goes.

Hanna Moellenhoff is a Psychology major at Missouri State with plans to get her master's degree in Social Work. Eventually, she'd like to get her doctorate in Clinical Psychology and practice as a professional psychologist.

Ledis Molina is a native Spanish speaker from El Salvador and a senior majoring in Computer Animation at Missouri State. In addition to preparing for her senior exhibition, she is a contributor to the *Digital Auto de Fe of 1601* Project.

Lynn Morrow, MSU alumnus, has published in Missouri and Ozarks history since 1978. He is the founding director of the Local Records Preservation Program, Missouri State Archives.

Andrea Flores Navarrete is a graphic artist and contributor to the *Digital Auto de Fe of 1601* Project.

Rhy Norton has served as Senior Scientist and Lab Manager at Dynamic DNA Laboratories since its founding in 2015. He received his master's degree in Biology from Missouri State University and has over twelve years' experience working in micro- and molecular biology laboratories.

Mike O'Brien is a retired reporter, editor, and columnist with the *Springfield News-Leader*. For twenty-five years he taught Journalism courses at Missouri State and Drury universities.

Austin O'Reilly is an entrepreneur and experienced DNA scientist with professional history in the biotech and genetic industries. After obtaining his master's degree in Molecular Biology from Texas Christian University, he founded Dynamic DNA Laboratories in 2015 and is currently serving as its CEO.

Robert S. Patterson is Professor of Astronomy at Missouri State University. His studies of yellow supergiant and cepheid variable stars make use of the laboratory and equipment at Baker Observatory.

Thomas A. Peters serves as Dean of Library Services at Missouri State University. He is finishing a book about the *Ozark Jubilee*, a live nationally broadcast TV show that originated in Springfield from 1955 to 1960.

Sarah Powell is a B.F.A. student in Art and Design at Missouri State University. She is serving as a historical illustrator and character concept designer for the *Digital Auto de Fe of 1601* Project.

Alex T. Primm is a Missourian and inveterate storyteller whose interests include environmentalism, folk culture, and oral history.

Jennifer Probst is a graduate student at Mizzou working towards her doctorate in Biological Sciences. She graduated from Missouri State University with a degree in Biology and minor in Chemistry. Her background is in farming.

Susanna B. Reichling works in the Music and Media area of Meyer Library. She is part of a team of library faculty and staff who provide instruction in new technologies, including 3D design and printing, robotics, and virtual and augmented reality.

Dulce Alejandra Martínez Roldán is a student of Architecture at the *Universidad Anáhuac Mayab* in Mérida, Yucatán, México. In 2018-2019, she was an International Research Exchange Student in the MSU Honors College and contributor to the *Digital Auto de Fe of 1601* Project.

Elizabeth J. Rozell is Associate Dean and Professor of Management at Missouri State University and has taught for more than twenty years. Dr. Rozell has authored over sixty publications which have appeared in such journals as *Personnel Psychology, Group and Organization Studies*, and the *Journal of Psychology*.

Austin Sams is a Cell and Molecular Biology major and Chemistry minor at Missouri State University, and he plans to attend medical school after graduation. He is a member of Phi Mu Alpha Sinfonia Music Fraternity of America and an avid supporter of the arts.

Richard Schur is Professor of English at Drury University, where he teaches courses in African American literature and culture, American literature, and the history of the English language.

Dan Scott AIA, MBA is an architect, re-developer and resident in downtown Springfield. His extensive community involvement has earned him the elite designation of Citizen Architect from the American Institute of Architects.

Michaela Šimonová is a doctoral candidate in the Religious Studies Department at Comenius University of Bratislava, Slovak Republic. She has contributed concept art and design to the *Digital Auto de Fe of 1601* Project.

Clifton M. Smart III is President of Missouri State University. He is finishing his eighth year in this role, having previously served as the university's general counsel.

Lloyd A. Smith is Professor and Graduate Coordinator of Computer Science at Missouri State University. His research and teaching interests include multimedia digital libraries, speech-driven and multimodal user interfaces, music information retrieval, and computer-aided music education.

Nicholas Stoll is a Psychology major and Healthcare Management minor at Missouri State University. At this time, he plans on attending graduate school for Clinical Psychology.

Grace Sullentrup is a Sociology and Psychology major at Missouri State University. She plans to further her education after graduation but loves learning too much to pigeonhole herself. She teaches color guard and serves as a resident assistant at Missouri Scholars Academy each summer.

Emma Sullivan is a Global Studies and Spanish double-major at Missouri State University. She is passionate about studying political movements in Latin American countries and how public expression enables the political. She hopes to continue traveling and engaging in linguistics and culture.

Thomas E. Tomasi is Professor of Biology at Missouri State University. His teaching includes physiological ecology, energetics, and mammals; his personal interests include woodworking, wrestling (which he referees locally), recycling, and bats.

Frederick Ulam is a clinical neuropsychologist whose current interest lies in translating recent developments in neuroscience into practical treatments for neurological and neuropsychiatric conditions. He has researched and applied quantitative electroencephalography in assessing brain function.

María del Carmen Rodríguez Viesca is a student of Architecture at the *Universidad Anáhuac Mayab* in Mérida, Yucatán, México. In 2018-2019, she was an International Research Exchange Student in the MSU Honors College and contributor to the *Digital Auto de Fe of* 1601 Project.

Kenneth R. Vollmar is Professor in the Department of Computer Science at Missouri State University. His teaching and research interests include algorithms, FPGA implementations, and special integer sequences.

Nick Vollmar is a recent graduate of Rice University in Computer Science and Math. He is a software professional in beautiful Austin, Texas.

Jesse Walker-McGraw is an Anthropology major at Missouri State University with minors in History and Child and Family Development. In the future, she hopes to become an archaeologist and do research in Europe and South America.